# FORGOTTEN CALCULUS:
# A Refresher Course
# For Business Applications

BARBARA LEE BLEAU, Ph.D.
Jefferson-Pilot Professor
of Management Science
Division Head
Business Administration and Economics
Pfeiffer College
Misenheimer, North Carolina

BARRON'S

Barron's Educational Series, Inc.

To MAЯY LOU
Your concern, support, and encouragement
made my writing hours more enjoyable.

© Copyright 1988 by Barron's Educational Series, Inc.

*All inquiries should be addressed to:*
Barron's Educational Series, Inc.
250 Wireless Boulevard
Hauppage, New York 11788

*Library of Congress Catalog Card No. 87-33279*
International Standard Book No. 0-8120-3958-0

**Library of Congress Cataloging-in-Publication Data**

Bleau, Barbara Lee.
    Forgotten calculus.

    Bibliography
    Includes index.
    1. Calculus.   I. Title.
QA303.B663   1988       515       87-33279
ISBN 0-8120-3958-0

PRINTED IN THE UNITED STATES OF AMERICA

1   100   9 8 7 6 5 4 3

# Contents

## OTHER BOOKS CITED IN TEXT

Bleau, Barbara Lee, *Forgotten Algebra: A Refresher Course*, Barron's Educational Series, Inc.

Barnett, Raymond A., and Michael R. Ziegler, *Calculus for Management, Life and Social Sciences*, Fourth Edition, Dellen Publishing Company.

Budnick, Frank, S., *Applied Mathematics for Business, Economics, and the Social Sciences*, Second Edition, McGraw-Hill Book Company.

Hoffmann, Laurence D., *Calculus for Business, Economic, and the Social Sciences*, Third Edition, McGraw-Hill Book Company.

Piascik, Chester, *Calculus with Applications to Management, Economics, and the Social and Natural Sciences*, Merrill Publishing Company.

# Preface

If you want to develop an intuitive understanding of calculus as well as an appreciation of the usefulness of calculus to solve managerial, business, economic, and social science problems, then this book was written for you.

*Forgotten Calculus* is an introductory book in differential and integral calculus with the emphasis to applying it in areas such as managerial business applications and problems in economics and other social science areas.

The book differs from the usual calculus text in one major way. There is no attempt to develop, derive, or prove any of the formulas, principles or operational techniques of the calculus. These are simply presented, explained in great detail, and then applied in examples. In other words, this text is aimed at the uses of the calculus, not its development as a subject in mathematics.

Although the title, *Forgotten Calculus*, implies that this is not the reader's first experience with the subject, such need not be the case. The book is suitable as a self-teaching guide for anyone without the slightest previous knowledge of what calculus is all about. Or the book may be used by a person wishing to "brush up" their calculus skills, whether this need results from inadequate learning in the first place or from the passing of time.

*Forgotten Calculus* was designed as a self-pacing workbook. Each unit of the workbook provides explanations and includes numerous examples, problems, and exercises with detailed solutions to facilitate self-study. Special emphasis has been placed on the techniques you will need in solving practical problems in the social and managerial sciences.

I am indebted to Lester Schlumpf who checked the entire manuscript for accuracy and worked all the problems. His suggestions during the preparation of this book were most helpful.

October, 1987                                                                                  Barbara Lee Bleau

# UNIT 1

## Functions

In this first unit of your workbook, you will learn the meaning of a function, a term that we will be using throughout the book. When you have completed the unit, you will be able to read and to use functional notation and identify the domain.

We are going to start with a picture.

This is a picture of a machine. We can define our machine to do most anything. Since this is a workbook about calculus, our machine is going to do calculations with numbers. Suppose the numbers 1, 2, and 3 are located as shown and that the machine "takes a number, squares it, and adds 1."

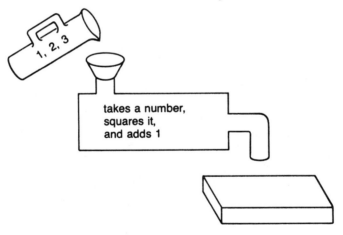

It should be clear what happens as each number is put into the machine.

If 1 is put into the machine, the machine takes the number, squares it, adds 1, and out comes 2.

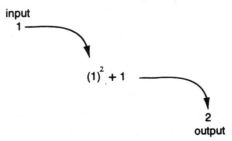

In a similar fashion, if 2 had been put in, the output would have been 5. If 3 had been put in, the output would have been 10. In the machine drawing, write the numbers 2, 5, and 10 in the box.

The values that can be put into the machine are referred to as the domain. The **domain** is the set of all possible values that can be used as input. In this example, the domain is the set of numbers 1, 2, and 3, or $D = \{1, 2, 3\}$. Only those three numbers can be put into this particular machine. Why? Because that is the way we created it on page one.

The values that come out are referred to as the range. The **range** is the set of all possible values that are output. In the picture, the range is the set of numbers 2, 5, and 10, or $R = \{2, 5, 10\}$.

Both the domain and range in this example are **finite**, meaning there are a countable number of elements or values in each. The domain has three elements and the range has three elements. An **infinite** domain or range would have an uncountable number of elements, such as the domain being the set of whole numbers or $D = \{0, 1, 2, 3 \ldots\}$. The three dots mean that the numbers continue on in the same pattern.

The "takes a number, squares it, and adds 1" is referred to as the **rule**.

Notice our picture has three parts: the domain, the range, and the rule.

The domain, often denoted by $x$, is the **independent variable**, whereas the range, often denoted by $y$, is the **dependent variable**. A way to remember the distinction between the two is that we may select any value from the domain to put into the machine; but once a specific value of $x$ is selected, such as 3, the output or the $y$-value is dependent upon it.

By using $x$ to denote the independent variable, it is now possible to rewrite the rule algebraically, $x^2 + 1$. At the same time, we will name the machine, $f$. Typically $f$ is the most common name used, although $g$ and $h$ are popular, followed by some of the Greek letters; e.g., $\alpha$, $\beta$, and $\gamma$.

The picture has been drawn again to include all the terminology.

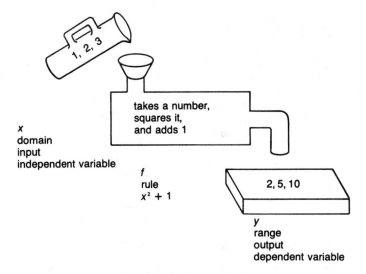

$x$
domain
input
independent variable

takes a number,
squares it,
and adds 1

$f$
rule
$x^2 + 1$

2, 5, 10

$y$
range
output
dependent variable

---

> **Definition:**    A **function** is a rule that assigns to each element in the domain one and only one element in the range.

---

As with the picture, the definition refers to three parts: domain, range, and rule. The key concept, though, is the one and only one phrase. For each value put in, only a single value may come out if the expression is to be a function. Otherwise the expression is classified as **a relation**.

The machine named $f$ is a function because for each value put in only a single value came out.

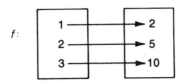

Rather than having to draw pictures each time, all of the above information can be combined into a single statement using functional notation.

$$y \; = \; f(x) \; = \; x^2 + 1 \text{ with } D \; = \; \{1, 2, 3\}$$

dependent name independent rule      domain
variable     variable

The above statement is read "$y$ equals $f$ of $x$ equals $x$ squared plus one with domain, $D$, equal to the set of numbers 1, 2, and 3."

Caution: $f(x)$ is read "$f$ of $x$" and does not mean multiplication. It is used to indicate that the name of the function is $f$ and that the variable inside the parentheses, in this example $x$, is to be the independent variable.

A variation on the above notation is:

$$f \; = \; \{(x, y) \; : \; y = x^2 + 1, \; x \in \{1, 2, 3\}\}$$

name   independent dependent    rule     domain
variable   variable

read "such that"   read "belongs to"

The above statement is read "$f$ equals the set of all ordered pairs $x$ and $y$, such that $y$ equals $x$ squared plus one and $x$ belongs to the set of numbers 1, 2, and 3." With this notation, the name of the function is first and the dependent variable is always the last one of the grouped variables listed inside the parentheses, in this case $y$.

It is time to stop and try some questions. Using $f$ as defined by the picture, answer the following questions.

1.  If $x = 2$, what is $y$?                                    Answer:    5

2.  If $x = 3$, what is $y$?                                    Answer:    10

3.  If $y = 2$, what is $x$?                                    Answer:    1

The confusing part is when the same questions are asked using functional notation. Here are some examples. Continue to use $f$ as defined earlier.

## EXAMPLE 1

Find $f(2)$.

Solution:    $f(2)$, read "$f$ of 2," is asking if $x = 2$, what is $y$? It is exactly the same as question one above and the answer is 5, or $f(2) = 5$.

## EXAMPLE 2

Find $f(3)$.

Solution:    $f(3)$ is asking the same thing as question two above, but using function notation. The answer is 10, or $f(3) = 10$.

## EXAMPLE 3

Find $f^{-1}(2)$.

Solution:    $f^{-1}(2)$ is read "$f$ inverse of 2." The question is asking if the dependent value is 2, what is $x$? As in question three above, the answer is 1. $f^{-1}(2) = 1$

Here are a few questions for you to try. Use $f$ as defined in the picture to answer each of the following.

## Problem 1

Find $f(1)$, $f^{-1}(10)$, and $f^{-1}(5)$.

Solution:

Answers:   2, 3, 2

## Problem 2

Find $f(4)$. Be careful; the answer is not 17.

Solution:

Answer:    $f(4)$ is undefined, because 4 is not an element of the domain. $f$ is defined only for the numbers 1, 2, and 3.

## Problem 3

Imagine that you are in a class with other students named Smith, Jones, and Howard. The machine is defined to assign grades at the conclusion of the course. If the choice were yours, what would the range be?

Solution:

Answer:  There are many answers possible. The most popular answer is that the range would contain only the letter grade $A$. It would mean that every one in the class will receive an $A$.

## Problem 4

Is the example used in problem 3 a function?

Solution:

Answer:  Yes. For each name entered, only one grade is given. It is a function even if everyone might be assigned the same grade.

## Problem 5

If the example in problem 3 were reversed, would it be a function? In other words, when the $A$ is entered, the output would be the names of those students receiving an $A$.

Solution:

Answer:  No. For the letter grade of an $A$, four names would be printed. That violates the concept of "for each $x$ there is a single $y$ value."

Unless specified, the domain of a function is all values for which the dependent variable is defined and real. This is usually the set of real numbers with two exceptions that will be discussed later. Remember the real numbers are all the counting numbers, zero, whole numbers, positive and negative integers, fractions, rationals, and irrational numbers.

## EXAMPLE 4

Given:   $y = g(x) = 3x + 1$.
Identify each part of the expression and
find:     $g(2)$, $g(-5)$, $g(a)$, $g(3x)$, $g(a + b)$, and $g^{-1}(7)$.

Solution:   $y$ is the dependent variable.

$x$ is the independent variable.

$g$ is the name of the function.

The rule is: 3 times the number plus 1.

The domain is assumed to be the set of all real numbers since nothing to the contrary is specified.

$g(x) \quad = 3x \qquad + 1$

$g(2) \quad = 3(2) \qquad + 1 = \qquad 6 + 1 = \qquad 7$

$g(-5) \quad = 3(-5) \quad + 1 = -15 + 1 = -14$

$g(a) \quad = 3(a) \qquad + 1 = 3a + 1$

$g(3x) \quad = 3(3x) \qquad + 1 = 9x + 1$

$g(a + b) = 3(a + b) + 1 = 3a + 3b + 1$

$g^{-1}(7) = 2$, because from above when $y = 7$, the $x$ value was 2.

Try the following problem before continuing.

## Problem 6

Given:   $y = f(x) = 5 + 2x$.

Find:     $f(0)$, $f(2)$, $f(1/2)$, $f(-9)$, $f(x - 1)$, and $f^{-1}(9)$.

Solution:   $f(x) = 5 + 2x$

$f(0) =$

$f(2) =$

$f(1/2) =$

$f(-9) =$

$f(x - 1) =$

$f^{-1}(9) =$

Answers: 5, 9, 6, $-13$, $3 + 2x$, and 2

The following is an example of a function with a **split domain**. In this example the rule has three parts. The choice of which one to be used is dependent upon the value of $x$.

## EXAMPLE 5

Given: $h(x) = \begin{cases} x^2 + 1, & \text{if } x > 0 \\ 10, & \text{if } x = 0 \\ 2x, & \text{if } x < 0. \end{cases}$

Find: $h(2)$, $h(3)$, $h(0)$, and $h(-1)$.

Solution: $h(2) = (2)^2 + 1 = 4 + 1 = 5$   The top rule was used because 2, the $x$ value, was greater than 0.

$h(3) = (3)^2 + 1 = 9 + 1 = 10$

$h(0) = 10$   The middle rule was used because $x = 0$.

$h(-1) = 2(-1) = -2$   The bottom rule was used because $x = -1$, which is negative.

Although functions commonly are expressed as algebraic expressions, sets and graphs also can be used.

## EXAMPLE 6

Given: $F = \{(1, 2), (1, 3), (2, 4)\}$.

Find:  a.  The domain.

b.  The range.

c.  $F(2)$.

d.  Is $F$ a function?

e.  $F^{-1}(2)$.

Solution:  $F$ is the name. The ordered pair $(1, 2)$ means when $x$ is 1, $y$ is 2. Similarly $(1, 3)$ means when $x = 1$, $y = 3$ and $(2, 4)$ means when $x = 2$, $y = 4$.

a.  The domain is the set of all $x$ values: $D = \{1, 2\}$. Note: there is no need to write the 1 twice.

b.  The range is the set of $y$ values: $R = \{2, 3, 4\}$.

c.  $F(2) = 4$. The notation is asking when $x$ is 2, what is $y$?

d.  No, $F$ is not a function. When $x$ is 1, there are two different values for $y$, 2 and 3.

e.  $F^{-1}(2) = 1$. The notation is asking when the $y$ value is 2, what is $x$?

## Problem 7

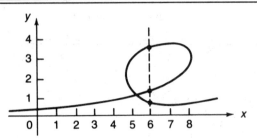

Does the graph above specify a function?

Solution:

Answer:   No, for example when $x = 6$, there are three values for $y$.

Notice that if you are able to draw a vertical line that intersects the graph at more than one point, then the graph does not represent a function. This is referred to as the **vertical line test**.

Thus far we have looked at functions with a single independent variable. A **bivariate function** has two independent variables. Think of a bivariate function as having two separate sets of input values rather than just one.

## EXAMPLE 7

Given:   $z = h(x, y) = 3x + y^2$.

Identify each part of the expression and

find:   $h(2, 3)$ and $h(5, -1)$.

Solution:      $z$    $=$    $h(x, y) = 3x + y^2$

dependent        two              rule: 3 times the first number plus the second number squared
variable         independent
        name     variables

$$z = h(2, 3)    = 3(2) + (3)^2   = 6 + 9 = 15$$
$$z = h(5, -1) = 3(5) + (-1)^2 = 15 + 1 = 16$$

A **multivariate function** has more than two independent variables. The following example has four independent variables. Subscripts can be used rather than different letters. $x_1$ is read "$x$ sub 1." $x_1, x_2, x_3,$ and $x_4$ denote different variables.

## EXAMPLE 8

Given:   $y = f(x_1, x_2, x_3, x_4) = 2x_1 + x_2 x_3 x_4$.

Find:   $y = f(23, -1, 10, 0)$.

Solution:   $y = f(x_1, x_2, x_3, x_4) = 2x_1 + x_2 x_3 x_4$

$$f(23, -1, 10, 0) = 2(23) + (-1)(10)(0) = 46 + 0 = 46$$

# LIMITATIONS ON THE DOMAIN

When functions are expressed by means of an algebraic expression, the domain is seldom mentioned. The domain, remember, is the set of values of the independent variable for which a function can be evaluated. In most cases, this is the set of all real numbers. There are two exceptions.

First Exception

If the function contains a quotient, all values of the independent variable that would result in the denominator equaling a value of 0 must be excluded from the domain. The reason is because division by zero is undefined.

## EXAMPLE 9

Given: $f(x) = \dfrac{3}{x-2}$, find the domain.

Solution: The domain is the set of all real numbers except 2 because 2 would result in the denominator equaling 0.

## EXAMPLE 10

Given: $g(x) = 1/x$, find the domain.

Solution: The domain is the set of all real numbers except 0.

## EXAMPLE 11

Given: $h(z) = \dfrac{3-z}{z+5}$, find the domain.

Solution: The domain is the set of all real numbers except $-5$ because $-5$ would result in the denominator assuming a value of 0.

## EXAMPLE 12

Given: $G(y) = \dfrac{y-7}{3}$, find the domain.

Solution: The domain is the set of all real numbers. There is no number that must be excluded because the denominator will never be 0.

## EXAMPLE 13

Given:  $H(x) = \dfrac{3}{(x-2)(x-1)(x+6)}$, find the domain.

Solution:  The domain is the set of all real numbers except 2, 1, and $-6$ because each of those would result in the denominator assuming a value of zero.

## Second Exception

If taking an even root, the independent variable must be defined so that the quantity under the radical sign is zero or positive; i.e., non-negative.

## EXAMPLE 14

Given:  $f(x) = \sqrt{x}$, find the domain.

Solution:  The quantity under the radical sign must be zero or positive, therefore the domain is the set of all $x$ such that $x \geq 0$.

## EXAMPLE 15

Given:  $g(x) = \sqrt{x-5}$, find the domain.

Solution:  To say that the quantity under the radical sign must be zero or positive is the same as saying      $x - 5 \geq 0$
or      $x \geq 5$; therefore the domain is the set of all $x$ such that $x \geq 5$ or, using set notation: $D = \{x \mid x \geq 5\}$.

## EXAMPLE 16

Given:  $h(x) = \sqrt{x^2 + 1}$, find the domain.

Solution:  The rule states that the quantity under the radical sign must be zero or positive (non-negative). $x^2 + 1$ is always positive. Therefore no numbers need be excluded. The domain is the set of reals.

## EXAMPLE 17

Given:  $t(x) = \sqrt[3]{x}$, find the domain.

Solution:  This example is to find the cube root, which is an odd root. The second exception does not apply. The domain is the set of reals.

You should now understand the basic concept of a function. Given a function you should be able to identify the domain, range, rule, name, independent, and dependent variables. These basic terms will be used continually throughout the following units.

You should also be able to read and evaluate questions written in functional notation, including those for bivariate and multivariate functions as well as functions with split domains.

Remember also that if the domain for a function $f$ is not specified, it is assumed to be the set of real numbers with two exceptions:

1.  if $f$ contains a quotient, the domain must exclude any numbers that result in the denominator being zero.

2.  if $f$ contains a radical sign for an even root, the domain must be only those values that result in the quantity under the radical sign being zero or positive (non-negative).

Before beginning the next unit you should evaluate the following functions and answer all the questions. When you have completed them, check your answers against those at the back of the book.

## EXERCISES

Given:    $f(x) = x^2 - 2x + 3$, find each of the following:

1.  $f(0)$

2.  $f(10)$

3.  $f(-2)$

4.  $f(a)$

5.  $f^{-1}(3)$

6.  $f(-x)$

7.  The domain for $f$.

Given:    $g(x) = 1/(x - 3)$, find each of the following:

8.  $g(5)$

9.  $g(c)$

10.  $g(a + 3)$

11.  The domain for $g$.

Given: $q = h(p) = \sqrt{13 - p}$, find each of the following:

12. Which letter denotes the independent variable?

13. The domain for $h$.

14. $h(12)$

15. $h(13)$

16. $h(-3)$

17. $h(2)$

18. Given $\alpha(x, y, z) = x - 3z + xy$, find $\alpha(5, 3, -2)$.

Given: $H(x) = \begin{cases} x + 1, & \text{if } x \le 3 \\ x - 1, & \text{if } x > 3, \text{ find:} \end{cases}$

19. $H(7)$

20. $H(0)$

21. $H(-4)$

22. $H(3)$

Given: $G = \{(5, 6), (5, 7), (8, 5)\}$

23. State the domain for $G$.

24. State the range for $G$.

25. Is $G$ a function?

---

If additional practice is needed:

Barnett and Ziegler, pages 55–57, problems 1–6, 13–22, 57–62, 75, 76
Budnick, pages 56–57, problems 1–14, 28–31
Hoffmann, pages 8–10, problems 1–32
Piascik, pages 17–20, problems 1–5, 18–25

# UNIT 2

## Functions: Explicit/Implicit, Composite, and Applications with Restricted Domains

In this second unit several more topics related to functions will be explained. You will learn the difference between implicit and explicit functions and what is meant by a composite function. When you have completed the unit, you will be able to evaluate composite functions. You also will be able to interpret functional notation and specify restricted domains for application problems.

## EXPLICIT AND IMPLICIT FUNCTIONS

For a function of the form, $y = f(x)$, $y$ is said to be an **explicit** function of the independent variable $x$.

For example, if we are given $y = x + 2$, $y$ is said to be an explicit function of the variable $x$, or $y = f(x)$. Stated another way, if you are given a value for $x$, you could calculate the value for $y$. Stated still another way, the equation has been solved for $y$ in terms of the other variables, meaning that $y$ is on one side of the equal sign and all other variables and constants are on the other side.

If the same equation is solved for $x$, $\qquad y = x + 2$

$$y - 2 = x$$

or, $\qquad x = y - 2$

Now $x$ can be said to be an explicit function of $y$, or $x = g(y)$. The equation has been solved for $x$ in terms of $y$ and $y$ is to be considered the independent variable.

If the same equation is rewritten as $y - x - 2 = 0$, it is termed an **implicit** equation. It is an equation that does *not* directly express one variable in terms of the others.

## EXAMPLE 1

Given the implicit equation:   $3x + y - 15 = 0$.

Find, if possible, the explicit functions $y = f(x)$ and $x = g(y)$.

Solution:   To find $y = f(x)$ means to solve, if possible, the given equation for $y$ in terms of the variable $x$.

$$3x + y - 15 = 0 \qquad \text{implicit}$$

$$y = -3x + 15 \qquad \text{explicit because } y = f(x)$$

To find $x = g(y)$ means to solve, if possible, the given equation for $x$ in terms of the variable $y$.

$$3x + y - 15 = 0 \qquad \text{implicit}$$

$$3x = -y + 15$$

$$x = -\frac{1}{3}y + 5 \qquad \text{explicit because } x = g(y)$$

## EXAMPLE 2

Given the implicit equation:   $2x^3 - x + 3y = 10$.

Find, if possible, the explicit functions $y = f(x)$ and $x = g(y)$.

Solution:   $2x^3 - x + 3y = 10 \qquad \text{implicit}$

$$3y = 10 + x - 2x^3$$

$$y = \frac{10 + x - 2x^3}{3} \qquad \begin{array}{l}\text{explicit because } y \text{ is expressed} \\ \text{in terms of } x, \text{ or } y = f(x).\end{array}$$

$2x^3 - x + 3y = 10$

$\quad 2x^3 - x = 10 - 3y$

Unless you know something I don't, there is no way to solve this cubic equation for $x$. Therefore, it is not possible to write an explicit function for $x$.   $x = g(y)$ does not exist.

Try solving this problem yourself.

## Problem 1

Given the implicit equation:   $2q + 11p = 33{,}000$.

Find, if possible, the explicit functions $p = f(q)$ and $q = g(p)$.

Solution:

$$\text{Answer:} \quad p = f(q) = 3{,}000 - \frac{2}{11} q$$

$$q = g(p) = 16{,}500 - \frac{11}{2} p$$

# COMPOSITE FUNCTIONS

If two functions are combined to form a third function, the third function sometimes is referred to as a **composite** function.

We will look at some easy examples first.

## EXAMPLE 3

Given:   $f(x) = x^2 + 1$ and $g(x) = x/2$.

Find:    $f(7) + g(10)$.

Solution:   Now the reason for naming functions should be apparent. The name is used to indicate which rule is to be used. Obviously the functions must have different names.

The rule for $f$: takes a number, squares it, and adds 1.

The rule for $g$: takes a number, divides it by 2.

$f(7)$ indicates that 7 is to be put into the function $f$, and $g(10)$ indicates that 10 is to be put into the function $g$.

$$
\begin{aligned}
f(7) + g(10) &= \quad f(7) \quad + g(10) \\
&= [(7)^2 + 1] + [10/2] \\
&= \quad 50 \quad + 5 \\
&= \quad 55
\end{aligned}
$$

## EXAMPLE 4

Given:   $f(x) = x^2 + x - 5$ and $g(x) = \sqrt{x}$.

Find:    $g(0)/f(3)$.

Solution:   $g(0) = \sqrt{0} = 0$ and $f(3) = (3)^2 + 3 - 5 = 7$.

Therefore                           $\dfrac{g(0)}{f(3)} = \dfrac{0}{7} = 0$

# EXAMPLE 5

Given:   $f(x) = x^2 + 2x - 1$ and $g(x) = 2x^2 + 5x$.

Find:     $h(x) = g(x) - 2f(x)$.

---

Solution:   $h(x) = g(x) - 2f(x)$        means 2 multiplied times the function $f$

$$= (2x^2 + 5x) - 2(x^2 + 2x - 1)$$

$$= 2x^2 + 5x - 2x^2 - 4x + 2$$

$$= x + 2$$

Note that $h$ is a new function, created by combining $g$ and $f$.

The rule for $h$: takes the number and adds 2.

---

Before we look at more complicated composite functions, solve the next few problems for practice.

# Problem 2

Given:   $f(x) = x + 1$, $g(x) = x - 2$, and $h(x) = 2x$.

Find:     a.   $3h(5) + f(0)$.

            b.   $H(x) = h(x) - f(x) - g(x)$.

            c.   $G(x) = 2h(x) + g(x)$.

---

Solution:

Answers:   a. 31,   b. 1,   c. $5x - 2$

---

   Thus far our composite functions have been formed by adding, subtracting, multiplying, or dividing functions. A *composite* function also exists when one function is considered a function of another function. This is the more common interpretation of a composite function. To illustrate this idea, we will draw another picture, this time with two function machines.

The first function, $g$, takes the number and doubles it.

The second function, $f$, takes the number and adds five.

Find the value of $y$ if first 10 is put into the function $g$, and then that output is put into the function $f$.

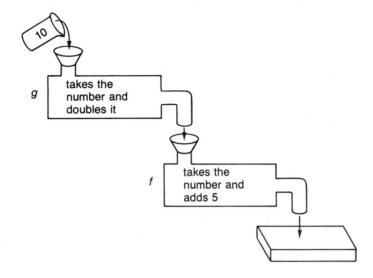

With the help of the drawing, the answer should be straightforward to calculate. The answer is 25. It is when the same question is asked using functional notation that the issue becomes confused. Here is the same question using functional notation.

$$\text{Given } y = f(x) = x + 5 \text{ and } y = g(x) = 2x, \text{ find } y = f(g(10)).$$

It will help if you recall that the rule for removing groups of parentheses is to work from the inside out The same rule applies here.

To find $f(g(10))$:

> first 10 is to be put into the function $g$,
> then the result of that is put into the function $f$.
> In case you wondered, $f(g(10))$ is read "$f$ of $g$ of 10."

# EXAMPLE 6

Given $y = f(x) = x + 5$ and $y = g(x) = 2x$, find $y = g(f(10))$.

Solution:   $y = g(f(10))$

10 first is to be put into the function $f$.

$$f(10) = 10 + 5 = 15$$

$= g(15)$

15 then is to be put into the function $g$.

$= 2(15)$

$= 30$

# EXAMPLE 7

Given $y = f(x) = x + 5$, find $y = f(f(3))$.

Solution:   $y = f(f(3))$

3 first is to be put into the function $f$.

$$f(3) = 3 + 5 = 8$$

$= f(8)$

8 then is to be put into the function $f$.

$= 8 + 5$

$= 13$

As you can see, composite functions of this type are really not as bad as at first glance. Try a few yourself with some new functions.

# Problem 3

Given:   $f(x) = x^2 + 1$  and  $g(x) = x/2$.

Find:      a.   $g(f(3))$.

        b.   $g(f(5))$.

        c.   $f(g(8))$.

Solution:

Answers:   a. 5,   b. 13,   c. 17

Now that you got all of those correct, here are a few more that use the same functions. These are for fun.

# Problem 4

Given:   $f(x) = x^2 + 1$ and $g(x) = x/2$.

Find:      a.   $f(f(f(0)))$.

        b.   $g(g(g(16)))$.

        c.   $f(g(f(1)))$.

        d.   $g(f(g(g(20))))$.

---

Solution:

Answers:   a. 5,   b. 2,   c. 2,   d. 13

---

So far all of the composite functions that were a function of a function have been numerical. That is to say we have been finding such values as $f(g(\text{some number}))$. The next examples explain how to find a new function, say $h$, that is a function of a function of a variable.

===

## EXAMPLE 8

Given:   $g(x) = 2x$ and $f(x) = x + 5$.

Find:   $h(x) = f(g(x))$.

---

Solution:   $h(x) = f(g(x))$

$x$ first is put into the function $g$.

$$g(x) = 2x$$

$= f(2x)$

$2x$ then is put into the function $f$.

$= 2x + 5$

The functions $g$ and $f$ are the same functions used in the drawing on page 17.   $h(x)$ should be thought of as a new function that combines $f$ and $g$ into a single machine with the rule, takes a number, doubles it, and adds 5.

---

===

## EXAMPLE 9

Using $f$, $g$, and $h$ as defined in Example 8, find $f(g(7))$.

---

Solution:   There are two approaches to solving this problem.

1.   $f(g(7)) = f(g(7))$

$= f(14)$   because $g(7) = 2(7) = 14$

$= 14 + 5$

$= 19$

or using the rule from Example 8,

2.  $f(g(x)) = 2x + 5$

$f(g(7)) = 2(7) + 5$

$= 19$

Use whichever method you are most comfortable with; obviously the answer should be the same either way.

## EXAMPLE 10

Given:   $y = H(x) = x^2 - 5$ and $y = G(x) = x + 3$.

Find:     $y = H(G(x))$ and simplify.

Solution:   $y = H(G(x))$

$= H(x + 3)$               Working from the inside out, $G(x) = x + 3$.

$= (x + 3)^2 - 5$          H takes the number, squares it, and subtracts 5.

$= (x^2 + 6x + 9) - 5$     Reminder:   $(x + 3)^2 = (x + 3)(x + 3)$

$= x^2 + 6x + 4$

$= x(x + 3) + 3(x + 3)$

$= x^2 + 3x + 3x + 9$

$= x^2 + 6x + 9$

# RESTRICTED DOMAINS

In Unit 1, the discussion on domains and ranges was from a purely mathematical point of view. When considering application problems, maybe you call them word problems, the domain and range often need to be restricted so that the values of the variables "make sense."

## Problem 5

Given:   $y = f(x) = 5x$.

Find the domain for $f$.

Solution:

Answer:   The domain is the set of reals, because there is no quotient or radical to an even power requiring that any numbers must be excluded.

# EXAMPLE 11

Given:   $y = f(x) = 5x$

where $x$ = the number of hours worked per day

and   $y$ = wages earned per day in dollars.

Find the restricted domain for $f$.

Solution:   This is the same function as given in Problem 5, but now $x$ has been defined as the number of hours worked per day. We should reason that since there are 24 hours in a day and negative numbers would be meaningless in this situation, the only values that "make sense" for $x$ are from 0 to 24, inclusive.

The restricted domain is $0 \leq x \leq 24$.

# Problem 6

Given:   $y = f(x) = 5x$   with restricted domain of $0 \leq x \leq 24$

where $x$ = the number of hours worked per day

and   $y$ = wages earned per day in dollars.

Calculate and interpret:   $f(3)$, $f(0)$, and $f(24)$.

Solution:

Answers:   $f(3) = 15$; $15 are the wages earned per day for working 3 hours.

$f(0) = 0$; $0 are the wages earned per day for working 0 hours.

$f(24) = 120$; $120 are the wages earned per day for working 24 hours.

# EXAMPLE 12

Find the restricted range for Problem 6 and interpret.

Solution:   Since the restricted domain is $0 \leq x \leq 24$

and $f(0) = 0$

and $f(24) = 120$

the restricted range is $0 \leq y \leq 120$.

In other words, the minimum amount of wages that can be earned per day is $0 for working no hours. The maximum amount of wages that can be earned per day is $120 for working 24 hours. The actual wages earned will be anywhere from $0 to $120, inclusive.

---

The last example, another application problem, combines the concepts of restricted domains and composite functions.

## EXAMPLE 13

H. Jones is an excellent, but persnickety, gardener who enjoys warm weather. In fact, the warmer the weather, the more hours he is willing to work. Each morning at 8 A.M., he checks the temperature and makes a decision as to the number of hours he will work that day, which in turn determines his salary for the day. His decision rules are:

$$y = f(x) = 4x$$

$$x = g(t) = \frac{t - 30}{2}$$

where $x$ = the number of hours worked per day

$y$ = salary earned per day by H. Jones

$t$ = temperature in $F°$ at 8 A.M.

a.  Interpret $y = f(x)$.

b.  Interpret $x = g(t)$.

---

Sol·tion:   a.   Reading from left to right:

$$y = f(x)$$

The salary earned per | is  a function of the number of hours
day by H. Jones                              worked per day

b.   $x = g(t)$   The number of hours worked per day is a function
of temperature in $F°$ at 8 A.M.

---

## EXAMPLE 13 (Continued)

c.  Calculate and interpret $y = f(5)$.

d.  Calculate and interpret $x = g(32)$.

e.  Calculate and interpret $y = f(g(32))$.

---

Solution:   c.   $y = f(5)$

$= 4(5)$

$= 20$       If the gardener works five hours he will earn $20 for the day.

   d.  $x = g(32)$

   $= (32 - 30)/2$

   $= 1$          If the temperature is 32 degrees at 8 A.M., H. Jones will work one hour.

   e.  $y = f(g(32))$

   $= f(1)$

   $= 4(1)$

   $= 4$          If the temperature is 32 degrees at 8 A.M., H. Jones will earn $4 for the day.

## Problem 7

Use the information provided in Example 13.

Calculate the interpret $y = f(g(44))$.

Solution:

   Answer:   28. If the temperature is 44 degrees at 8 A.M., the gardener's salary for the day will be $28.

## EXAMPLE 13 (Continued)

f.  Find and interpret $y = h(t) = f(g(t))$.

Solution:   f.   $y = h(t) = f(g(t))$

$$= f\left(\frac{t - 30}{2}\right)$$

$$= \frac{4(t - 30)}{2}$$

$$= 2t - 60$$

The salary earned per day by H. Jones is a function of the temperature in $F°$ at 8 A.M.

## EXAMPLE 13 (Continued)

g.  Find and interpret $y = h(50)$, where $h$ is the function from part f.

Solution:   g.   $y = h(50)$

$= 2(50) - 60$

$= 100 - 60$

$= 40$

If the temperature is 50 degrees at 8 A.M., the salary earned for the day by H. Jones will be $40.

Note that by using function $h$ the salary was found directly from the given temperature without going through the intermediate step of calculating the number of hours worked per day.

---

## Problem 8

Use the information provided in Example 13.

Name the independent variable for function $f$.

Find the restricted domain for $f$.

---

Solution:

Answer:    The independent variable for $f$ is $x$.
           Restricted domain for $f$: $0 \leq x \leq 24$.

---

## EXAMPLE 13 (Continued)

h.   Name the independent variable for function $g$.

i.   Find the restricted domain for $g$.

---

Solution:    For $x = g(t) = \dfrac{t - 30}{2}$, the independent variable is $t$.

Regarding the restricted domain for $g$, the following question must be answered: are there any limitations on the values that can be used for $t$?

From Problem 8, we know that the smallest value for $x$ is 0,

$$\text{and since } x = \frac{t - 30}{2}$$

$$\text{then} \qquad 0 = \frac{t - 30}{2}$$

$$0 = t - 30$$

$$\text{or} \qquad t = 30 \text{ is the smallest value for } t.$$

Any number less than 30 would result in a negative number for $x$, which would be meaningless for this problem. A gardener cannot work a negative number of hours.

From Problem 8, we know that the largest value for $x$ is 24,

$$\text{and since } x = \frac{t - 30}{2}$$

$$\text{then} \quad 24 = \frac{t - 30}{2}$$

$$48 = t - 30$$

$$\text{or} \quad t = 78 \text{ is the largest value for } t.$$

Any number larger than 78 would result in $x$ being larger than 24.

If you don't believe me, try 80.

Restricted domain for $g$: $30 \leq t \leq 78$.

---

In summary we distinguished between implicit and explicit functions. For some implicit functions, you should be able to rewrite the function as an explicit function of one of the variables.

Also you should be able to read, interpret, and evaluate composite functions, both numerical and algebraic. When two or more functions are combined as a function of a function, such as $y = f(g(x))$, the composite function is evaluated from the inside out.

Application problems were introduced to illustrate the necessity for sometimes specifying a restricted domain to ensure that the value of the variable "makes sense."

Now try to work the following problems. In addition, some of the application problems contain formulas that will be used again in later units.

# EXERCISES

Given: $f(x) = x + 1$

$g(x) = x - 2$

$h(x) = 2x + 3$

$H(x) = x^2 + x$

$F(x) = x^2 - 3x$

$G(x) = 3x - 5$

Find and simplify:

1. $f(g(5))$

2. $f(H(5))$

3. $G(F(2))$

4. $g(F(G(2)))$

5. $F(x + 1)$    Be careful with this one.

6. $f(h(x))$

7. $2f(x) + g(x) - h(x)$

8. $h(f(x))$

9. $H(2x) - h(x)$

10. $G(F(x))$

11. Given the equation $5x + 6y = 72$, find (if possible) the explicit functions $y = f(x)$ and $x = g(y)$.

12. It is estimated that $t$ years from now the population of an urban community will be $N(t) = -t^2 + 400t + 50,000$.

    a. What is the current population of the community?

    b. What will the population be 10 years from now?

13. The Alexander Company produces hand-crocheted rugs. The overhead cost for such items as insurance and rent is $6,452. In addition, each rug costs $72 to make, including all labor and materials. Let $C(x)$ represent the total cost of producing $x$ hand-crocheted rugs; then

$$C(x) = 6,452 + 72x \text{ with } x \geq 0.$$

The function $C$ is called a **total cost function**.

    a. Determine the total cost of producing 10 rugs.

    b. Calculate and interpret $C(100)$.

    c. Find $C(0)$.
       $C(0)$ is called the **fixed cost**. It is the cost to the company when no rugs are produced. The cost of producing each additional rug, $72, is called the **variable cost per unit**.

---

Cost Function

Total Cost = fixed cost + (variable cost/unit)(number of units)
           or if the *total* variable cost is given,
Total Cost = fixed cost + total variable cost

---

14. At another location, the Alexander Company has a fixed cost of $13,200 and a variable cost per unit of $43 on braided rugs. Let $x$ represent the number of braided rugs produced and $C(x)$ equal the total cost of producing $x$ braided rugs.

    a. Write the equation that defines the cost function.

    b. Calculate and interpret $C(0)$, $C(10)$, and $C(100)$.

    c. State the restricted domain for $C$.

15. Crocheted rugs are popular at the present time and the Alexander Company is able to sell each rug for $153. If $R(x)$ represents the total sales revenue gained from selling $x$ crocheted rugs, then

$$R(x) = 153x \text{ with } x \geq 0.$$

The function $R$ is called a **total revenue function**.

    a. Find the total revenue gained from selling 50 rugs.

    b. Calculate and interpret $R(100)$ and $R(0)$.

> Revenue Function
>
> Total Revenue = (selling price per unit)(number of units sold)
> which is often shortened to price · quantity.

16. The Company's braided rugs sell for only $78, but the company is able to sell quite a few. Let $R(x)$ represent the total revenue gained from selling $x$ braided rugs.

   a. Write the total revenue equation relating $R(x)$ and $x$.

   b. Calculate and interpret $R(1,000)$.

   c. State the restricted domain for $R$.

17. Profit for an organization is the difference between total revenue and total cost. If $P(x)$ represents the total profit from the production and sale of $x$ crocheted rugs, then

$$P(x) = R(x) - C(x)$$
$$= 153x - (6,452 + 72x)$$
$$= 81x - 6,452 \quad \text{with } x \geq 0$$

   a. Find the profit (or loss) gained from the production and sale of 200 rugs.

   b. Calculate and interpret $P(50)$.

> Profit Function
>
> Profit = Total Revenue − Total Cost

18. $C(x_1, x_2, x_3)$ is the cost function in dollars for Mary Lou's Toy Company with

$$C(x_1, x_2, x_3) = 1,750 + 7.5x_1 + 6x_2 + 10.6x_3$$

where $x_1$ = number of toy soldiers,

$x_2$ = number of teddy bears, and

$x_3$ = number of dolls manufactured.

   a. Calculate and intepret (10, 25, 100)

   b. Interpret each of the numbers in the cost function $C$.

---

If additional practice is needed:

Barnett and Ziegler, pages 56–57, problems 13–30, 37–56
Budnick, page 57, problems 21–27, page 65, problems 22–31
Hoffmann, pages 11–12, problems 34–58
Piascik, pages 17–18, problems 6–12

# UNIT 3

---

# Graphing Linear Functions

The purpose of this unit is to provide you with an understanding of how to graph a linear function of one variable. When you have finished the unit, you will be able to recognize a linear function and be able to graph it quickly and accurately.

---

Definition:   The **graph of a function** $f$ is the set of all points whose coordinates $(x, y)$ satisfy the function $y = f(x)$.

---

Traditionally the horizontal axis is labeled as the independent variable.

The vertical axis is labeled as the dependent variable.

## LINEAR FUNCTIONS

---

Definition:   $y = f(x) = mx + b$, with $m$ and $b$ being real numbers, $m$ and $b$ not both zero, is called a **linear function** involving one independent variable $x$ and a dependent variable $y$. Sometimes this is shortened to read a linear function of one variable, meaning there is only one independent variable.

---

In other words, a linear function of one variable has:

1.   One independent and one dependent variable.

2.   Neither the independent nor dependent variable raised to a power greater than 1.

3.   Neither the independent nor dependent variable appearing in any denominator.

4.   No term containing a product of the independent and dependent variables.

Here are some examples of linear functions of one variable:

$$y = f(x) = 3x - 1$$

$$y = g(x) = 1 + 5x$$

$$y = h(x) = -4x$$

$$q = F(p) = 1{,}743 + 0.03p$$

$$y = G(x) = 25$$

Here are some examples that are *not* linear functions of one variable:

$$y = f(x) = x^{-4} + 3x$$

$$z = g(x, y) = 5xy$$

$$y = h(x) = x^2 - x + 1$$

$$y = N(x) = 1/(x - 2)$$

$$w = G(x, y, z) = 3x + 2y - z$$

Before proceeding, determine why each of the above is *not* a linear function of one variable.

# GRAPHING

The graph of a linear function is a straight line.

There are three basic procedures for graphing linear functions: plotting points, using intercepts, or graphing with slope-intercept.

## Plotting Points

To graph a linear function by plotting points, locate three points whose coordinates satisfy the equation and connect them with a straight line. Actually only two points are needed; the third point serves as a check.

To locate each point:

1. Select some convenient value for $x$. Any value will do because the domain is the set of all real numbers.

2. Substitute this value into the function.

3. Solve for $y$.

## EXAMPLE 1

Graph:  $y = f(x) = -\dfrac{1}{2}x + 6.$

Solution:   $y = f(x) = -\dfrac{1}{2}x + 6$ is a linear function; therefore its graph will be a straight line.

Locate the first point:

1.  Select a convenient value for $x$.   Let $x = 2$.

2.  Substitute into the function.             $y = f(2)$

3.  Solve for $y$.                                    $= -\dfrac{1}{2}(2) + 6$

$= -1 + 6$

$= 5$

Thus $(2, 5)$ is a point on the graph of the function because it satisfies the equation:

$$5 \underset{\underline{?}}{} -\frac{1}{2}(2) + 6$$

$$5 = 5$$

Repeat the procedure to find a second point.

1.  Select a value for $x$.             Let $x = 0$.

2.  Substitute in the function.       $y = f(0)$

3.  Solve for $y$.                              $= -\dfrac{1}{2}(0) + 6$

$= 6$

Thus $(0, 6)$ is a second point on the graph because $(0,6)$ satisfies the equation:

$$6 \underset{\underline{?}}{} -\frac{1}{2}(0) + 6$$

$$6 = 6$$

Repeat the procedure to find a third point.

1.  Select a value for $x$.                   Let $x = 4$.

2.  Substitute into the function.          $y = f(4)$

3.  Solve for $y$.                                 $= -\dfrac{1}{2}(4) + 6$

$= 4$

Thus $(4, 4)$ is a third point on the graph.

Plot the three points and connect with a straight line.

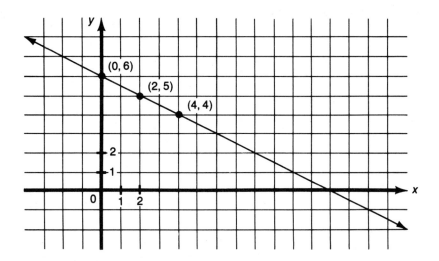

Answer:

Note that there was nothing special about the three values selected for $x$. The domain of the function was the set of all reals so any values could have been selected. It often happens that $x = 0$ is a convenient value to use. The other two were selected so that $y$ would result in an integral value.

The point where the line crosses the $y$-axis is called the **$y$-intercept**. In other words, the $y$-intercept is the value of $y$ when $x = 0$. In the above example the $y$-intercept is 6.

The point where the line crosses the $x$-axis is called the **$x$-intercept**. The $x$-intercept is the value of $x$ when $y = 0$.

# EXAMPLE 2

Find the $x$-intercept for $y = f(x) = -\dfrac{1}{2}x + 6$.

Solution:   The $x$-intercept is the value of $x$ when $y = 0$.

The equation is      $y = -\dfrac{1}{2}x + 6$

By substitution:      $0 = -\dfrac{1}{2}x + 6$

Clear of fractions:   $0 = -x + 12$

$x = 12$

The $x$-intercept is 12.

Refer to the graph of Example 1 to verify that the line crosses the $x$-axis at 12.

Do not let the words linear function confuse you.

Example 1 was the explicit function $\qquad y = f(x) = -\dfrac{1}{2}x + 6.$

If it is rewritten as an implicit equation $\qquad y = -\dfrac{1}{2}x + 6$

$$2y = -x + 12$$

we have $\quad x + 2y = 12,$

which is a linear equation in two variables. A linear equation in two variables is the equation of a straight line. So all along we have been talking about graphing a line by plotting three points.

Try graphing the next linear function yourself.

## Problem 1

Graph: $\quad y = f(x) = -4x + 5.$

Solution:    It is a linear function; the graph will be a straight line.
Locate three points whose coordinates satisfy the equation.

For example, let $x = 0$; then $y = f(0) =$ _____

let $x = 2$; then $y =$ _____

let $x =$ _____

Plot the three points on the graph provided, and connect them with a straight line.

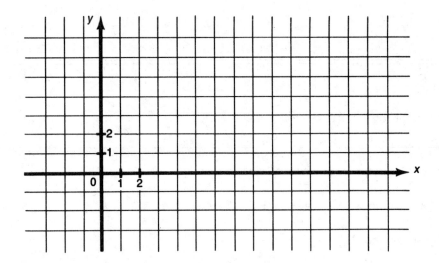

Reminder:    Be sure to put arrows at each end to indicate that the line continues on in both directions.

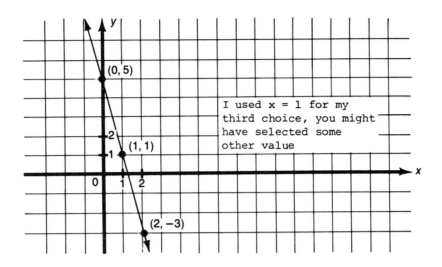

Answer:

---

If you need additional practice or explanation on plotting points to graph a line, refer to Unit 22, Graphing Linear Equations in Two Variables, of FORGOTTEN ALGEBRA.

## Using Intercepts

To graph a linear function using intercepts, locate the $x$- and $y$-intercepts and connect them with a straight line.

To locate each point:

1.  Rewrite the function as an implicit equation. Whenever possible an implicit equation should be simplified before locating the points used to graph it.
2.  Find the $x$-intercept, which is the value of $x$ when $y = 0$.
3.  Find the $y$-intercept, which is the value of $y$ when $x = 0$.

## EXAMPLE 3

Graph:   $y = f(x) = \dfrac{3}{5}x + 3$ using intercepts.

---

Solution:   1.   Rewrite the function as an implicit equation.

$$y - \frac{3}{5}x = 3$$

Simplify the equation by clearing of fractions first.

$$5y - 3x = 15$$

2.  Find the $x$-intercept, which is the value of $x$ when $y = 0$.

    By substitution:                        $5(0) - 3x = 15$

                                                $-3x = 15$

                                                  $x = -5$

    The $x$-intercept is $-5$ or the point $(-5, 0)$.

3.  Find the $y$-intercept, which is the value of $y$ when $x = 0$.

    By substitution:                        $5y - 3(0) = 15$

                                                  $5y = 15$

                                                   $y = 3$

    The $y$-intercept is $3$ or the point $(0, 3)$.

    Plot the two points and connect with a straight line.

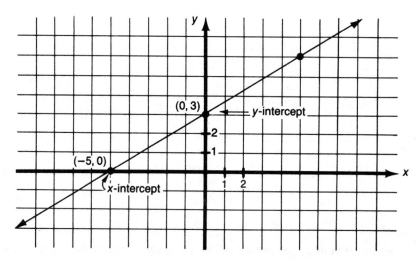

Answer:

---

The danger with using intercepts to graph a line is that you are only locating two points. There is no third point serving as a check. It is easy to make a mistake. I recommend finding a third point just to be safe.

# SLOPE

Before continuing on with the third procedure for graphing linear functions, we need to introduce the word slope. One of the most important characteristics of a line is its slope.

> Definition:  **Slope of a line** $= m = \dfrac{\text{change in } y}{\text{change in } x} = \dfrac{\Delta y}{\Delta x}$

$\Delta$ is the Greek letter delta. $\frac{\Delta y}{\Delta x}$ is read "delta $y$ over delta $x$". It means the change in $y$ over the change in $x$.

Think of the slope of a line as a rate of change. It is the rate at which $y$ changes with respect to $x$.

A fundamental property of a straight line is that the slope is constant.

It can be shown that if $y$ is written as a function of $x$, the coefficient of $x$ is the slope of the line and the constant term is the $y$-intercept.

---

Definition: $y = f(x) = mx + b$ is said to be in **slope-intercept form**. $m$ is the slope of the line and $b$ is the $y$-intercept.

---

# EXAMPLE 4

Identify the slope and $y$-intercept for $y = f(x) = 2x + 3$.

---

Solution: $y = f(x) = mx + b$ is slope-intercept form.

$y = f(x) = 2x + 3$ is written in slope-intercept form. Therefore the coefficient of $x$ is the slope and the constant term is the $y$-intercept. $y = f(x) = 2x + 3$

slope    $y$-intercept

The slope of the line is 2.

The $y$-intercept is 3.

---

You are probably wondering, so the slope is 2, what does that mean?

It might be clearer if we rewrite 2 as $\frac{2}{1}$ and think of the slope as a rate of change, $m = \frac{\Delta y}{\Delta x} = \frac{2}{1}$. Usually the independent variable is read first. The interpretation of this slope would be: if $x$ increases by 1, the $y$ value will increase by 2.

Graphically an interpretation of the slope $m = \frac{\Delta y}{\Delta x} = \frac{2}{1}$ is that you can start at any point on the line and as $x$ increases by 1 (moves one unit to the right), and $y$ increases by 2 (moves two units up), the new point also is located on the line. This ratio remains constant all along the line. It makes no difference where you start on the line. Each time you let $x$ increase by 1 and $y$ increase by 2, the new point also will be located on the line. Several examples are provided in the following solution.

---

## EXAMPLE 5

Graph:   $y = f(x) = 2x + 3$.

Solution:

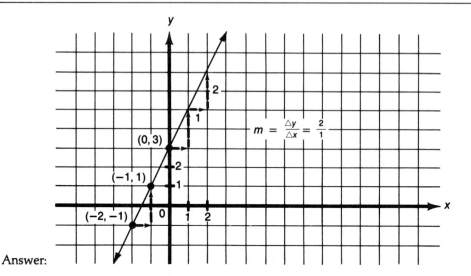

Answer:

In this example the slope was a positive number. A **positive** slope means that as $x$ increases, $y$ increases. Stated another way, as you move from left to right, the line is rising.

Before proceeding, we need to comment on the sign of a fraction.

$$\frac{-a}{b} = \frac{a}{-b} = -\frac{a}{b}$$

These three are equivalent fractions. The negative sign may be placed either in the numerator, in the denominator, or in front of the entire fraction. My personal choice is to always put the negative sign in the numerator.

## EXAMPLE 6

Identify the slope and interpret for $y = f(x) = -\frac{1}{2}x + 4$.

Solution:   $y = f(x) = -\frac{1}{2}x + 4$ is written in slope-intercept form.

Thus the slope is the coefficient of $x$.

The slope of the line is $\frac{-1}{2}$ or $\frac{1}{-2}$ or $-\frac{1}{2}$, whichever you prefer.

There are three interpretations depending on which form you used. Each one is correct.

$\dfrac{\Delta y}{\Delta x} = \dfrac{-1}{2}$   If $x$ increases by 2, the $y$ value will *decrease* by 1.

$\dfrac{\Delta y}{\Delta x} = \dfrac{1}{-2}$   If $x$ *decreases* by 2, the $y$ value will increase by 1.

$$\frac{\Delta y}{\Delta x} = -\frac{1}{2}$$ If $x$ increases by 1, the $y$ value will *decrease* by $\frac{1}{2}$.

For the remainder of the book, my answers will be written with the negative sign being put in the numerator.

A graph will illustrate that all three of the interpretations are correct. Each one will result in points located on the line.

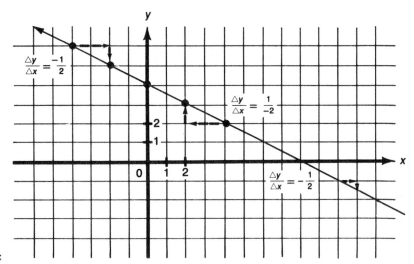

Answer:

In Example 6 the slope was a negative number. A **negative** slope means that as $x$ increases, $y$ decreases. Stated another way, as you move from left to right, the line is falling.

## Problem 2

Identify the slope and $y$-intercept for $g(x) = .53x - 2.17$.

Solution:

Answer:   The slope is .53 and the $y$-intercept is $-2.17$.

Now we are ready to use the slope and $y$-intercept for graphing linear functions.

## Graphing with Slope-intercept

To graph a linear function using slope-intercept, plot the $y$-intercept and locate two additional points using the slope. Connect the points with a straight line.

To locate the points:

1.  Rewrite the equation in slope-intercept form, if necessary.

2.  Identify the slope and $y$-intercept.

3.  Plot the $y$-intercept.

4.  **Start at the $y$-intercept**; use the slope to locate a second point.

5.  Continue from the second point; use the slope to locate a third point.

## EXAMPLE 7

Graph $y = f(x) = 3x + 1$ using the slope and $y$-intercept.

Solution:   1.   $y = f(x) = 3x + 1$ is in slope-intercept form.

slope        $y$-intercept

2.   The $y$-intercept is 1.

The slope is 3.

$$m = \frac{\Delta y}{\Delta x} = 3 = \frac{3}{1}$$

If $x$ increases by 1, the $y$ value increases by 3.

3.   Plot the $y$-intercept at 1, the first point.

4.   **Start at the $y$-intercept of 1.** Let $x$ increase by 1 (move one unit to the right), then $y$ increases by 3 (move three units up); that is your second point. Put a dot at the second point.

5.   Continue from the second point. Let $x$ increase by 1 (move one unit to the right), then $y$ increases by 3 (move three units up); that is your third point. Put a dot at the third point.

Connect the three points with a straight line.

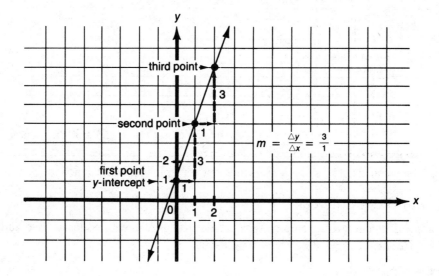

Answer:

# EXAMPLE 8

Graph $y = f(x) = -\dfrac{2}{3}x + 5$ using the slope and $y$-intercept.

Solution: 1. $y = f(x) = -\dfrac{2}{3}x + 5$ is in slope-intercept form.

slope    $y$-intercept

2. The $y$-intercept is 5.

The slope is $-2/3$.

$$m = \frac{\Delta y}{\Delta x} = \frac{-2}{3}$$

If $x$ increases by 3, the $y$ value will *decrease* by 2.

3. Plot the $y$-intercept of 5, the first point.

4. **Start at the y-intercept of 5.** Let $x$ increase by 3 (move three units to the right), then $y$ decreases by 2 (move two units down). Put a dot at the second point.

5. Continue from the second point. Let $x$ increase by 3 (move three units to the right), then $y$ decreases by 2 (move two units down). Put a dot at the third point.

Connect the three points with a straight line.

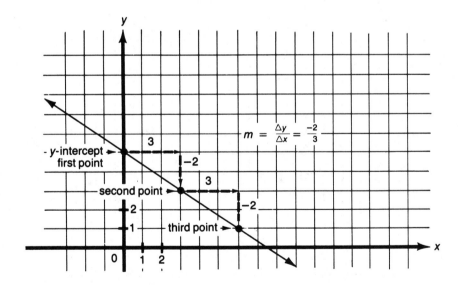

Answer:

# EXAMPLE 9

Graph $y = f(x) = \dfrac{x}{5} - 2$ using the slope and $y$-intercept.

Solution:   Rewrite $y = f(x) = \dfrac{x}{5} - 2$ in slope-intercept form.

$$y = f(x) = \frac{1}{5}x - 2$$

slope      $y$-intercept

The $y$-intercept is $-2$ and the slope is $1/5$.

$$m = \frac{\Delta y}{\Delta x} = \frac{1}{5}$$

If $x$ increases by 5, the $y$ value will increase by 1.

Plot the $y$-intercept.

**Start at the $y$-intercept.** Use the slope to locate a second point. Continue from the second point. Use the slope to locate a third.

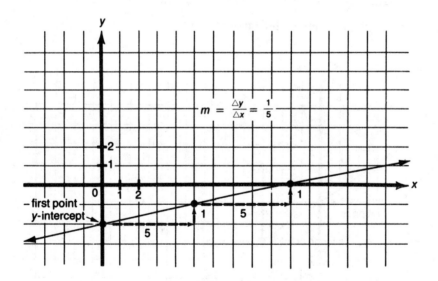

Answer:

Now try to graph a linear function yourself using this procedure.

# Problem 3

Graph $y = f(x) = 5x - 4$ using the slope and $y$-intercept.

Solution:   $y = f(x) = 5x - 4$ is in slope-intercept form.

The $y$-intercept is _____.

The slope is

$$m = \frac{\Delta y}{\Delta x} = \underline{\hspace{3cm}}.$$

If $x$ increases by 1, the $y$ value will _____.

Plot the $y$-intercept on the graph provided, and find two additional points using the slope. Connect the points with a straight line.

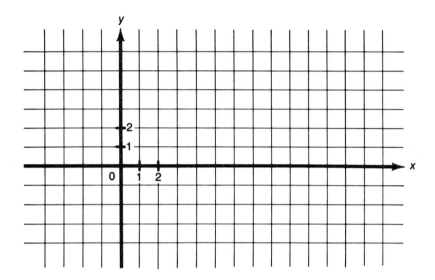

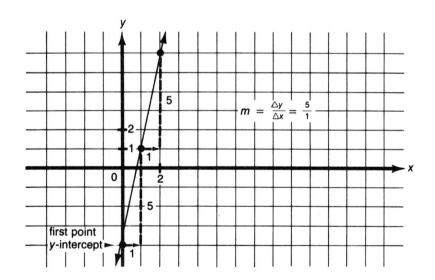

Answer:

Try the next graph without any clues.

## Problem 4

Graph $y = f(x) = -\dfrac{4}{3}x + 3$ using the slope and $y$-intercept.

Solution:

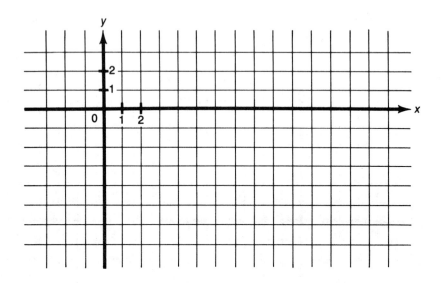

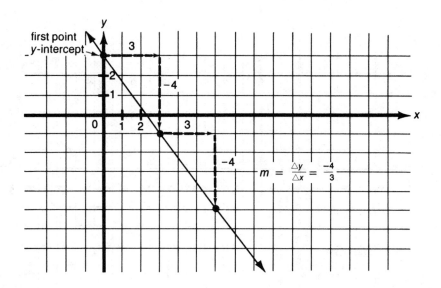

Answer:

The next two examples illustrate what happens when either $m$ or $b$ are equal to zero

# EXAMPLE 10

Graph $y = f(x) = 2x$ using the slope and $y$-intercept.

Solution:   Recall that a linear function is of the form $y = f(x) = mx + b$ with $m$ and $b$ not both zero.

Compare: $y = f(x) = \boxed{m}\ x + \boxed{b}$

$y = f(x) = \boxed{2}\ x\quad \boxed{\phantom{0}}$   so $m = 2$ and $b = 0$.

Therefore $y = f(x) = 2x$ is classified as a linear function; its graph will be a straight line with slope of 2 and $y$-intercept of 0.

$$m = \frac{\Delta y}{\Delta x} = \frac{2}{1}$$

Plot the $y$-intercept at 0.   Start at 0 (the $y$-intercept), locate two additional points using the slope of 2.

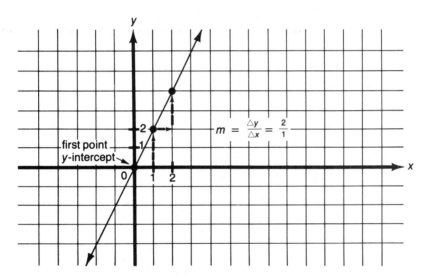

Answer:

# EXAMPLE 11

Graph $y = f(x) = 7$ using the slope and $y$-intercept.

Solution:   By a similar line of reasoning,

compare:   $y = f(x) = \boxed{m}\ x + \boxed{b}$

$y = f(x) = \boxed{\phantom{0}}\quad \boxed{7}$   so $m = 0$ and $b = 7$.

Thus $y = f(x) = 7$ is also a linear function; its graph will be a straight line with slope equal to 0 and $y$-intercept of 7.

$$m = \frac{\Delta y}{\Delta x} = 0 = \frac{0}{1}$$

Think of a slope of 0 as meaning if $x$ increases by 1, the $y$ value does *not* change.

The $y$-intercept is plotted at 7.

**Start at the $y$-intercept of 7**. If $x$ increases by 1, the $y$ value does *not* change and remains at 7. There is your second point.

Locate the third point yourself.

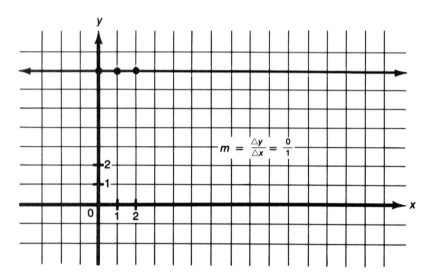

Answer:

Observe:    The graph of $y = f(x) = 7$ is a horizontal line.
Each point on the line has a $y$-coordinate of 7.

The $y$-intercept is 7.

There is no $x$-intercept.

---

In this example the slope was zero. A **zero** slope means that as $x$ increases, $y$ remains the same. Stated another way, the graph of the line with a slope of zero is horizontal.

> The graph of a linear function $y = f(x) = p$, where $p$ is a real number, is a **horizontal line** with $y$-intercept of $p$. Its slope is 0.

## EXAMPLE 12

Graph $y = f(x) = -2$.

---

Solution:   The graph of the linear function $y = f(x) = -2$ is a horizontal line with $y$-intercept of $-2$.

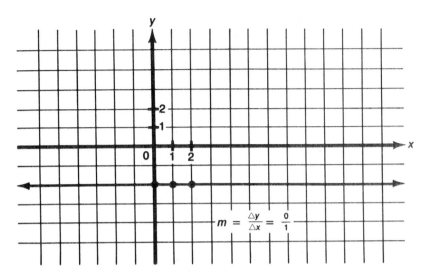

Answer:

---

The advantage of graphing linear functions using the slope and $y$-intercept should be obvious. Most linear functions are written in slope-intercept form to start with thus making the slope and $y$-intercept easily identified. Starting with the $y$-intercept, three points can then be quickly plotted with a minimum amount of work and the line drawn.

The last example is a function in terms of $p$ and $q$.

# EXAMPLE 13

Graph $q = D(p) = -2p + 8$.

Identify the slope and interpret.

---

Solution:   Recall if $q = D(p) = -2p + 8$

dependent     independent
variable       variable

Since the horizontal axis is always the independent variable, the horizontal axis will be labeled as $p$, with the vertical axis being labeled $q$.

Compare:   $y = f(x) = \boxed{m}\, x + \boxed{b}$

$q = D(p) = \boxed{-2}\, p + \boxed{8}$   so $m = -2$ and $b = 8$.

Thus $q = D(p) = -2p + 8$ is a linear function; its graph will be a straight line.

The $y$-intercept is 8. It is easier to refer to 8 as the $y$-intercept even though the vertical axis is labeled $q$ in this situation.

The slope is $-2$ or $m = \dfrac{\Delta y}{\Delta x} = \dfrac{-2}{1} = \dfrac{\Delta q}{\Delta p}$

To interpret a slope of $-2$, remember that the independent variable is read first; if $p$ increases by 1, the $q$ value will *decrease* by 2.

Answer:

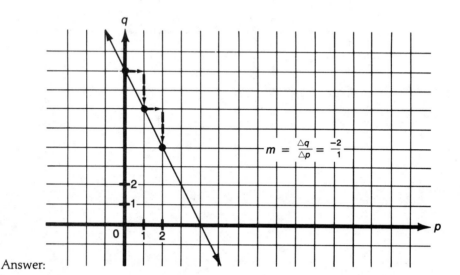

You should now be able to recognize a linear function of one variable and graph it using one of three methods: plotting points, locating the $x$- and $y$-intercepts, or using the slope and $y$-intercept.

To plot points:

   select three convenient values for $x$;

   substitute these values into the function and solve for $y$;

   connect the three points with a straight line.

To graph using the $x$- and $y$-intercepts:

   locate the $y$-intercept (the value of $y$ when $x = 0$);

   locate the $x$-intercept (the value of $x$ when $y = 0$);

   connect the two points with a straight line.

To graph using the slope and $y$-intercept:

   locate the $y$-intercept;

   start at the $y$-intercept, use the slope to locate two additional points;

   connect the three points with a straight line.

Also, given any linear function of one variable, you should be able to identify the slope and both the $x$- and $y$-intercepts, if they exist.

Before beginning the next unit you should graph the following functions.

# EXERCISES

Identify the slope and $y$-intercept and graph each function.

1. $y = f(x) = 2x + 5$

2. $y = f(x) = \dfrac{7}{3}x - 1$

3. $y = \dfrac{2}{5}x + 3$

4. $f(x) = x - 4$

5. $g(x) = -\dfrac{3}{2}x + 7$

6. $h(x) = -x + 2$

7. $y = f(x) = \dfrac{x}{3} + 1$

8. $y = f(x) = .25x + 1.50$  Hint: $.25 = \dfrac{25}{100} = \dfrac{1}{4}$

9. $q = D(p) = -3p + 11$

10. Given: $x - 3y = -12$

    a. Rewrite in slope-intercept form, $y = f(x)$.

    b. Identify and interpret the slope.

    c. Find both the $x$- and $y$-intercepts.

    d. Graph.

---

If additional practice is needed:

Barnett and Ziegler, page 56, problems 31–36
Hoffmann, page 34, problems 6–15
Piascik, page 48, problem 2

# UNIT 4

## Writing Linear Functions

The purpose of this unit is simply the reverse of Unit 3. Given the graph of a linear function, we want to be able to write its equation. When you have finished the unit, you will be able to write a linear function that describes the relationship between the independent and dependent variables.

There are several approaches for writing linear functions. We will use only one, the point-slope formula.

---

Point-slope Formula

$y - y_1 = m(x - x_1)$ is the **point-slope formula** for a line with slope $m$ and containing the point $(x_1, y_1)$.

---

The point-slope is used to write the equation of a line. In order to use the formula, two items of information are required: the slope of the line and a given point $(x_1, y_1)$ on the line.

## EXAMPLE 1

Write the linear function containing the point (1, 2) with slope 3.

---

Solution:  The graph of a linear function is a straight line.

To write the equation of a line, use the point-slope formula.

In order to use the formula, we need two items, the slope and a given point on the line. The slope was given as 3, so $m = 3$. The line was to contain the point (1, 2), so $(x_1, y_1) = (1, 2)$.

Use point-slope formula.    $y - y_1 = m(x - x_1)$

Substitute.    $y - 2 = 3(x - 1)$

Remove parentheses.    $y - 2 = 3x - 3$

Solve for $y$.    $y = 3x - 1$

So the answer is:    $y = f(x) = 3x - 1$

## Problem 1

Graph $y = f(x) = 3x - 1$. Verify that the slope is 3 and that the line contains the point $(1, 2)$.

Solution:

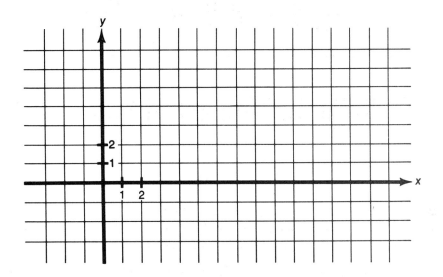

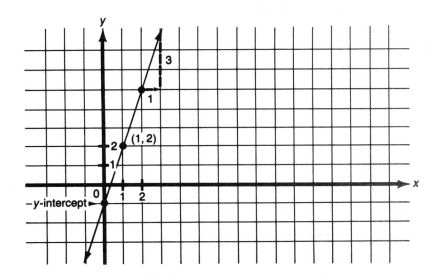

Answer:

## EXAMPLE 2

Write the linear function containing the point $(6, -1)$ with slope 2/3.

Solution:   The graph of a linear function is a straight line.

To write the equation of a line, use the point-slope formula. This time the slope is 2/3 and the given point is $(6, -1)$.

Use the point-slope formula.     $y - y_1 = m(x - x_1)$

Substitute.                             $y - (-1) = \dfrac{2}{3}(x - 6)$

Remove parentheses.              $y + 1 = \dfrac{2}{3}x - 4$

Solve for $y$.                            $y = \dfrac{2}{3}x - 5$

So the answer is:                      $y = f(x) = \dfrac{2}{3}x - 5$

Before continuing, let's make sure you are able to write a linear function yourself.

## Problem 2

Write the linear function containing the point $(-10, 2)$ with slope 1/2.

Solution:   Use the point-slope formula.     $y - y_1 = m(x - x_1)$

Answer:   $y = f(x) = \dfrac{1}{2}x + 7$

To summarize what has been said so far:

$y - y_1 = m(x - x_1)$ is the point-slope formula. It is used to write the equation of a line, given the slope and a point $(x_1, y_1)$ on the line.

The point-slope formula is not always the fastest method for writing linear functions, but I like it because it works for all situations. That way you need to remember only the one formula. It is the only method we will use in the book.

# CALCULATING THE SLOPE

Thus far we always have been given the slope. What happens when that is not the case? First, the definition of the slope of a line needs to be expanded.

Given two points on a line such as $P(x_1, y_1)$ and $Q(x_2, y_2)$, the slope is found by taking the difference of the $y$-coordinates divided by the difference of the $x$-coordinates. In symbols it would look like this.

$$\text{Definition:} \quad \text{Slope of a line} = m = \frac{\Delta y}{\Delta x} = \frac{y_2 - y_1}{x_2 - x_1}$$

## EXAMPLE 3

Use the definition to find the slope of the line connecting the points $(0, 1)$ and $(2, 5)$.

Solution:   Let $(0, 1)$ be the first point, therefore $x_1 = 0$ and $y_1 = 1$.

Then $(2, 5)$ is the second point and $x_2 = 2$ and $y_2 = 5$.

By substituting into the definition,

$$m = \frac{\Delta y}{\Delta x} = \frac{y_2 - y_1}{x_2 - x_1} = \frac{5 - 1}{2 - 0} = \frac{4}{2} = 2$$

The slope of the line is 2.

The points $(0, 1)$ and $(2, 5)$ have been plotted on the grid below and the line drawn connecting them. From the previous unit, remember that if the slope is 2, $m = \frac{\Delta y}{\Delta x} = 2 = \frac{2}{1}$. If we start at any point on the line and as $x$ increases by 1 (moves one unit to the right), and $y$ increases by 2 (moves two units up), the new point is located on the line. This ratio remains constant all along the line. It makes no difference where you start. Each time $x$ increases by 1 and $y$ increases by 2, another point will be located on the line. Several examples are provided.

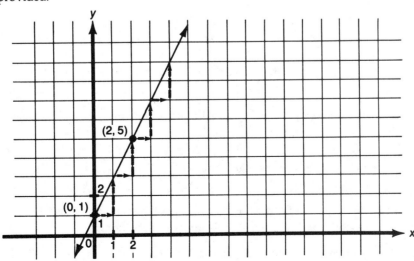

## EXAMPLE 4

Find the slope of the line connecting the points (2, 2) and (0, 3).

Solution:   Let (2, 2) be the first point, therefore $x_1 = 2$ and $y_1 = 2$.

Then (0, 3) is the second point with $x_2 = 0$ and $y_2 = 3$.

By substituting into the definition,

$$m = \frac{\Delta y}{\Delta x} = \frac{y_2 - y_1}{x_2 - x_1} = \frac{3 - 2}{0 - 2} = \frac{1}{-2}$$

The slope of the line is $1/-2$, $-\dfrac{1}{2}$, or as I prefer to write it, $-1/2$.

One of the troublesome areas about finding the slope of a line is remembering to subtract the coefficients in the same order. It makes no difference which of the two points is specified as the first and which is the second, but the same designation must be used for calculating both $\Delta x$ and $\Delta y$.

## EXAMPLE 5

Find the slope of the line connecting the points $(-3, 5)$ and $(7, -2)$.

Solution:   Let $(-3, 5)$ be the first point and $(7, -2)$ be the second point.

Then

$$m = \frac{\Delta y}{\Delta x} = \frac{y_2 - y_1}{x_2 - x_1} = \frac{(-2) - 5}{7 - (-3)} = \frac{-2 - 5}{7 + 3} = \frac{-7}{10}$$

The slope of the line is $-7/10$.

Had we let $(7, -2)$ be the first point and $(-3, 5)$ be the second point, the result would have been the same.

$$m = \frac{\Delta y}{\Delta x} = \frac{y_2 - y_1}{x_2 - x_1} = \frac{5 - (-2)}{(-3) - 7} = \frac{5 + 2}{-3 - 7} = \frac{7}{-10}$$

This way the slope is calculated to be $7/-10$, but that equals $-7/10$.

Unfortunately, students often want to simply take the larger number minus the smaller number. Using the numbers from Example 5, a common error is to take $5 - (-2) = 7$ and $7 - (-3) = 10$ and conclude that the slope is a positive $7/10$. Doing so disregards which point is considered the first and all too frequently yields the wrong answer as it would in this case.

An alternative method for finding the slope is to actually subtract the coordinates of the points as they are written as ordered pairs. This eliminates the necessity for remembering which is to be the first point and which is to be the second point for substituting into the definition.

The next example will be worked using this alternative method. I think you will find it much easier.

# EXAMPLE 6

Find the slope of the line connecting the points $(0, -1)$ and $(-5, 3)$.

Solution:  Subtract the coordinates of the points as written. Remember the rule for subtracting integers is to change the sign of the bottom number and then treat the problem as addition.

$$(0, -1)$$
$$(-5, \quad 3)$$

$$5, -4$$

This is the change in $x$.        This is the change in $y$.

$$m = \frac{\Delta y}{\Delta x} = \frac{-4}{5}$$

Isn't that a neat way to find the slope?

It is time to return to our original purpose, that of writing linear functions.

# EXAMPLE 7

Write the linear function containing the points $(0, -1)$ and $(-5, 3)$.

Solution:  The graph of a linear function is a straight line.

To write the equation of a line, we use the point-slope formula.

In order to use the formula, we need to know the slope and a given point on the line.

From Example 6, the slope was found to be $-4/5$.

Two points on the line are given; select one to use in the formula. I chose to use $(0, -1)$, but it would not make any difference which one was selected.

Use formula.        $y - y_1 = m(x - x_1)$

Substitute.          $y - (-1) = \frac{-4}{5}(x - 0)$

Remove parentheses.    $y + 1 = \frac{-4}{5}x$

Solve for $y$.          $y = \frac{-4}{5}x - 1$

So the answer is:      $y = f(x) = \frac{-4}{5}x - 1$

Suppose we had selected the second point $(-5, 3)$ to use in the formula. Verify that the answer would be the same.

## Problem 3

Write the linear function containing the point $(-5, 3)$ with slope of $-4/5$.

Solution:

Answer:   $y = f(x) = \dfrac{-4}{5}x - 1$

The procedure then for writing a linear function when given two points on the line is:

1.  Use the two points to first calculate the slope.

2.  Select one of the points to be used as the given point on the line. Either of the two points may be used; select the simplest.

3.  Use the point-slope formula to write the equation of the line.

Now try a few problems yourself.

## Problem 4

Find the slope of the line connecting the points $(5, 8)$ and $(4, 3)$.

Solution:

Answer:   $m = 5$

## Problem 5

Write the linear function containing the points $(5, 8)$ and $(4, 3)$.

Solution:

Answer:   $y = f(x) = 5x - 17$

## Problem 6

Find the slope of the line connecting the points (5, 1) and (−3, −4).

Solution:

Answer:   $m = 5/8$

## Problem 7

Write the linear function containing the points (5, 1) and (−3, −4).

Solution:

Answer:   $y = f(x) = \dfrac{5}{8}x - \dfrac{17}{8}$

## EXAMPLE 8

The lines connecting the points from Problems 4 and 6 are shown on the graph. Verify, by starting at any point and moving to a new point, that the slope of the first line is 5 and the slope of the second line is 5/8.

Solution:

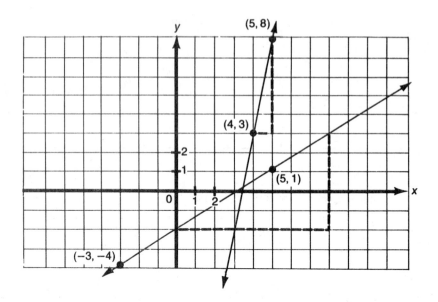

Answer:

Observe that both lines have a positive slope. Graphically a positive slope means that as you move left to right, the line is rising. Stated in symbols, if $m > 0$, the line is rising.

Also notice that the first line, with a slope of 5, is rising much faster than the second line with a slope of 5/8. Think of the comparison this way: if $x$ increases by 1, the $y$ value of the first line would increase by 5 whereas the $y$ value of the second line would only increase by 5/8.

The larger the value of $m$, the faster the line will rise. Visualize what a line would look like going through the origin with a slope of 3 and a second line, also going through the origin, but with a slope of 100. Which line will be rising faster or, stated another way, which is the steeper line?

Let's continue with writing linear functions.

## Problem 8

Write the linear function containing the points $(3, -2)$ and $(1, 1)$.

Solution:

Answer:  $y = f(x) = \dfrac{-3}{2} x + \dfrac{5}{2}$

## Problem 9

Write the linear function containing the points $(-4, 5)$ and $(-1, 2)$.

Solution:

Answer: $y = f(x) = -x + 1$

## Problem 10

The lines connecting the points from Problems 8 and 9 are shown on the graph. Verify, by starting at any point and moving to the next point, that the slope of the line from Problem 8 is $-3/2$ and the slope of the line from Problem 9 is $-1$.

Solution:

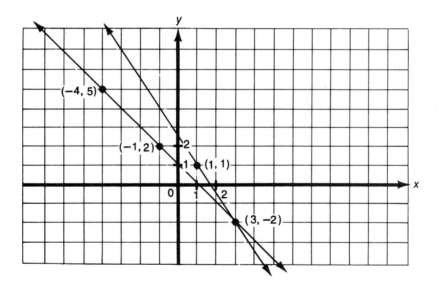

This time both lines have a negative slope or, as you move from left to right, the lines are falling. Stated in symbols, if $m < 0$, the line is falling.

Think what a line would look like with a slope of $-50$. The more negative the slope, the faster the line falls.

The next problem is easy, but be careful.

## Problem 11

Write the linear function containing the points (3, 4) and ( −2, 4).

Solution:

Answer:   $y = f(x) = 4$

## EXAMPLE 9

Verify by graphing, that the slope of the line connecting the points (3, 4) and ( −2, 4) is 0.

Solution:

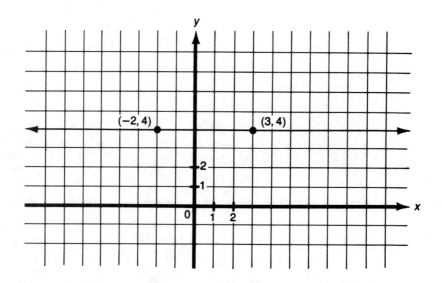

Starting at any point on the line, as $x$ increases by 1, the $y$ value does not change, but remains at 4.

In symbols, if $m = 0$, the line is horizontal.

The next example is a special situation.

# EXAMPLE 10

Write the linear function containing the points (3, 7) and (3, 5) and graph.

Solution:   First use the two points to calculate the slope.    (3, 7)

$$\underline{(3, 5)}$$

0, 2

$$\text{Then } m = \frac{\Delta y}{\Delta x} = \frac{2}{0} = \text{undefined}$$

As I am sure you remember, division by zero is undefined. Therefore since the slope, $m$, in this example is undefined, the point-slope formula cannot be used.

Maybe the graph will help.

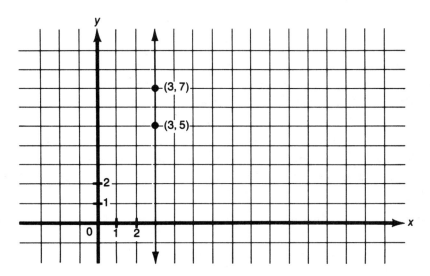

Observe:   It is not a function.

The graph is a vertical line.

The $x$-intercept is 3.

There is no $y$-intercept.

Each point on the line has an $x$-coordinate of 3.

There is no linear function that can be written containing the points (3, 7) and (3, 5).

In symbols, if $m$ is undefined, the line is vertical.

However, even though the line in Example 10 is not a function, it is possible to write an equation for the line.

> The equation of a **vertical line** through a given point $(r, p)$ is $x = r$.

The graph of a linear equation $x = r$, where $r$ is a real number, is a vertical line with $x$-intercept of $r$. Its slope is undefined.

# EXAMPLE 11

Write the equation of a vertical line through the point (3, 5).

Solution:   The equation of a vertical line through the given point, (3, 5), is $x = 3$.

# EXAMPLE 12

Write the equation of the line connecting the points $(-2, 5)$ and $(-2, 13)$.

Solution:   Use the two points to calculate the slope.

$$\begin{array}{rr} (-2, & 5) \\ (-2, & 13) \\ \hline 0, & -8 \end{array}$$

$$m = \frac{\Delta y}{\Delta x} = \frac{-8}{0} = \text{undefined}$$

Since the slope of the line connecting the points is undefined, the line must be vertical. The equation of the vertical line through the point $(-2, 5)$ is $x = -2$.

You should now be able to write a linear function to describe the relationship between variables if you are given either the slope and one point on the line or two points on the line. If you are given the two points, the slope must be calculated first. In either situation, the point-slope formula would be used to write the linear function.

To generalize the relationship between the slope and the graph of a line, always moving from left to right,

if $m > 0$, the line is rising,

if $m < 0$, the line is falling,

if $m = 0$, the line is horizontal,

if $m$ is undefined, the line is vertical.

A vertical line is a special case. It is not a function, but the equation of the line can be written as $x = r$, where the line contains the point $(r, p)$.

Before beginning the next unit you should write a linear function, given the information in each problem.

## EXERCISES

Find a linear function, if possible, relating $x$ and $y$, with the given properties. Write the final answer in slope-intercept form, $y = mx + b$.

1.  Line passes through $(4, -3)$ with slope 5.

2.  Line passes through $(6, 7)$ with slope $-\dfrac{1}{6}$.

3.  Line passes through $(7, 5)$ and $(-3, 5)$.

4.  Slope equals 0.15 and the point $(2.35, 100)$ lies on the line.

5.  $(-5, -2)$ and $(-3, -8)$ lie on the line.

6.  The line contains the points $(400, 200)$ and $(3,600, 1,000)$.

7.  A vertical line passing through $(-9, 5)$.

8.  The line through $(2.75, 300)$ and $(3.95, 435)$.

9.  A line passing through $(6, 4)$ and parallel to $y = 5x + 2$. Hint: recall that if two nonvertical lines are *parallel*, then they have the same slope. Or, if $m$ is the slope of one line, any line parallel to it must also have a slope of $m$.

10. The line through $(-7, 2)$ and perpendicular to $y = 3x - 1$. Hint: Recall that if two nonvertical lines are *perpendicular*, then their slopes are negative reciprocals of each other. Stated another way, if $m$ is the slope of one line, any line perpendicular to it would have a slope of $-1/m$.

If additional practice is needed:

Barnett and Ziegler, page 37, problems 23–50
Budnick, page 103, problems 1–23
Hoffmann, page 35, problems 1–5, 17–27
Piascik, page 48, problems 3–12

# UNIT 5

## Linear Applications

In this unit we will consider several applications of linear functions. The emphasis will be not only on the mathematical answer, but also the interpretation of the answer in terms of the problem.

## DEMAND FUNCTIONS

**Demand functions** describe the relationship between the demand for a product and its price. Suppose that $p$ denotes the price of a certain product. Let $D$ denote the demand function. Then $q = D(p)$ is the number of units of the product that consumers are willing to buy when the market price is $p$, or quantity demanded $= D$(market price).

## EXAMPLE 1

In a small farm town, the demand function for Margaret's homemade cakes is defined by

$$q = D(p) = -2p + 10 \qquad \text{with } 0 \leq p \leq 5$$

$$\text{where } p = \text{price in dollars per cake}$$

$$\text{and} \quad q = \text{number of cakes sold at price } p.$$

a. Calculate the demand for Margaret's cakes if the price is $3 per cake.

b. Calculate and interpret $D(1)$.

Solution:   a.   $q = D(p) = -2p + 10$

$q = D(3) = -2(3) + 10$

$= -6 + 10$

$= 4$

If the price of a cake is $3, Margaret will sell 4 cakes.

b.   $q = D(p) = -2p + 10$

$q = D(1) = -2(1) + 10$

$= -2 + 10$

$= 8$

If the price of a cake is $1, Margaret will sell 8 cakes.

# EXAMPLE 2

Graph:   $q = D(p) = -2p + 10$ with $0 \leq p \leq 5$.

Solution:   $q = D(p) = -2p + 10$ is a linear function and usually its graph is a straight line. But this function has a restricted domain; so, instead of being the entire line, its graph will be only a line segment. The line segment will start at the lower limit of the restricted domain, $p = 0$, and end at the upper limit of the restricted domain, $p = 5$.

Lower limit:   Let $p = 0$, then $q = D(0) = -2(0) + 10 = 10$

The starting point of the line segment is $(0, 10)$.

Upper limit:   Let $p = 5$, then $q = D(5) = -2(5) + 10 = 0$.

The endpoint of the line segment is $(5, 0)$.

The notation $D(p)$ indicates that $p$ is to be the independent variable; therefore, the horizontal axis is labeled $p$ and $q$ is on the vertical axis.

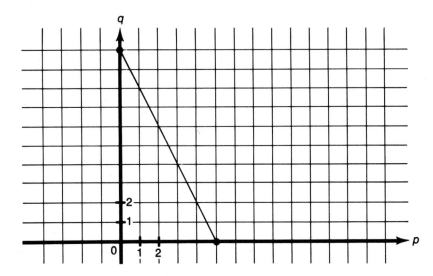

Answer:

## EXAMPLE 3

Interpret the ordered pair (0, 10) from Example 1.

Solution:    (0, 10) corresponds to $p = 0$ and $q = 10$.

If the price of a cake is $0, Margaret will sell 10; or, stated in more common terms, if the cakes are free, Margaret can give away only 10. Remember that this is a small town.

## EXAMPLE 4

Identify and interpret the slope from Example 1.

Solution:    The slope is $-2$ or $m = \dfrac{\Delta y}{\Delta x} = \dfrac{\Delta q}{\Delta p} = -2 = \dfrac{-2}{1}$

Remember that the independent variable is read first; if $p$ increases by 1, the $q$ value will decrease by 2. Although that is the mathematical interpretation of the slope, since this is an application problem, the definitions of $p$ and $q$ as stated in the problem should be used.

If the price in dollars per cake increases by $1, the number of cakes sold at that price will decrease by 2. Or, stated still another way, for each additional dollar increase in price, the number of cakes sold will decrease by 2.

A typical demand function will have a negative slope because as the price increases for a product, the demand for that product usually decreases. Of course there are exceptions.

Here is a problem for you to do.

## Problem 1

Use $D(p)$ from Example 1 to calculate and interpret $D(4)$, $D(5)$, and $D(15)$.

Solution:

Answers: $D(4) = 2$. If the price per cake is $4, Margaret will sell 2 cakes.

$D(5) = 0$. If the price per cake is $5, Margaret will sell 0 cakes.

From this information we can conclude that if the price per cake is $5 or more, Margaret won't sell any cakes.

$D(15)$ is undefined. At a price of $15 per cake, she won't be able to sell any cakes.

# SUPPLY FUNCTIONS

**Supply functions** relate the market price to the quantities that suppliers are willing to produce and sell. Suppose that $p$ is the price of a certain product and $S$ denotes the supply function. Then $q = S(p)$ is the number of units of this product that suppliers are willing to produce and sell when the market price is $p$, or quantity supplied $= S($market price$)$.

   In most cases, the supply function has a positive slope because as the market price increases suppliers usually are willing to produce and sell more.

## EXAMPLE 5

Based on her costs and other interests, Margaret has determined that the supply function for her cakes is given by

$$q = S(p) = 2p - 2 \qquad \text{with } p \geq 1$$

$$\text{where } p = \text{price in dollars per cake}$$

$$\text{and} \quad q = \text{number of cakes she is willing}$$
$$\text{to bake and sell at price } p.$$

How many cakes is Margaret willing to bake if she sells them for $4 each?

Solution:   The supply function is $q = S(p) = 2p - 2$

If $p = \$4$, then $\qquad q = S(4) = 2(4) - 2$

$$= 6$$

   If the market price of cakes is $4, Margaret would be willing to bake and sell 6 cakes.

## Problem 2

Calculate the supply of Margaret's cakes at $p = \$2$ per cake.

Solution:

Answer:   2

## Problem 3

Calculate and interpret $S(15)$.

Solution:

Answer:   If the market price of cakes is $15 per cake, Margaret would be willing to bake and sell 28 cakes.

# EQUILIBRIUM

**Equilibrium** occurs if there is a price at which the supply and demand are equal. Stated in symbols, equilibrium occurs where $S(p) = D(p)$.

## EXAMPLE 6

Find the equilibrium price for Margaret's cakes.

Solution:   Equilibrium occurs where $S(p) = D(p)$.

By substitution
$$2p - 2 = -2p + 10$$
$$4p = 12$$
$$p = 3$$

At a price of $3 per cake, the number of cakes supplied and demanded will be equal.

Although this was not asked for, either of the two functions can be used to find $q$:

$$q = S(p) = 2p - 2$$
$$S(3) = 2(3) - 2$$
$$= 4$$

At a price of $3 per cake, the number of cakes supplied *and* sold will be 4.

## Problem 4

Graph the supply and demand functions on the same set of axes and observe where the equilibrium point occurs.

Solution:

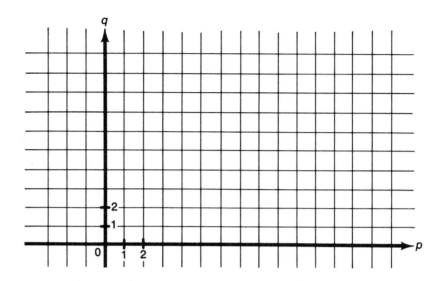

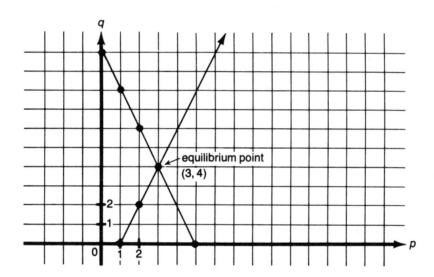

Answer:

# EXAMPLE 7

The demand function for cellular telephones at the Wygand Manufacturing Company is defined by

$$q = D(p) = 15,000 - 30p \quad \text{with } 0 \le p \le 500$$

where $p$ = the price in dollars per telephone,
and $q$ = the quantity demanded at price $p$.

Calculate and interpret $D(100)$.

Solution:   $q = D(p)$     $= 15{,}000 - 30p$

$q = D(100) = 15{,}000 - 30(100)$

$= 15{,}000 - 3{,}000$

$= 12{,}000$

If the price is $100 per telephone, the demand will be for 12,000 telephones.

## EXAMPLE 8

Identify and interpret the slope from Example 7.

Solution:   The slope is $-30$. If the price increases by $1, the demand will decrease by 30 units; or, for every dollar increase in price, the demand for the telephones will decrease by 30 telephones.

## Problem 5

If the price of a telephone *decreases* by $1, demand will increase by how many telephones?

Solution:

Answer:   30

## EXAMPLE 9

If cellular telephones are priced at $200 each, the Wygand Manufacturing Company is willing to produce 400 telephones. If the telephone's price drops to $150, the company is willing to produce only 200 telephones. If $p$ is the price per telephone and $q$ corresponds to supply, use the concept of slope to find the rate of change of supply with respect to price. Assume the relationship is linear.

Solution:   At price of $p = \$200$, quantity supplied is $q = 400$ or   (200, 400)

At price of $p = \$150$, quantity supplied is $q = 200$ or $\underline{(150, 200)}$

50, 200

The slope $= m = \dfrac{\Delta y}{\Delta x} = \dfrac{\Delta q}{\Delta p} = \dfrac{200}{50} = 4$

The rate of change of supply with respect to price is 4.

## Problem 6

If the price of a cellular telephone increases by \$1, the supply will decrease by how many telephones?

Solution:

Answer:   4

## EXAMPLE 10

Determine the supply function for cellular telephones with $q = S(p)$.

Solution:   What do we know about this supply function?

It is assumed to be linear; therefore its graph is a line.

When $p = 200$, $q = 400$ or $(200, 400)$ is a point on the line.

When $p = 150$, $q = 200$ or $(150, 200)$ is also on the line.

The slope has been calculated to be 4.

I draw pictures whenever possible. It usually makes the problem easier for me to comprehend. The graph below is a quick sketch of the information provided in the problem.

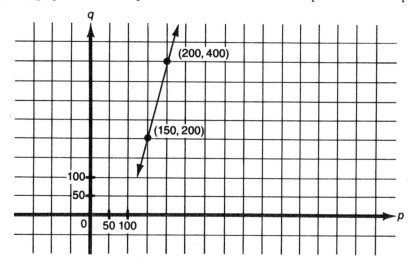

To determine the supply function, we are actually being asked to write the equation of the line. To write the equation of a line, use the point-slope formula.

$$\text{Point-slope formula:} \qquad y - y_1 = m(x - x_1)$$

$$\text{Use one point and slope:} \quad y - 400 = 4(x - 200)$$

$$y - 400 = 4x - 800$$

$$y = 4x - 400$$

Be careful; this is not the answer. The problem asked for $q = S(p)$. In other words, the answer is to be written in terms of $p$ and $q$ rather than $x$ and $y$. But that is easy. $S(p)$ indicates that $p$ is to be the independent variable. By exchanging variables the correct answer is:

$$q = S(p) = 4p - 400$$

## EXAMPLE 11

Where does the supply function from Example 10 cross the horizontal axis?

What is the economic significance of this point?

Solution:   The point where the supply function crosses the horizontal axis is the $x$-intercept. The $x$-intercept is where $y = 0$. Actually it is the $p$-intercept, but it is easier to think of the $x$-intercept.

$$y = 4x - 400$$

$$0 = 4x - 400$$

$$4x = 400$$

$$x = 100$$

The point where the supply function crosses the horizontal axis is at 100. Mathematically that means the restricted domain for the supply function is $p \geq 100$; because if $p$ were less than 100, $q$ would be negative and meaningless. The company cannot produce a negative number of telephones. The economic significance of this point is that at a price of $100 or less, the Wygand Manufacturing Company would be unwilling to produce any cellular telephones.

## EXAMPLE 12

Find the equilibrium price and the corresponding number of telephones supplied and demanded.

Solution:   Equilibrium occurs where   $S(p) = D(p)$.

By substitution        $4p - 400 = 15,000 - 30p$

$$34p = 15,400$$

$$p = 452.94$$

Use either function to determine $q$.

$$q = S(p) = 4p - 400$$

$$q = S(452.94) = 4(452.94) - 400$$

$$= 14,11.76$$

$$\approx 1,412$$

If the price of Wygand Manufacturing's cellular telephones is $452.94, the number of telephones supplied and demanded will be approximately 1,412.

Ready to try one yourself?

## Problem 7

In an attempt to forecast the demand for a particular style of running shoes, Bill noted that if the price per pair was $50, he sold only 30 pairs. But when he had a special and lowered the price to $45, 45 pairs were sold. Bill assumes there is a linear relationship between the price and the number of pairs sold.

a.  Determine $q = D(p)$ where $p$ is the price per pair of running shoes and $q$ is the number of pairs of running shoes demanded at price $p$.

A quick sketch of the information might help.

b.  Use your function to predict how many pairs of running shoes will be sold if Bill has a half-price sale and the price of the shoes is lowered to $25.

c.  Graph the function.

d.  What is the restricted domain?

e.  What is the restricted range?

Solution:

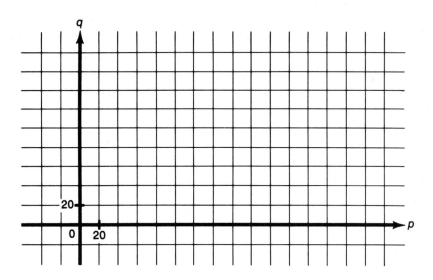

Answers:   a.   $q = D(p) = -3p + 180$

b.   If the price of shoes is $25, it is estimated that Bill will sell 105 pairs.

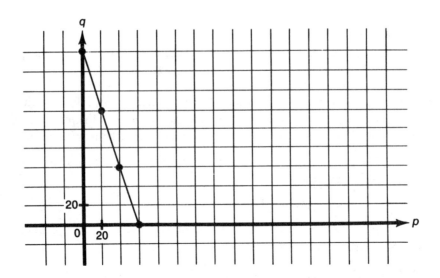

c.

d.   $0 \le p \le 60$

e.   $0 \le q \le 180$

---

Let's leave supply and demand functions for the time being and look at a few other types of applications. The objective here will be to interpret the meaning of points, slope, and intercepts in the context of the application.

---

## EXAMPLE 13

The police in a major city have released crime figures indicating reported crime was up 11.4% in May from April of this year. One police consultant has estimated that the number of reported crimes can be predicted by the function $n = f(t) = 1{,}850 + 322t$, where $n =$ the number of reported crimes per month and $t =$ time measured in months *since* January ($t = 0$ corresponds to January of this year).

a.   Which is the independent variable? Do a quick *sketch* of information.

b.   Identify the $y$-intercept and interpret the meaning in this application.

c.   Identify the slope and interpret the meaning in this application.

d.   Predict the number of reported crimes that will occur this coming October.

Solution: a. $f(t)$ denotes that $t$ is the independent variable.

Compare: $y = mx + b$

$n = 322t + 1{,}850$

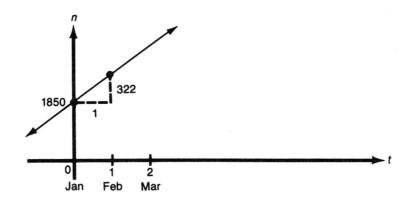

b. The $y$-intercept is 1,850. In January of this year, the number of reported crimes was estimated to be 1,850.

c. The slope is 322. For each additional month starting from January of this year, the number of reported crimes is estimated to increase by 322.

d. $n = f(t) = 322t + 1{,}850$ and October corresponds to $t = 9$.

$n = f(9) = 322(9) + 1{,}850$

$= 4{,}748$

It is estimated that there will be 4,748 reported crimes this coming October.

---

The following functions were defined in the Exercises in Unit 2 and are restated here for convenience.

Revenue Function = (selling price/unit)(number of units sold)

Cost Function = fixed cost + (variable cost/unit)(number of units)
or if the *total* variable cost is given,

= fixed cost + total variable cost

Profit = Total Revenue − Total Cost

## EXAMPLE 14

From Unit 2, the following functions were determined for the Alexander Company:

$C(x) = 13,200 + 43x$

$R(x) = 78x$

        where $x =$ the number of braided rugs produced by the Company,

        $C(x) =$ the total cost in dollars of producing $x$ braided rugs,

        and $R(x) =$ the total revenue in dollars of selling $x$ braided rugs.

a.   Identify and interpret the slope of the total cost function.

b.   Identify and interpret the slope of the total revenue function.

Solution:   a.   The slope of $C(x)$ is 43 or $m = \dfrac{\Delta y}{\Delta x}$ or $\dfrac{\Delta C}{\Delta x} = 43 = \dfrac{43}{1}$.

              Notice that the variable cost per unit is the slope.

              For each additional braided rug produced, total cost will increase by \$43.

        b.   The slope of $R(x)$ is 78 or $m = \dfrac{\Delta y}{\Delta x}$ or $\dfrac{\Delta R}{\Delta x} = 78 = \dfrac{78}{1}$.

              For each additional braided rug sold, total revenue will increase by \$78.

# BREAK-EVEN ANALYSIS

Break-even analysis identifies the level of operation or level of output where the manufacturer is just able to break even on operations, neither incurring a loss nor earning a profit. The **break-even point** represents the level of output at which total revenue equals total cost, $R(x) = C(x)$.

## EXAMPLE 15

Determine how many braided rugs the Alexander Company must sell in order to break even.

Solution:   Break even is where total revenue equals total cost,

$$R(x) = C(x)$$

By substitution
$$78x = 13,200 + 43x$$

$$35x = 13,200$$

$$x = 377.14$$

        The Alexander Company must sell approximately 377 braided rugs in order to break even.

As previously stated, the break-even point is the level of output where the producer is just able to break even, neither incurring a loss nor earning a profit. Any changes from the break-even level of operation will result in either a profit if the output is increased, or a loss if the output is decreased.

## EXAMPLE 16

Verify that if the Alexander Company produced and sold 450 braided rugs, they would make a profit.

Solution:   First determine the profit function for braided rugs.

$$\text{Profit} = \text{Total Revenue} - \text{Total Cost}$$

$$P(x) = R(x) - C(x)$$

$$= 78x - (13{,}200 + 43x)$$

$$= 78x - 13{,}200 - 43x$$

$$= 35x - 13{,}200$$

Use the profit function to determine the profit if 450 rugs are sold.

$$P(450) = 35(450) - 13{,}200$$

$$= 15{,}750 - 13{,}200$$

$$= 2{,}550$$

If 450 braided rugs are produced and sold, the Alexander Company will make $2,550 profit.

## Problem 8

Verify, by selecting some number less than 450, that if the number of braided rugs produced and sold is less than the break-even level the Company would experience a loss.

Solution:

Answers will vary depending upon the number you selected.

## EXAMPLE 17

A manufacturer of T-shirts has a fixed cost of $30,000 and a variable cost per shirt of $9. The selling price per shirt is $12. Determine how many shirts must be sold to break even.

Solution:   Let $x$ be the number of T-shirts, determine $R(x)$ and $C(x)$.

Total Revenue = (selling price/unit)(number of units sold)

$$R(x) = 12x$$

Total Cost    = fixed cost + (variable cost/unit)(number of units)

$$C(x) = 30{,}000 \quad + 9x$$

Break even is where total revenue equals total cost or,

$$R(x) = C(x)$$

By substitution    $12x = 30{,}000 + 9x$

$$3x = 30{,}000$$

$$x = 10{,}000$$

The manufacturer must sell 10,000 T-shirts to break even.

---

# EXAMPLE 18

Continuing with Example 17, how many T-shirts must be sold to make a profit of $6,000?

---

Solution:   First determine the profit function for T-shirts.

$$\text{Profit} = \text{Total Revenue} - \text{Total Cost}$$

$$P(x) = R(x) - C(x)$$

$$= 12x - (30{,}000 + 9x)$$

$$= 3x - 30{,}000$$

Now the question becomes that of finding $x$ so that profit will be $6,000.

$$P(x) = 3x - 30{,}000$$

By substitution    $6{,}000 = 3x - 30{,}000$

$$-3x = -30{,}000 - 6{,}000$$

$$-3x = -36{,}000$$

$$x = 12{,}000$$

The manufacturer must sell 12,000 T-shirts to make a profit of $6,000.

---

# EXAMPLE 19

Mary Lou's Toy Company can produce 50 teddy bears at a total cost (fixed plus variable) of $625, while 150 such teddy bears cost $1,225 to produce. Mary Lou assumes there is a linear relationship between cost and the number of teddy bears produced. Determine total cost = $f$(number of teddy bears produced).

Solution:    What do we know about this function?

Neither fixed costs nor variable costs are given, so we cannot simply substitute into the total cost function.

The way the question has been phrased, determine total cost = $f$(number of teddy bears produced), indicates that the number of teddy bears is to be the independent variable while total cost is meant to be the dependent variable.

So, let $x$ = the number of teddy bears produced

and $y$ = total cost.

The function is assumed to be linear; the graph is a line.

When $x = 50$, $y = 625$ or $(50, 625)$ must be a point on the line.

When $x = 150$, $y = 1{,}225$ or $(150, 1{,}225)$ must also be on the line.

Does all of this sound familiar?

Again a quick sketch of the information might be helpful.

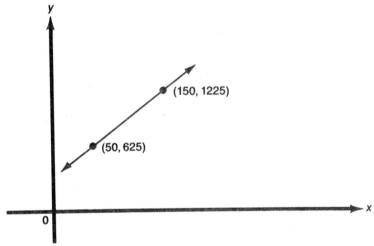

Actually we are being asked to do nothing more than write the equation of a line given two points on it.

Calculate the slope:      $(150, 1{,}225)$
                          $(\ 50,\ \ \ 625)$
                          $\overline{\phantom{xx}100,\ \ \ 600\phantom{xx}}$     $m = \dfrac{\Delta y}{\Delta x} = \dfrac{600}{100} = 6$

Point-slope formula:              $y - y_1 = m(x - x_1)$

Use one point and slope:          $y - 625 = 6(x - 50)$

                                  $y - 625 = 6x - 300$

                                  $y = 6x + 325$

Answer:    $y = f(x) = 6x + 325$ with $x \geq 0$

where $x$ = the number of teddy bears produced

and   $y$ = the total cost in dollars of producing $x$ teddy bears.

## Problem 9

Use the information from Example 19 to answer the following questions.

a.   What will the cost be if 300 teddy bears are produced?

b.   Find the variable cost per unit.

c.   Find the fixed cost.

Solution:

Answers:   a. $2,125;   b. $6;   c. $325

The last example looks at an application problem with a split domain.

## EXAMPLE 20

My friend Rosalie is a free lance writer and is quite good at her job. She specializes in writing annual reports for corporations. Annual reports average between twelve and sixteen pages and her fee, which is a function of the number of pages she writes, includes the cost of research, on-site visits, and all necessary rewrites. Rosalie's schedule of fees is given by

$$F(x) = \begin{cases} 7,000, & \text{if } 1 \leq x \leq 5 \\ 5,000 + 500x, & \text{if } 5 < x \leq 10 \\ 1,000x, & \text{if } 10 < x \end{cases}$$

where $F(x) =$ Rosalie's fee in dollars for writing $x$ pages.

a.   Graph $F(x)$.

b.   Interpret $F(x)$ in this application.

c.   What would Rosalie's fee be for a 15-page annual report?

d.   Determine her fee for writing an 8-page annual report.

Solution:   a.   The function has a split domain that has three parts.

One part is:

$$F(x) = 7,000, \quad \text{if } 1 \leq x \leq 5$$

This part of the function is linear. The graph will be a line segment.

Lower limit:      Let $x = 1$ and find $F(1)$.

$$F(x) = 7,000$$

$$F(1) = 7,000$$

One endpoint of the line segment is (1, 7,000).

Upper limit:    Let $x = 5$ and find $F(5)$.

$$F(x) = 7,000$$

$$F(5) = 7,000$$

The other endpoint of the line segment is $(5, 7,000)$.

The second part is:

$$F(x) = 5,000 + 500x, \quad \text{if } 5 < x \leq 10$$

This part of the function is linear. The graph will be a line segment too.

Lower limit:    Let $x = 5$ and find $F(5)$.

$$F(x) = 5,000 + 500x$$

$$F(5) = 5,000 + 500(5)$$

$$= 5,000 + 2,500$$

$$= 7,500$$

One endpoint of the line segment is $(5, 7,500)$.

This endpoint is graphed as an open circle because the 5 was not included in the interval $5 < x \leq 10$.

Upper limit:    Let $x = 10$ and find $F(10)$.

$$F(x) = 5,000 + 500x$$

$$F(10) = 5,000 + 500(10)$$

$$= 5,000 + 5,000$$

$$= 10,000$$

The other endpoint of the line segment is $(10, 10,000)$.

The third part is:

$$F(x) = 1,000x, \quad \text{if } 10 < x$$

This part of the function is also linear but the graph will be what is called a **half line**. The half line has one endpoint at the lower limit but continues on since it has no upper limit.

Lower limit:    Let $x = 10$ and find $F(10)$.

$$F(x) = 1,000x$$

$$F(10) = 1,000(10)$$

$$= 10,000$$

The one endpoint of the half line is $(10, 10,000)$.

This endpoint also would be graphed as an open circle because the 10 is not included in the interval $10 < x$. But notice that the open circle will be "filled in" by one of the endpoints of the second part.

Since there is no upper limit, we need to either find the coordinates of a second point or use the slope to locate a second point. In this case it is just as easy to pick another value for $x$. But remember it must be greater than 10.

Suppose we let $x = 11$. Find $F(11)$.

$$F(x) = 1,000x$$
$$F(11) = 1,000(11)$$
$$= 11,000$$

(11, 11,000) is a second point on the half line.

The graph of $F(x)$ is shown on the grid. Do you see why the third part is referred to as a half line?

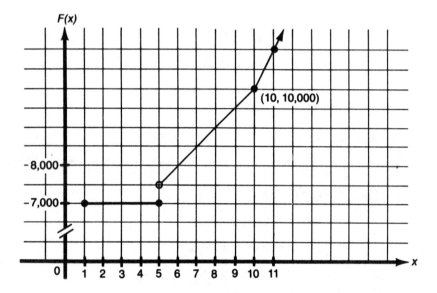

b.   Try doing the interpretation of $F(x)$ yourself before reading the answer. It might help to look at the graph.

Answer:   Rosalie charges a flat fee of $7,000 for writing an annual report anywhere from one page to five pages in length. Anything over five pages, even a fraction of a page, up to and including ten pages is charged at the rate of $5,000 plus $500 per page. Any report in excess of ten pages is charged at the rate of $1,000 per page.

c. and d.   I will leave the last two questions for you.

Answers:   c. $15,000;   d. $9,000

You should now be able to interpret points, slopes, and intercepts for various application problems, especially those involving cost and revenue functions and break-even points. In some instances, you should be able to write the linear function that relates the variables as described by the given information. Often a quick sketch proves helpful.

In addition you should have a reasonably good sense of what is meant by supply and demand functions and how to find the equilibrium price.

Before beginning the next unit you should work the following problems. These are probably some of the more difficult ones that you will encounter in this book, but try them. There are only a few. If you need help, remember the answers are given in the back.

## EXERCISES

Be sure to *interpret* each answer in terms of the application problem.

1.  A street vendor in Harrisburg has noticed that the amount of lemonade he sells during the lunch hour is closely related to the temperature. He estimates that

    $$L(t) = 1.5t - 60$$

    > where $t =$ the temperature in Fahrenheit
    >
    > and $L(t) =$ the number of lemonades sold at lunch

    a.  Find $L(86)$.

    b.  Determine the significance of $L(40)$.

    c.  Identify the slope.

2.  Marguerite is attending school and earning extra money by providing a cleaning service. For $10 an hour she will come and clean your house. Marguerite brings her own cleaning supplies and often has a helper, Christine. She estimates that her weekly costs and revenue are given by

    $$C(x) = 38.50 + 4.5x$$

    $$R(x) = 10x$$

    > where $x =$ the number of hours worked per week,
    >
    > $C(x) =$ Marguerite's weekly costs in dollars, and
    >
    > $R(x) =$ Marguerite's total revenue in dollars.

    a.  How many hours must Marguerite work to break even?

    b.  Marguerite would like to attend an upcoming summer concert. Tickets are priced at $22.00 and go on sale in a week. How many hours must she work to have sufficient money for a ticket?

3.  Consider the relation $p + 2q - 5000 = 0$, where $p$ equals the price of a product in dollars and $q$ equals the quantity demanded in units.

    a.  Determine the corresponding explicit demand function $q = D(p)$ and answer the following questions.

    b.  Determine $D(2,000)$.

    c.  Identify the $y$-intercept.

    d.  Comment about $D(5,000)$.

    e.  Identify the slope.

4.  If city maps are priced at $5 each, suppliers are willing to produce 46 maps. If the price of maps increases to $9 each, suppliers are willing to produce 86. It is assumed the relationship is constant; i.e., linear.

    a.   Determine quantity supplied $= S(\text{price})$.

    b.   Calculate $S(15)$.

    c.   Find the $x$-intercept.

    d.   What is the restricted domain for $S$?

5.   If city maps are priced at $5 each, there will be a demand for 65 maps. If they are priced at $10 each, the demand drops to 40. Assume the relationship is linear.

    a.   Determine quantity demanded $= D(\text{price})$.

    b.   Find both intercepts.

    c.   What is the restricted domain for $D$?

    d.   Determine the equilibrium price and corresponding quantity for city maps.

6.   The average cost of a gallon of gasoline has been declining at a constant rate in recent years. In 1981, the average cost was $1.40 while in 1985 it was only $1.20.

    a.   Express the average cost as a function of time where $t = 0$ corresponds to 1980.

    b.   Identify the slope.

    c.   If the trend continues, when will the average cost of a gallon of gasoline be 86.5 cents?

7.   Graph: $f(x) = \begin{cases} 2x + 1, & \text{if } 0 \le x \le 4 \\ 3, & \text{if } 4 < x < 9 \\ -\dfrac{2}{3}x + 12, & \text{if } 9 \le x \end{cases}$

---

If additional practice is needed:

Barnett and Ziegler, page 38, problems 51–57; page 57, problems 81–83
Budnick, page 65, problems 30–31; page 75, problems 15–16
Hoffmann; page 35, problems 28–40; page 46, problems 18–28
Piascik, page 49, problems 13–18; pages 17–18, problems 7–12

# UNIT 6

## Quadratic Functions

In this unit you will learn to recognize quadratic functions of one variable and to graph them using a minimum number of well-chosen points.

---

Definition: $y = f(x) = ax^2 + bx + c$, with $a$, $b$, and $c$ being real numbers, $a \neq 0$, is called a **quadratic** or **second-degree function of one variable**.

---

In other words, a second-degree function of one variable must contain a squared term, $x^2$, and no higher powered term.

In this unit we will deal only with second-degree or quadratic functions of one variable.

There are basically two ways to graph quadratic functions:

1.  Find and plot a large number of points—this can be time consuming, and the key points missed.

2.  Plot a few well-chosen points based on knowledge of quadratic functions—the approach we will use.

---

Definition: $y = f(x) = ax^2 + bx + c$ is called the **standard form** of a quadratic function of one variable.

Note: *All* terms are on the right side of the equal sign with *only* $y$ or $f(x)$ on the left.

---

Like linear functions, the graph of a quadratic function is very predictable. The following are some basic facts that can be used when graphing a quadratic function of one variable.

# BASIC FACTS ABOUT THE GRAPH OF A QUADRATIC FUNCTION IN STANDARD FORM:
## $y = f(x) = ax^2 + bx + c$

1.  The graph of a quadratic function is a smooth, ⌣ ⌣ -shaped curve called a parabola.

2.  If $a$, the coefficient of the squared term, is positive, the curve opens up: ⌣ .

    If $a$ is negative, the curve opens down: ⌢ .

3.  The $y$-intercept is $c$, the constant. Remember that the $y$-intercept is the value of $y$ when $x = 0$; thus $y = a(0)^2 + b(0) + c = c$.

4.  There are **at most two $x$-intercepts**, "at most" meaning there can be two, one, or none. The $x$-intercepts are the values of $x$ when $y = 0$; thus the solution to $0 = ax^2 + bx + c$ yields the $x$-intercepts.

5.  The low point on the curve (or the high point if the curve opens down) is called the **vertex**.

6.  The vertex is located at the point

$$\left( \frac{-b}{2a}, \frac{4ac - b^2}{4a} \right)$$

7.  The curve is symmetric to a vertical line through the vertex. The vertical line is called the **axis of symmetry**.

Based on the above information about quadratic functions of one variable, it is now possible to graph such functions using only a few well-chosen points. By "few" I mean no more than seven, but at least three.

---

# SUGGESTED APPROACH FOR GRAPHING QUADRATIC FUNCTIONS OF ONE VARIABLE

1.  Write the quadratic function in standard form.

2.  Determine whether the parabola opens up or down.

3.  Find the $y$-intercept at $c$.

4.  Find the $x$-intercepts, if any, by solving $0 = ax^2 + bx + c$.

    Note:   For a review of factoring and solving second-degree equations of this type, see Units 16–21 of FORGOTTEN ALGEBRA.

5. Find the coordinates of the vertex:

$$\left(\frac{-b}{2a}, \frac{4ac - b^2}{4a}\right)$$

6. Locate one point on either side of the vertex. The $x$-values should be within one unit of the vertex.

7. Plot the above points, and connect them with a smooth, $\cup$-shaped curve.

The following examples will illustrate this approach.

# EXAMPLE 1

Graph: $y = f(x) = x^2 - 8x + 7$.

Solution:   1. Write in standard form and compare:

$$y = f(x) = \boxed{a}x^2 \boxed{+b}x \boxed{+c}$$
$$y = f(x) = \boxed{\phantom{0}}x^2 \boxed{-8}x \boxed{+7}$$

Thus $a = 1$, $b = -8$, $c = 7$.

Graph will be a parabola.

2. Since $a$ is positive, curve opens up: $\cup$.

3. The $y$-intercept is $c = 7$.

4. Find the $x$-intercepts by solving:

$$0 = x^2 - 8x + 7$$
$$0 = (x - 1)(x - 7)$$
$$x = 1, \qquad x = 7$$

The $x$-intercepts are 1 and 7.

5. Find the coordinates of the vertex.

$$\frac{-b}{2a} = \frac{-(-8)}{2(1)} = \frac{8}{2} = 4$$

$$\frac{4ac - b^2}{4a} = \frac{4(1)(7) - (-8)^2}{4(1)} = \frac{28 - 64}{4} = \frac{-36}{4} = -9$$

The vertex is located at $(4, -9)$.

6. Locate one point on either side of the vertex. Since the vertex is at $x = 4$:

let $x = 3$; then $y = f(3) = (3)^2 - 8(3) + 7 = 9 - 24 + 7 = -8$

let $x = 5$; then $y = f(5) = (5)^2 - 8(5) + 7 = 25 - 40 + 7 = -8$

7. Plot the above points, and connect them with a smooth curve.

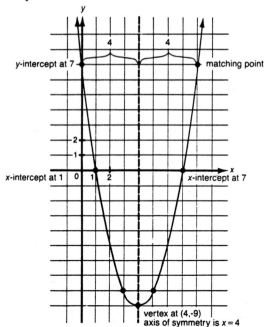

Answer:

vertex at (4,-9)
axis of symmetry is $x = 4$

---

Before going on to another example, let's mention a few points about the above graph. Be sure to notice that every ordered pair whose coordinates satisfy the function, $f(x) = x^2 - 8x + 7$, must correspond to some point on the curve. And every point on the curve must have coordinates that satisfy the function.

The axis of symmetry is indicated with a dashed line. Because the curve is symmetrical about this line, there must be a point 4 spaces away on the right that corresponds to the $y$-intercept of 7. The actual coordinates of the point are not important—only its location, which is shown, need concern us.

The arrows on each end of the parabola indicate that the curve continues upward.

# EXAMPLE 2

Graph:   $y = f(x) = 3 + x^2$.

---

Solution:   1.   Write in standard form and compare:

$$y = f(x) = \boxed{\phantom{a}}x^2 \boxed{\phantom{+b}}\phantom{x} \boxed{+3}$$
$$y = f(x) = \boxed{a}x^2 \boxed{+b}x \boxed{+c}$$

Thus $a = 1, b = 0, c = 3$.

Graph will be a parabolic curve.

2.   Since $a$ is positive, curve will open up: ∪ .

3.   The $y$-intercept is at $c = 3$.

4.   Find the $x$-intercepts, at most two, by solving:

$$0 = x^2 + 3$$

$$-3 = x^2 \text{ has no solution}$$

Therefore there are no $x$-intercepts.

5. Find the coordinates of the vertex:

$$\frac{-b}{2a} = \frac{0}{2(+1)} = 0$$

To find the $y$-coordinate for the vertex, we can, of course, use the formula: $(4ac - b^2)/4a$, but there is an alternative method.

Once the $x$-coordinate is determined, the $y$-coordinate can be calculated by substituting into the original function:

if $x = 0$, then $f(0) = 3 - (0)^2 = 3$.

The vertex is located at $(0, 3)$.

For this particular example, using the original function was an easier way to determine $y$ than using the formula for the vertex, but the choice is yours.

6. Locate one point on either side of the vertex:

$$\text{let } x = \quad 1; \text{ then } y = f(1) = 3 + (1)^2 = 3 + 1 = 4$$

$$\text{let } x = -1; \text{ then } y = f(-1) = 3 + (-1)^2 = 3 + 1 = 4$$

7. Plot the above points, and connect them with a smooth curve.

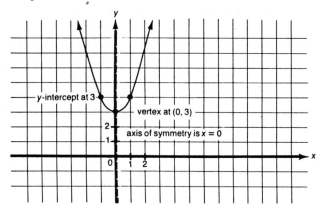

Answer:
If more detail is needed, plot additional points.

Now it is time for you to try a few problems yourself.

# Problem 1

Graph:  $y = f(x) = x^2 - 2x - 8$.

Solution:  1. Write in standard form and compare.

2. Curve will open _____.

3. The $y$-intercept is at _____.

4. The $x$-intercepts, at most two, are _____ and _____.

5.  The vertex is located at _____.

6.  Locate one point on either side of the vertex.

    Step 6 may be omitted because we have already a point on either side of the vertex, the $y$-intercept and its matching point.

7.  Plot the above points on the grid provided and connect them with a smooth curve.

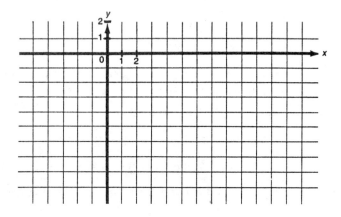

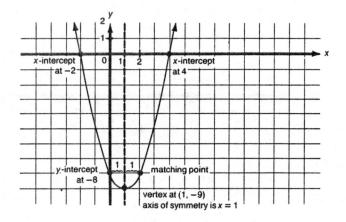

Answer:

# Problem 2

Graph:  $y = f(x) = 2x^2$.

Solution:

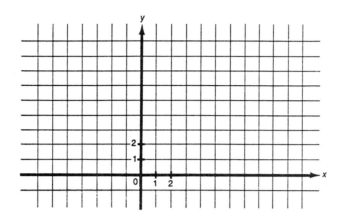

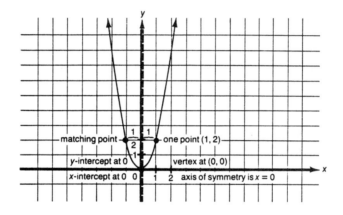

Answer:

---

The next example illustrates still another method for locating the vertex.

## EXAMPLE 3

Graph:  $y = f(x) = x^2 - 6x + 5$.

---

Solution:  1.  This is a quadratic function with $a = 1$, $b = -6$, and $c = 5$.

2.  Parabolic curve opens up.

3.  The $y$-intercept is at 5.

4.  The $x$-intercepts are:

$$0 = x^2 - 6x + 5$$

$$0 = (x - 1)(x - 5)$$

$$x = 1, \quad x = 5$$

5.  As before, we could locate the vertex using the formula, but in this example there is an easier method. Recall that the axis of symmetry is a vertical line through the vertex that cuts the curve in the middle; the two sides of the curve must match. Since in this example the $x$-intercepts are at 1 and 5, where must the axis of symmetry be? It must be midway between the two points. Thus the $x$-coordinate of the vertex is 3, the value in the middle of 1 and 5.

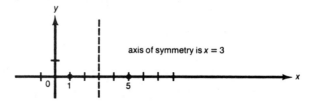

The $y$-coordinate may be calculated by either method discussed earlier. In this case I will substitute into the original function:

$$\text{if } x = 3, \ f(x) = (3)^2 - 6(3) + 5 = 9 - 18 + 5 = -4$$

The vertex is located at $(3, -4)$.

Step 6 may be omitted because we have already a point on either side of the vertex, the two $x$-intercepts.

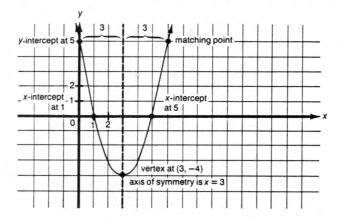

Answer:

---

As you probably suspected, all quadratic functions are not as easy to work with as the preceding ones. Consider the next example.

# EXAMPLE 4

Graph:  $y = f(x) = x^2 + 2x - 6$.

---

Solution:   1.  This is a quadratic function with $a = 1$, $b = 2$, and $c = -6$.

2.  Parabolic curve opens up.

3.  The $y$-intercept is at $-6$.

4.  Find the $x$-intercepts, at most two, by solving:

$$0 = x^2 + 2x - 6$$

Since we are unable to factor, use the quadratic formula:

$$x = \frac{-b \pm \sqrt{b^2 - 4ac}}{2a}$$

$$x = \frac{-2 \pm \sqrt{(2)^2 - 4(1)(-6)}}{2(1)}$$

$$x = \frac{-2 \pm \sqrt{4 + 24}}{2}$$

$$x = \frac{-2 \pm \sqrt{28}}{2} \quad \text{are the } x\text{-intercepts.}$$

If you have a calculator or a set of tables, $\sqrt{28} \approx 5.29$

$$x = \frac{-2 + 5.29}{2} \quad \text{and} \quad x = \frac{-2 - 5.29}{2}$$

$$x = 1.6 \qquad\qquad x = -3.6$$

The $x$-intercepts are approximately 1.6 and $-3.6$.

5. The vertex is located at $(-1, -7)$ because:

$$\frac{-b}{2a} = \frac{-2}{2(1)} = -1 \quad \text{and} \quad f(-1) = (-1)^2 + 2(-1) - 6 = 1 - 2 - 6 = -7$$

Step 6 may be omitted because again we have a point on either side of the vertex, the two $x$-intercepts.

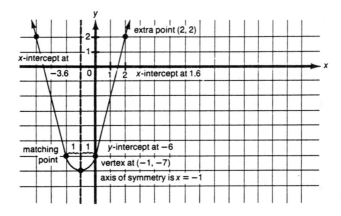

Answer:

Comments: To get a better sense of where the curve actually crosses the $x$-axis, I plotted a convenient extra point in quadrant I, $(2, 2)$. There would be a corresponding point in quadrant II. Then by connecting the points with a smooth curve, the location of the $x$-intercepts is well determined.

## Problem 3

Graph:   $y = f(x) = x^2 + 6x + 9$.

Solution:

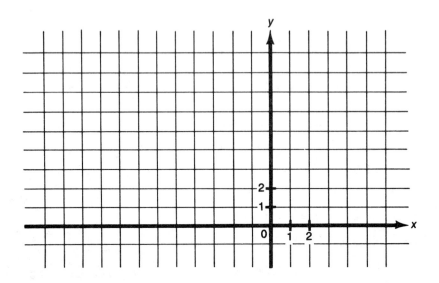

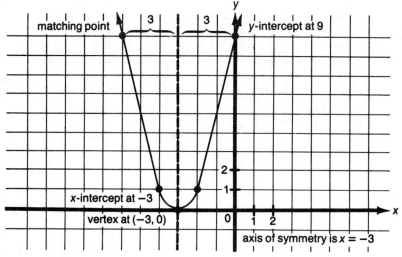

Answer:

# Problem 4

Graph:   $y = f(x) = -x^2 + 4x.$

Solution:

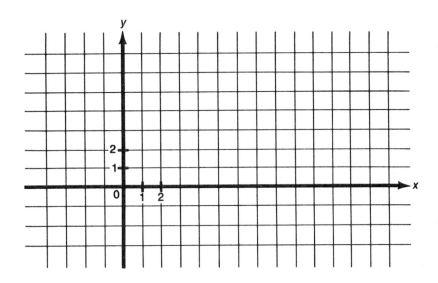

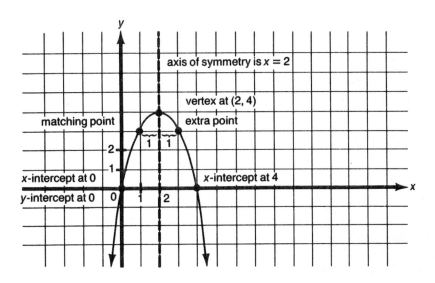

Answer:

You should now be able to recognize a quadratic function of one variable, of the type $y = f(x) = ax^2 + bx + c$ with $a \neq 0$. Also you should be able to graph it using a minimum number of well-chosen points based on your knowledge of quadratic functions.

Briefly, the graphing approach is as follows:

1.   Write the function in standard form and compare.

2.   Determine whether the parabola opens up or down.

3.   The $y$-intercept is at $c$.

4.   Solve $0 = ax^2 + bx + c$ to find the $x$-intercepts, if any.

5.   The vertex is at

$$\left( \frac{-b}{2a}, \frac{4ac - b^2}{4a} \right)$$

6.   If needed, locate a point on either side of the vertex.

7.   Plot the points, and connect them with a smooth, $\cup$-shaped curve.

Now try the following exercises.

## EXERCISES

Graph each of these by using your knowledge of quadratic functions of one variable to plot a few well-chosen points.

1.   $y = f(x) = x^2 + 2x + 1$

2.   $y = f(x) = -x^2 + 2x + 3$

3.   $y = f(x) = -2x^2 + 4x$

4.   $y = f(x) = 2x^2 - x - 10$

5.   $y = f(x) = -x^2 - 7$

6.   $y = f(x) = x^2 - 4x + 2$

7.   $y = f(x) = 3x^2 - 9x$

If additional practice is needed:

Barnett and Ziegler, page 56, problems 63–80
Bleau, page 185, problems 1–10
Budnick, page 393, problems 1–16
Piascik, page 50, problems 19–20

# UNIT 7

## Quadratic Applications

Now that you have mastered the mechanics of graphing quadratic functions, we will look at some applications. These will be similar in nature to those discussed in Unit 5, except the functions will be quadratics.

In the following examples the emphasis will be on doing sketches rather than detailed graphs. This is in part because of the large numbers that frequently occur in application problems. However, even with a sketch, the key points—the vertex and intercepts—must still be shown.

Recall the suggested approach for graphing quadratic functions is:

1. Write the quadratic function in standard form.

2. Determine whether the parabola opens up or down.

3. Find the $y$-intercept at $c$.

4. Find the $x$-intercepts, if any, by solving $0 = ax^2 + bx + c$.

5. Find the coordinates of the vertex:
$$\left( \frac{-b}{2a}, \frac{4ac - b^2}{4a} \right)$$

6. Locate one point on either side of the vertex. The $x$-values should be within one unit of the vertex.

   Note: Often this step may be omitted with application problems unless a detailed graph is required.

7. Plot the above points and connect them with a smooth, ∪-shaped curve.

Notice in the following examples that the scales on the $x$- and $y$-axis have had to be adjusted to accommodate the larger numbers.

## EXAMPLE 1

Graph:   $y = f(x) = 2.5x^2 - 6{,}250$.

Solution:
1.  This is a quadratic function with $a = 2.5$, $b = 0$, and $c = -6{,}250$.

2.  Parabolic curve opens up.

3.  The $y$-intercept is at $-6{,}250$.

4.  Find the $x$-intercepts, at most two, by solving:

$$0 = 2.5x^2 - 6{,}250$$

$$-2.5x^2 = -6{,}250$$

$$x^2 = 2{,}500$$

$$x = \pm 50$$

The $x$-intercepts are 50 and $-50$.

5.  Recall as before, that the axis of symmetry is a vertical line through the vertex that cuts the curve in the middle. Since in this example the $x$-intercepts are at $-50$ and 50, the axis of symmetry must be midway between the two points. Thus the $x$-coordinate of the vertex is 0, the value in the middle of $-50$ and 50.

The $y$-coordinate will be calculated by using the original function:

$$\text{if } x = 0, \; f(0) = 2.5(0)^2 - 6{,}250 = -6{,}250.$$

The vertex is located at $(0, -6{,}250)$.

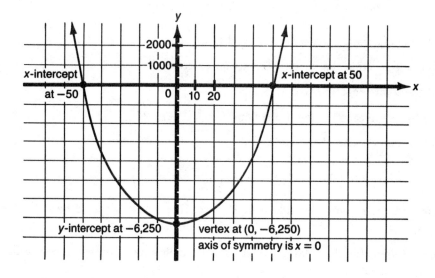

Answer:

You might be wondering why we are spending so much time graphing. Trust me when I say that all this practice will make life easier as you progress through the book.

## Problem 1

Graph:   $y = f(x) = x^2 - 600x$.

Solution:

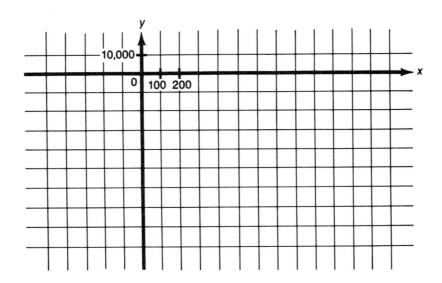

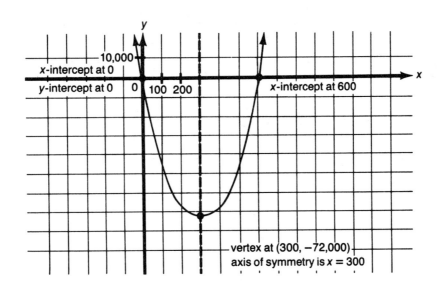

Answer:

Remember that supply functions relate the market price to the quantities that suppliers are willing to produce and sell.

## EXAMPLE 2

The supply function for ceiling hugger fans is defined by

$$S(p) = 2.5p^2 - 6{,}250$$

where $p$ = the price in dollars per fan

and $S(p)$ = the number of fans the manufacturer is willing to supply at price $p$.

a.  Sketch the function.

b.  Calculate and interpret $S(60)$.

c.  Is there any restriction on the domain?

Solution:    a.    $S(p) = 2.5p^2 - 6{,}250$ is the same function as Example 1 except with the independent variable changed to $p$. The sketch is repeated here for convenience.

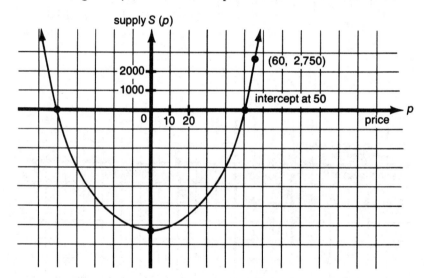

b.    $S(p) = 2.5p^2 - 6{,}250$

$S(60) = 2.5(60)^2 - 6{,}250$

$= 9{,}000 - 6{,}250$

$= 2{,}750$

At a price of $60 each, the manufacturer would be willing to supply 2,750 ceiling hugger fans.

c.    Since $p$ represents price, $p$ must be positive or zero; i.e., non-negative. From the graph, it should be apparent that if $p < 50$, then $q$ would be negative and meaningless. Remember $q$ is quantity and therefore could not be negative.

Restricted domain:    $50 \leq p$

## EXAMPLE 3

Suppose the total cost of manufacturing $x$ items is given by the function

$$y = C(x) = 3x^2 + x + 50$$

where $C(x)$ = the total cost in dollars

and      $x$ = the number of items manufactured.

Graph the function.

Solution:  1.  This is a quadratic function with $a = 3$, $b = 1$, and $c = 50$.

2.  The parabola opens up.

3.  The $y$-intercept is at 50.

4.  Find the $x$-intercepts, at most two, by solving:

$$0 = 3x^2 + x + 50$$

Since we are unable to factor, use the quadratic formula.

$$x = \frac{-b \pm \sqrt{b^2 - 4ac}}{2a}$$

$$x = \frac{-1 \pm \sqrt{(1)^2 - 4(3)(50)}}{2(3)}$$

$$x = \frac{-1 \pm \sqrt{1 - 600}}{6}$$

$$x = \frac{-1 \pm \sqrt{-599}}{6} \qquad \text{Cannot take the square root of a negative.}$$

There are no $x$-intercepts.

5.  Find the coordinates of the vertex.

$$\frac{-b}{2a} = \frac{-1}{2(3)} = \frac{-1}{6}$$

$$\frac{4ac - b^2}{4a} = \frac{4(3)(50) - (1)^2}{4(3)} = \frac{600 - 1}{12} = \frac{599}{12}$$

The vertex is located at $(-1/6, 599/12)$ or, if you prefer decimals, $(-.17, 49.92)$.

6.  I chose to locate an additional point on each side of the vertex due to the difficulty in graphing the vertex accurately.

let $x = 1$; then $f(x) = 3(1)^2 + 1 + 50 = 54$

let $x = -1$; then $f(x) = 3(-1)^2 + (-1) + 50 = 52$

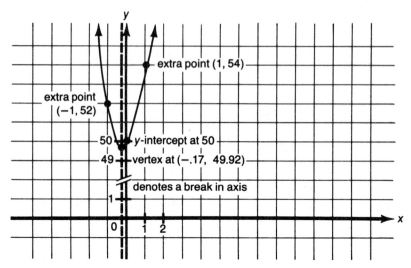

Answer:

## EXAMPLE 4

Use the information from Example 3 to answer the following questions.

a.   What is the cost of manufacturing 20 items?

b.   Interpret the $y$-intercept.

c.   Calculate and interpret $C(100)$.

Solution:   a.   $y = C(x)\quad = 3x^2 + x + 50$

$y = C(20) = 3(20)^2 + 20 + 50$

$= 3(400) + 20 + 50$

$= 1{,}270$

The cost of manufacturing 20 items will be $1,270.

b.   The $y$-intercept is 50.

If no units are manufactured, the cost will be $50. Or stated another way, fixed costs are $50.

c.   $y = C(x)\quad = 3x^2 + x + 50$

$y = C(100) = 3(100)^2 + 100 + 50$

$= 3(10{,}000) + 100 + 50$

$= 30{,}150$

The cost of manufacturing 100 units will be $30,150.

Recall demand functions describe the relationship between the demand for a product and its price.

# EXAMPLE 5

Betsy owns a boutique on an exclusive resort island. Over the past few years she has determined that the demand function for T-shirts with the resort's logo is defined by

$$q = D(p) = -p^2 - 10p + 5{,}600$$

where $p$ is the price in dollars for the T-shirts

and $D(p)$ is the quantity sold (demand) at price $p$.

Use the graph of the function to determine the restricted domain and range.

Solution:  1.  This is a quadratic with $a = -1$, $b = -10$, $c = 5{,}600$.

2. The parabolic curve opens down.

3. The $y$-intercept is at 5,600.

4. The $x$-intercepts are $-80$ and 70 because

$$0 = -p^2 - 10p + 5{,}600$$
$$0 = p^2 + 10p - 5{,}600$$
$$0 = (p + 80)(p - 70)$$
$$p = -80 \qquad p = 70$$

5. The vertex is located at $(-5, 5{,}625)$ because

$$\frac{-b}{2a} = \frac{-(-10)}{2(-1)} = \frac{10}{-2} = -5$$

and $D(-5) = -(-5)^2 - 10(-5) + 5{,}600 = -25 + 50 + 5{,}600 = 5{,}625$

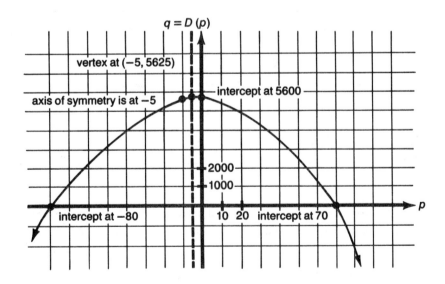

To determine the restricted domain and range, we should reason as follows. Since $p$ is price, any negative number would be meaningless. At the same time, $q$ is quantity and again any negative number would be meaningless. In other words, *both* $p$ and $q$ must be positive or zero; i.e., non-negative.

What restrictions are necessary to insure that both $p$ and $q$ are non-negative? Our graph from above should be of help. Both $p$ and $q$ are non-negative only in the first quadrant.

The restricted domain is the set of all possible values that can be used as input and "make sense" with reference to the problem. The domain always describes the independent variable, which in this application is $p$. The only relevant portion of the graph applicable to this problem is the part in the first quadrant.

The smallest value $p$ takes on in the first quadrant is 0.

The largest value $p$ takes on in the first quadrant is 70.

Therefore the restricted domain is:   $0 \le p \le 70$.

The restricted range describes the dependent variable, in this case $q$, and is the set of all possible values that are output and "make sense" in terms of the problem. Using the first quadrant, the smallest value $q$ takes on in the first quadrant is 0; the largest value $q$ takes on in the first quadrant is 5,600.

Therefore the restricted range is:   $0 \le q \le 5,600$.

The graph of $D(p)$ with the restricted domain is shown below. Verify for yourself that if the price of T-shirts is greater than $70, Betsy will not sell any.

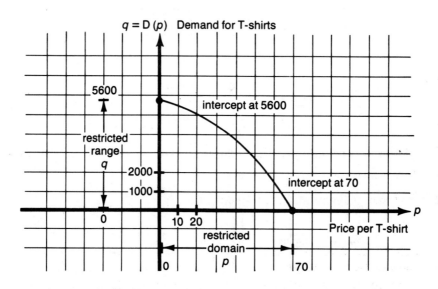

Answer:

# Problem 2

It is near the start of the season and Betsy has received a shipment of 4,400 T-shirts. She decides to price them at $50 each. How many T-shirts should she expect to sell at that price?

Solution:

Answer:   According to the demand function, at a price of $50 per shirt Betsy will sell only 2,600 shirts, leaving her with 1,800 unsold at the end of the season.

## Problem 3

Suppose Betsy does not want to carry any T-shirts over to the following season and a sale at the end of the season is impractical. What is the highest price she can put on the T-shirts and still be reasonably sure of selling all of them? I will get you started, but you are to finish the problem yourself.

Solution:

$$q = -p^2 - 10p + 5{,}600$$

$$4{,}400 = -p^2 - 10p + 5{,}600$$

$$p^2 + 10p - 1{,}200 = 0$$

Answer:    To sell all 4,400 T-shirts, she should price them at $30 each.

## EXAMPLE 6

The local country club, founded in 1970, has a membership estimated by the following function:

$$M(t) = -t^2 + 22t + 240$$

where $M(t)$ = the number of members in the country club at time $t$

and        $t$ = time measured in years since 1970.

Graph the function.

Solution:   1.   This is a quadratic function with $a = -1$, $b = 22$, and $c = 240$.

2.   Parabolic curve opens down.

3.   The $y$-intercept is at 240.

4.   Find the $x$-intercepts, at most two, by solving:

$$0 = -t^2 + 22t + 240$$

I am better at factoring if the squared term is positive; so I always multiply the entire equation by $-1$. Be careful when you do this. The entire equation must be multipled by $-1$; that means *both* sides. You should not attempt to do this with a function, as it usually complicates the expression rather than simplifies it.

$$0 = t^2 - 22t - 240$$

$$0 = (t + 8)(t - 30)$$

$$t = -8 \qquad t = 30$$

The $x$-intercepts are $-8$ and 30.

5.   The vertex is located at $(11, 361)$ because

$$\frac{-b}{2a} = \frac{-22}{2(-1)} = 11$$

and $f(11) = -(11)^2 + 22(11) + 240 = -121 + 242 + 240 = 361$

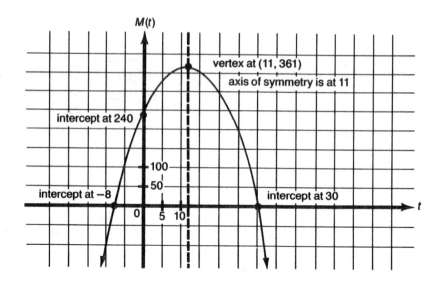

Answer:

---

Try answering the next questions yourself before looking at the answers.

===

## EXAMPLE 7

Use the information from Example 6 to answer the following questions.

a.  Calculate and interpret $M(5)$.

b.  Interpret the vertex.

c.  Interpret the intercepts.

---

Solution:   a.   $M(5)$ denotes $t = 5$ and $t = 5$ means five years since 1970.

$$M(t) = -t^2 + 22t + 240$$
$$M(5) = -(5)^2 + 22(5) + 240$$
$$= -25 + 110 + 240$$
$$= 325$$

In the year 1975, the country club had 325 members.

b.   The vertex is located at $(11, 361)$.

In 1981 $(1970 + 11)$, the membership of the country club was 361.

c.   The $y$-intercept is at 240 or the point $(0, 240)$.

In 1970 $(1970 + 0)$, the year the country club was founded, membership started at 240.

The $x$-intercept is at 30 or the point $(30, 0)$.

In the year 2000 $(1970 + 30)$, it is estimated that the membership of the country club will be 0.

# EXAMPLE 8

Use the information, especially the graph, from Example 6 to determine the restricted domain and range for $M(t)$.

---

Solution:   Again, *both* $t$ and $M(t)$ must be positive or zero because $t$ is time measured in years since the club's founding and $M(t)$ is the number of members. Neither could be a negative number.

Both $t$ and $M(t)$ are non-negative only in the first quadrant.

The domain always refers to the independent variable; in this application it is $t$. Using the graph from Example 6, the smallest value $t$ takes on in the first quadrant is 0; the largest value $t$ takes on in the first quadrant is 30.

Therefore the restricted domain is:   $0 \le t \le 30$.

To analyze it yet another way, if $t$ is less than 0, it would be prior to the founding of the club, and hence meaningless; if $t$ is greater than 30, the membership would be a negative number and again meaningless.

The range always describes the values of the dependent variable, in this case $M(t)$. Using the graph from Example 6, the smallest value $M(t)$ takes on in the first quadrant is 0; the largest value $M(t)$ takes on in the first quadrant is 361.

Therefore the restricted range is:   $0 \le M(t) \le 361$.

The relevant portion of the graph is shown below.

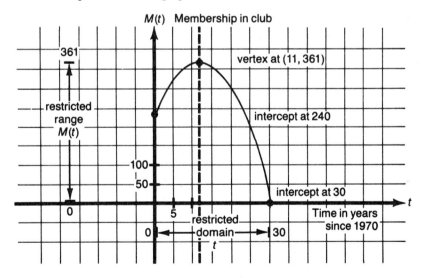

The graph of $M(t) = -t^2 + 22t + 240$ with $0 \le t \le 30$.

---

If you are able to answer the last few questions yourself, you have a good understanding of the material thus far.

## EXAMPLE 9

Use the graph of Example 8 to answer the following.

a.   At what year between 1970 and the present was the membership the largest?

b.   What was the largest membership?

c.   How would you describe the future prospects for the country club?

Solution:   a.   From the graph, the highest point on the parabola is the vertex that is located at $(11, 361)$. The interpretation of that point, as you probably recall, was that in the year 1981 the membership of the country club was 361.

Therefore the membership was the largest when $t = 11$, which corresponds to 1981.

b.   The largest membership was 361 which occurred in 1981.

c.   The club's prospects are poor; it is estimated that the country club will cease to have any members by the year 2000.

Are you starting to see the value of being able to graph quadratics?

You now should be able to interpret all points, especially the vertex and the intercepts, of a quadratic as they relate to various applications.

Here are a few comprehensive problems to try before continuing on to the next unit.

## EXERCISES

1.   Graph:   $y = f(x) = .01x^2 - 8x$.

2.   The number of pounds of seed supplied by farmers at a price of $p$ dollars per pound is given by $S(p) = p^2 - 100$.

   a.   Graph the supply function $S(p)$.

   b.   Calculate and interpret $S(18.50)$.

   c.   What would the price have to be in order for the farmers to be willing to meet a demand for 800 pounds?

3.   Suppose $D(p) = 2p^2 - 297p + 10,960$

   and        $S(p) = 2p^2 + 5p - 250$

                    where $D(p) =$ the demand function per unit at price $p$,

                            $S(p) =$ the supply function per unit at price $p$,

                    and        $p =$ the price per unit in dollars.

   Determine the equilibrium price and corresponding quantity.

4.   If a rubber ball is thrown vertically upward from the top of a building with an initial speed of 128 feet per second, its height (in feet) $t$ seconds later is given by $H(t) = -16t^2 + 128t + 320$.

   a.   Graph the function $H(t)$.

    b.   The point (2, 512) is on the graph of $H(t)$. Interpret its meaning with respect to the problem.

    c.   Compute and interpret $H(9)$.

    d.   Use the graph to determine when the ball will hit the ground.

    e.   Use the graph to determine how high the ball will rise.

    f.   How tall was the building?

5.   Jim recently opened a restaurant specializing in home-style cooking. His accountant has advised him that his profit is related to the number of customers as defined by

$$P(x) = -x^2 + 140x - 4{,}000$$

                where $P(x)$ = profit (or loss) in dollars per day

                and      $x$ = number of customers per day.

    a.   Graph $P(x)$.

    b.   Interpret the graph and all key points.

        If you can do this one, you've got it made!

# UNIT 8

## Polynomial Functions

In this unit you will learn to recognize polynomial functions of one variable and determine their degree. If the polynomial function is a cubic, you will be able to determine some, but not all, of the characteristics of its graph.

---

Definition:   $y = f(x) = ax^3 + bx^2 + cx + d$, with $a$, $b$, $c$, and $d$ being real numbers, $a \neq 0$, is called a **cubic** or **third-degree function of one variable**.

---

In other words, a third-degree function of one variable must contain a cubed term, $x^3$, and no higher powered term.

---

Definition:   $y = f(x) = ax^3 + bx^2 + cx + d$ is called the **standard form** of a cubic function of one variable.

Note:   *All* terms are on the right side of the equal sign with *only* $y$ or $f(x)$ on the left.

---

Recall that the graph of a function was defined as the set of all points, $(x, y)$, whose coordinates satisfy the function.

One way to graph cubic functions is to find and plot a large number of points. We are going to do one example together to illustrate this approach.

---

## Problem 1

Graph:   $y = f(x) = 2x^3 - 7x^2 - 7x + 5$.

Solution:  Find and plot a large number of points whose coordinates satisfy the function.

$$\text{Let } x = -2; \text{ then } y = 2(-2)^3 - 7(-2)^2 - 7(-2) + 5$$

$$= 2(-8) - 7(4) + 14 + 5$$

$$= -16 - 28 + 14 + 5$$

$$= -25, \text{ which gives the point } (-2, -25).$$

$$\text{Let } x = -1; \text{ then } y = 2(-1)^3 - 7(-1)^2 - 7(-1) + 5$$

$$= 2(-1) - 7(1) + 7 + 5$$

$$= -2 - 7 + 7 + 5$$

$$= 3, \text{ which gives the point } (-1, 3).$$

Find the remaining points yourself. Find all the points listed; otherwise you might miss the shape of the curve.

Let $x = 0$; then $y =$

Let $x = 1$; then $y =$

Let $x = 2$; then $y =$

Let $x = 3$; then $y =$

Let $x = 4$; then $y =$

Let $x = 5$; then $y =$

Plot the points on the grid provided and connect with a smooth curve. Locate as many more points as you need to be sure of its basic shape.

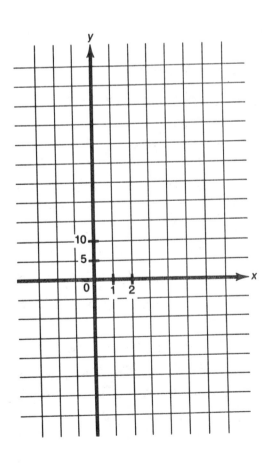

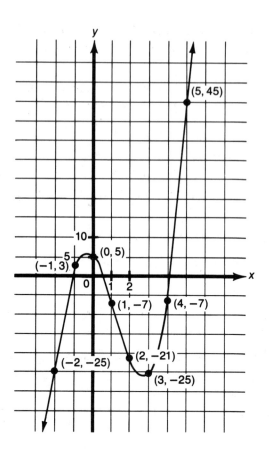

Answer:

---

Again notice that every ordered pair whose coordinates satisfy the function, $y = f(x) = 2x^3 - 7x^2 - 7x + 5$, must correspond to some point on the curve. And every point on the curve must have coordinates that satisfy the function.

As you probably observed, even with a calculator plotting points could become quite time consuming if the cubic function was more complicated, and the key points could be easily missed. So we will move to a second method.

Like linear and quadratic functions, the graph of a cubic function is very predictable. The following are some basic facts that can be used when graphing a cubic function of one variable.

# BASIC FACTS ABOUT THE GRAPH OF A CUBIC FUNCTION IN STANDARD FORM: $y = f(x) = ax^3 + bx^2 + cx + d$

1.  The graph of a cubic function is a smooth, ⌢⌣ -shaped curve.

2.  If $a$, the coefficient of the cubed term, is positive, the curve opens up; meaning, the right hand tail is going up: ⌢⌣ . If $a$ is negative, the curve opens down: ⌣⌢ .

3.  The $y$-intercept is $d$, the constant. Remember that the $y$-intercept is the value of $y$ when $x = 0$; thus $y = a(0)^3 + b(0)^2 + c(0) + d = d$.

4.  There are **at most three $x$-intercepts**, "at most" meaning there can be three or less. The $x$-intercepts are the values of $x$ when $y = 0$; thus the solution to $0 = ax^3 + bx^2 + cx + d$ yields the $x$-intercepts.

Based on the above information about cubic functions of one variable, it is possible to do nothing more than a rough sketch of the function at this time. In later units we will refine the approach so as to include more detail.

# SUGGESTED APPROACH FOR SKETCHING CUBIC FUNCTIONS OF ONE VARIABLE

1.  Write the cubic function in standard form.

2.  Determine whether the cubic opens up or down.

3.  Find the $y$-intercept at $d$.

4.  Find the $x$-intercepts, if any, by solving $0 = ax^3 + bx^2 + cx + d$. Unfortunately most cubic equations are not easily factored and we will be unable to find the $x$-intercepts.

5.  Plot the above points and connect them with a smooth,⌒⌣-shaped curve.

The following examples will illustrate this approach.

## EXAMPLE 1

Sketch:   $y = f(x) = x^3 - x^2 - 2x$.

Solution:   1.  Write in standard form and compare:

$$y = f(x) = ax^3 - bx^2 + cx + d$$
$$y = f(x) = \ \ x^3 - \ \ x^2 - 2x$$

Thus $a = 1$, $b = -1$, $c = -2$, $d = 0$.

Graph will be a cubic.

2.  Since $a$ is positive, curve opens up: ⌒⌣ .

3.  The $y$-intercept is $d = 0$.

4.  Find the $x$-intercepts by solving:

$$0 = x^3 - x^2 - 2x$$
$$0 = x(x^2 - x - 2)$$
$$0 = x(x - 2)(x + 1)$$
$$x = 0, \quad x = 2, \quad x = -1$$

The $x$-intercepts are $-1$, $0$, and $2$.

5.  Plot the above points and connect them with a smooth curve.

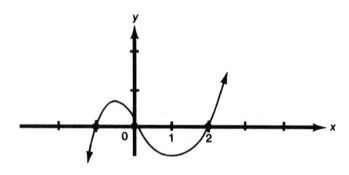

Answer:

Remember the intent is to do a rough sketch. We do not know how high the peak on the left goes up or how low the dip on the right goes down. That information will come later.

If greater detail was needed, we could plot more points.

# EXAMPLE 2

Sketch:  $y = f(x) = -x^3 - 5x^2 - 6x$.

Solution:   1.  Write in standard form and compare:

$$y = f(x) = ax^3 + bx^2 + cx + d$$

$$y = f(x) = -x^3 - 5x^2 - 6x$$

Thus $a = -1$, $b = -5$, $c = -6$, $d = 0$.

Graph will be a cubic.

2.  Since $a$ is negative, curve opens down: ⌣⌒↘ .

3.  The $y$-intercept is $d = 0$.

4.  Find the $x$-intercepts by solving:

$$0 = -x^3 - 5x^2 - 6x$$

$$0 = -x(x^2 + 5x + 6)$$

$$0 = -x(x + 2)(x + 3)$$

$$x = 0, \quad x = -2, \quad x = -3$$

The $x$-intercepts are $-3$, $-2$, and 0.

5.  Plot the above points and connect them with a smooth curve. Let me repeat, this is merely a sketch of the general shape. All we know at this point is that the curve is ⌣⌒↘ -shaped and goes through the three intercepts.

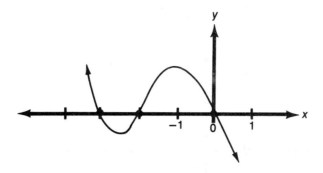

Answer:

# EXAMPLE 3

Sketch: $y = f(x) = (x - 3)(x + 2)(x - 5)$.

Solution: Believe it or not, this one is easier because it is factored already.

1. Write in standard form and compare.

As you probably have noticed, the only values that are actually used are $a$, the coefficient of the cubed term, and $d$, the constant. So rather than waste time multiplying everything out, determine only the cubed term and the constant.

$$y = f(x) = (x - 3)(x + 2)(x - 5)$$

$$= (x)(x)(x) + \quad \cdots \quad + (-3)(2)(-5)$$

$$= \quad x^3 + \quad \cdots \quad + 30$$

$$y = f(x) = a\,x^3 + bx^2 + cx + d$$

Thus $a = 1$ and $d = 30$.

Graph will be a cubic.

2. Since $a$ is positive, curve opens up: ⌢⌣.

3. The $y$-intercept is $d = 30$.

4. Find the $x$-intercepts by solving:

$$0 = (x - 3)(x + 2)(x - 5)$$

$$x = 3, \quad x = -2, \quad x = 5$$

The $x$-intercepts are 3, $-2$, and 5.

Wasn't that easy? But had the function been stated as $y = f(x) = x^3 - 6x^2 - x + 30$, it is unlikely that we would have been able to determine the $x$-intercepts unless you are far better than I at factoring cubics.

5.  Plot the above points and connect them with a smooth curve.

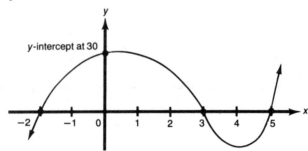

Answer:

# EXAMPLE 4

Sketch:   $y = f(x) = -x^3 + 1$.

Solution:  1.  Write in standard form and compare:

$$y = f(x) = \boxed{a}\, x^3 + bx^2 + cx \boxed{+d}$$
$$y = f(x) = \boxed{-}\, x^3 \qquad\qquad \boxed{+1}$$

Thus $a = -1$, $b = 0$, $c = 0$, $d = 1$.

Graph will be a cubic.

2.  Since $a$ is negative, curve opens down: ⤵.

3.  The $y$-intercept is at 1.

4.  Find the $x$-intercepts by solving:

$$0 = -x^3 + 1$$
$$x^3 = 1$$
$$x = 1$$

The $x$-intercept is 1.

5.  Several possibilities exist for the sketch, but they must all have three things in common. There is exactly one $x$-intercept and it is 1, the $y$-intercept is at 1, and the curve is shaped like ⤵ .

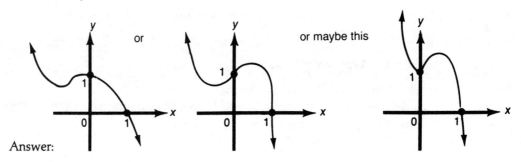

Answer:

Ready to try one?

## Problem 2

Sketch:  $y = f(x) = x^3 - 6x^2 + 9x$.

Solution:   1.   Write in standard form and compare.

2.   Curve will open _____.

3.   The $y$-intercept is at _____.

4.   The $x$-intercepts, at most three, are _____.

5.   Plot the points on the grid provided and connect them with a smooth curve.

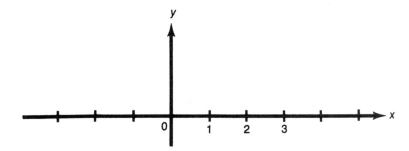

Answer:   Your sketch might have looked like either of the two below. There were only two $x$-intercepts. If your sketch showed more than two it was wrong.

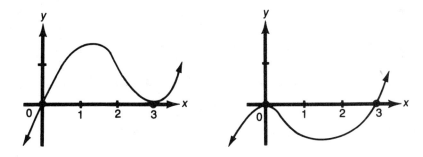

What happens if we are unable to factor the cubic equation to find the $x$-intercepts? There is no formula, such as the quadratic formula, that exists to help us. So without plotting points, there is little that can be done. The next example will illustrate what I mean.

## EXAMPLE 5

Sketch:  $y = f(x) = \frac{1}{3}x^3 - 3x^2 + 5x + 2$.

Solution:   1.   Write in standard form and compare:

$$y = f(x) = \boxed{a}x^3 \boxed{+ b}x^2 \boxed{+ c}x \boxed{+ d}$$

$$y = f(x) = \boxed{\frac{1}{3}}x^3 \boxed{- 3}x^2 \boxed{+ 5}x \boxed{+ 2}$$

Thus $a = \dfrac{1}{3}$, $b = -3$, $c = 5$, $d = 2$.

Graph will be a cubic.

2.   Since $a$ is positive, curve opens up.

3.   The $y$-intercept is at 2.

4.   Find the $x$-intercepts by solving:

$$0 = \frac{1}{3}x^3 - 3x^2 + 5x + 2$$

I could spend days and still never be able to factor, thus solve, the equation. It doesn't mean there aren't any $x$-intercepts; it only means I am unable to factor the cubic and find them.

The best that can be said is that the function has at most three $x$-intercepts.

5.   Not much can be done with a sketch other than to show that the curve is ⌢⌣ -shaped opening up and the $y$-intercept is at 2.

Here are some possibilities of what it might look like.

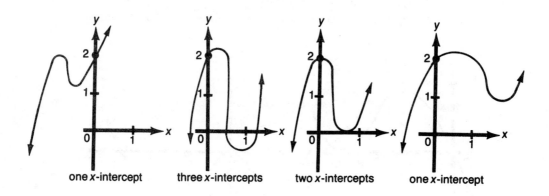

one $x$-intercept       three $x$-intercepts       two $x$-intercepts       one $x$-intercept

Of course, if we had nothing better to do, there is always plotting points.

A few units from now, we will come back and finish this one.

# POLYNOMIAL FUNCTIONS

Linear, quadratic, and cubic functions are all part of a larger group of functions called polynomial functions. Next we will consider what is meant by a polynomial function, how we can identify such functions, and what procedure we can use to graph them.

> **Definition:** A function $y = f(x) = a_n x^n + a_{n-1} x^{n-1} + \cdots + a_2 x^2 + a_1 x + a_0$, where $n$ is a counting number, $a_n \neq 0$, and $a_0, a_1, \ldots a_n$ are real numbers, is called an **$n$th degree polynomial function** involving one independent variable $x$ and a dependent variable $y$. Sometimes this is shortened to read a polynomial function of one variable, meaning there is only one independent variable.

In other words, an $n$th degree polynomial function of one variable has:

1. One independent variable and one dependent variable.

2. The dependent variable raised only to the first power.

3. One term containing the independent variable raised to the $n$th power.

4. The highest exponent being $n$.

5. All exponents being counting numbers.

6. Neither the independent nor dependent variable appearing in any denominator.

7. No term containing a product of the independent and dependent variables.

Here are some examples of polynomial functions of one variable and their degrees:

$$y = f(x) = x^4 + 3x - 1 \qquad \text{fourth degree}$$

$$y = g(x) = 1 - 5x^3 \qquad \text{third degree}$$

$$y = h(x) = 27x^{19} - 10x^7 + x \quad \text{nineteenth degree}$$

$$y = f(x) = x^2 - 3x + 4 \qquad \text{second degree}$$

$$y = F(x) = 2x - 7 \qquad \text{first degree}$$

Here are some examples that are not polynomial functions of one variable:

$$y = f(x) = \sqrt{x^2 + x - 1}$$

$$y = g(x) = \frac{3 - x}{x}$$

$$y^2 = x^3 + 3x^2 + 2x + 7$$

$$\frac{1}{y} = x^4 + 3x - 1$$

Before proceeding, determine why each of the above is not a polynomial function.

# GRAPHING POLYNOMIAL FUNCTIONS

First-, second-, and third-degree polynomial functions are nothing more than linear, quadratic, and cubic functions with the coefficients of the $x$-terms changed to different letters of the alphabet. And we know a lot about graphing linear, quadratic, and cubic functions.

To summarize what we know so far:

**First-degree polynomial functions:**   $y = f(x) = mx + b$

They also are called linear functions.

The graph is a straight line.

The slope is $m$.

The $y$-intercept is at $b$.

**Second-degree polynomial functions:**   $y = f(x) = ax^2 + bx + c$

They also are called quadratic functions.

The graph is a ⌣-shaped curve called a parabola.

If $a$ is positive, the curve opens up: ⌣.

If $a$ is negative, the curve opens down: ⌢.

The $y$-intercept is at $c$.

There are at most two $x$-intercepts.

The solution to $0 = ax^2 + bx + c$ yields the $x$-intercepts.

The low point on the curve (or the high point if the curve opens down) is called the vertex.

The vertex is located at the point

$$\left(\frac{-b}{2a}, \frac{4ac - b^2}{4a}\right).$$

The curve is symmetric to a vertical line (called the axis of symmetry) through the vertex.

**Third-degree polynomial functions:**   $y = f(x) = ax^3 + bx^2 + cx + d$

They also are called cubic functions.

The graph is a ⌒⌣-shaped curve.

If $a$ is positive, the curve opens up: ⌒⌣.

If $a$ is negative, the curve opens down: ⌣⌒ .

The $y$-intercept is at $d$.

There are at most three $x$-intercepts.

The solution to $0 = ax^3 + bx^2 + cx + d$ yields the $x$-intercepts.

**The predictability ends with third-degree polynomial functions.** Starting with fourth-degree polynomial functions, their graphs take on a variety of shapes. The best we can do at this stage is to offer the following generalizations.

# BASIC FACTS ABOUT THE GRAPHS
# OF POLYNOMIAL FUNCTIONS

1.   The graph is always a smooth curve without any breaks or sharp corners.

2.   The number of $x$-intercepts is at most equal to the degree of the polynomial function.

3.   The domain is the set of real numbers.

In other words, if you were to make up a fifth-degree polynomial function, there is no way to predict what its graph would look like other than to say it will be a nice smooth curve without any breaks or sharp corners, and it will have at most five $x$-intercepts. Graphing would require locating an adequate number of points and connecting them with a smooth curve.

The graph of your fifth-degree polynomial function might look like one of the following:

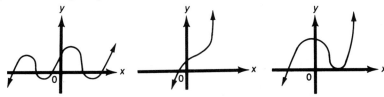

It could not look like this:

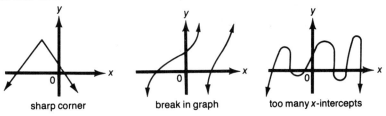

sharp corner             break in graph          too many $x$-intercepts

We will not attempt graphing any higher-order polynomial functions in this unit.

You should now be able to identify a polynomial function and to distinguish it from functions that are not polynomial functions. Also you should be able to describe the basic shape of the graphs for first-, second-, and third-degree polynomial functions. Remember that:

the graph of a first-degree polynomial function, also called a linear function, is a straight line;

the graph of a second-degree polynomial function, also called a quadratic function, is a ⌣ ⌢ -shaped curve called a parabola;

the graph of a third-degree polynomial function, also called a cubic function, is a ⌣⌢ -shaped curve.

Before beginning the next unit you should work the following problems.

## EXERCISES

By means of sketches, show the pertinent characteristics of each of the following functions, including *all* the $x$-intercepts.

1.  $f(x) = x^3 - 9x^2 + 18x$

2.  $f(x) = 2x^3 + 6x^2 - 20x$

3.  $f(x) = 12x^3 - 24x^2 + 12x$

4.  $f(x) = (x - 2)(x + 3)(x - 1)$

5.  $f(x) = -x^3 + 3x^2$

6.  $f(x) = x^3 - 4x$

7.  $f(x) = \begin{cases} 3, & \text{if } 5 < x \\ x^2, & \text{if } \quad x \le 2 \\ x + 1, & \text{if } 3 \le x \le 5 \end{cases}$

# UNIT 9

## The Derivative

This unit starts the beginning of what is traditionally called differential calculus. The purpose of this unit is to provide you with an understanding of a derivative. An intuitive approach will be used rather than formal, rigorous mathematical definitions. When you have finished the unit, you will be able to differentiate polynomial functions.

Recall the following definition from an earlier unit.

$$\text{Definition:} \quad \text{Slope of a line} = m = \frac{\Delta y}{\Delta x} = \frac{y_2 - y_1}{x_2 - x_1}$$

Recall that the slope of a line is the rate at which $y$ changes with respect to $x$. And a fundamental property of a straight line is that the slope is constant. As a review, consider the function $f(x) = \frac{2}{3}x + 5$, with slope $= m = \frac{\Delta y}{\Delta x} = \frac{2}{3}$. In other words, if $x$ increases by 3, the $y$ value will increase by 2 and this ratio will remain constant for the entire line.

**Differential calculus deals with finding the slope of a curve.**

When we speak of the slope of a curve, we really mean the slope of the tangent line to the curve at a point.

What do we mean by a tangent line? Without going into an elaborate explanation, think back to your days in geometry class. Remember a tangent to a circle? It was a line that touched the circle at only one point. That is close to what is meant here, except instead of it being a line, think of the tangent line as being a very short line segment.

Here we go with another picture.

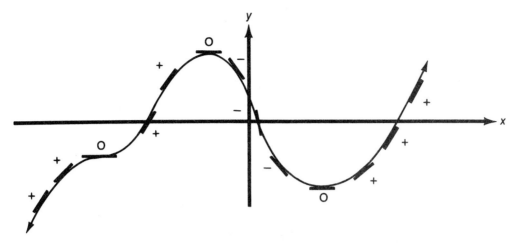

The short line segments represent tangent lines at various points along the curve.

Be sure to notice that the slopes of the tangent lines are changing. In some places the slopes are positive, in others negative, and at still other points, the slope is zero. Unlike a straight line with a constant slope, the slope of a curve is *not* constant.

---

Definition:   If $f(x)$ is a polynomial function, then $f'(x)$ is the symbol for the slope of the tangent line to the curve at any point.

---

Definition:   $f'(x)$ is called the **derivative**. $f'(x)$ is read "f prime of x."

---

# RULES FOR DETERMINING DERIVATIVES

Next follow four rules for determining derivatives. In this unit we will concentrate on the mechanics of finding the derivative and wait until the following unit before illustrating the relationship of the derivative to the slope of the curve.

Let me explain the notation to be used: $f(x)$, $u(x)$, and $v(x)$ denote functions of $x$, such as various polynomials, and $c$ denotes a constant, such as a number.

---

**Power Rule**

If $f(x) = x^n$, then $f'(x) = nx^{n-1}$ for all $n$, $n$ being a real number.

In words, if $x$ is raised to some exponent, the derivative is equal to the exponent times $x$ raised to the exponent minus 1.

---

Of all the rules for determining derivatives, the Power Rule is the one used most frequently. Memorize it!

We will do lots of examples. For the moment, leave Example 1 blank; we will come back to it later.

| Power Rule: | If $f(x) = x^n$, then $f'(x) = nx^{n-1}$ |
| --- | --- |

## EXAMPLE 1

## EXAMPLE 2
If $f(x) = x^2$, then $f'(x) = 2x^{2-1} = 2x$

## EXAMPLE 3
If $f(x) = x^3$, then $f'(x) = 3x^{3-1} = 3x^2$

## EXAMPLE 4
If $f(x) = x^4$, then $f'(x) = 4x^{4-1} = 4x^3$

## EXAMPLE 5
If $f(x) = x^5$, then $f'(x) = 5x^{5-1} = 5x^4$

Do you see the pattern? To find the derivative, you take the exponent times $x$ raised to the exponent minus one. Before returning to Example 1, try the next few yourself.

## Problem 1
If $f(x) = x^6$, then $f'(x) =$

## Problem 2
If $f(x) = x^{34}$, then $f'(x) =$

## Problem 3
If $f(x) = x^{88}$, then $f'(x) =$

## Problem 4
If $f(x) = x^{27}$, then $f'(x) =$

Answers:   $6x^5$, $34x^{33}$, $88x^{87}$, $27x^{26}$

By now you should be comfortable enough with the Power Rule that we can return to Example 1 and finish it. It should be obvious what function I want in there.

## EXAMPLE 1 (Continued)

If $f(x) = x$

$= x^1$        because if no exponent is written, the exponent is understood to be 1.

Then $f'(x) = 1x^{1-1}$

$= 1x^0$

$= x^0$

$= 1$        because the definition of a zero component is $b^0 = 1$, if $b \neq 0$.

We will refer to Examples 1–5 for the remainder of the unit, so go back and write in the following information in the blank space provided by Example 1: If $f(x) = x$, then $f'(x) = 1$.

---

**Derivative of a Constant**

If $f(x) = c$, then $f'(x) = 0$, where $c$ is a constant.

The derivative of a constant term is 0.

---

If $f(x) = c$, then $f'(x) = 0$.

EXAMPLE 6    If $f(x) = 5$, then $f'(x) = 0$.

EXAMPLE 7    If $f(x) = -10$, then $f'(x) = 0$.

EXAMPLE 8    If $f(x) = 2$, then $f'(x) = 0$.

---

**Derivative of a Constant Times a Function**

If $f(x) = c \cdot u(x)$, then $f'(x) = c \cdot u'(x)$.

The derivative of a constant times a function equals the constant times the derivative of the function.

---

## EXAMPLE 9

If $f(x) = 5x^2$, find $f'(x)$.

Solution:      If $f(x) = 5x^2$,      This is a constant times a function.

From the Power Rule,

then $f'(x) = 5(2x)$      the derivative of $x^2 = 2x$.

$= 10x$

---

## EXAMPLE 10

If $f(x) = -3x^4$, find $f'(x)$.

Solution:      If $f(x) = -3x^4$,      This is a constant times a function.

From the Power Rule,

then $f'(x) = -3(4x^3)$      the derivative of $x^4 = 4x^3$.

$= -12x^3$

---

## EXAMPLE 11

If $f(x) = 7x$, find $f'(x)$.

Solution:     If $f(x) = 7x$,

then $f'(x) = 7(1)$         From the Power Rule,

                             the derivative of $x$ is 1.

$= 7$

---

# EXAMPLE 12

If $f(x) = cx^n$, find $f'(x)$.

---

Solution:     If $f(x) = cx^n$,

                             From the Power Rule,

then $f'(x) = cnx^{n-1}$         the derivative of $x^n = nx^{n-1}$.

---

The last example is a special case of a constant times a function where the function is $x$ raised to some power, and can be used as a rule by itself. Stated in words, the derivative of a constant times $x$ raised to some exponent equals the constant times the exponent times $x$ raised to the exponent minus one.

Special Case:   If $f(x) = cx^n$, then $f'(x) = cnx^{n-1}$

The last rule of this unit combines all of the previous rules.

---

**Derivative of Sums and/or Differences**

If $f(x) = u(x) + v(x)$, then $f'(x) = u'(x) + v'(x)$.

If $f(x) = u(x) - v(x)$, then $f'(x) = u'(x) - v'(x)$.

The derivative of the sum (or difference) of two or more functions equals the sum (or difference) of their respective derivatives.

---

Until now, the first three rules dealt with taking the derivative of a single term. This rule defines how to take the derivative of several terms; the derivative is taken term by term.

---

# EXAMPLE 13

Find the derivative of $f(x) = x^3 + 7x^2 - 3x + 10$.

---

Solution:     If $f(x) = x^3 + 7x^2 - 3x + 10$,

then $f'(x) = 3x^2 + 7(2x) - 3(1) + 0$         from the Power Rule.

$= 3x^2 + 14x - 3$

# EXAMPLE 14

Find the derivative of $f(x) = 15x^4 - 2x^3 + 11x^2 + 3x - 1$.

Solution:   If $f(x) = 15x^4 - 2x^3 + 11x^2 + 3x - 1$,

then $f'(x) = 15(4x^3) - 2(3x^2) + 11(2x) + 3(1) - 0$

$= 60x^3 - 6x^2 + 22x + 3$

Not too bad is it?

Keep in mind what we are finding with all of this. The geometric interpretation is that the derivative is a formula, expressed in terms of $x$, for finding the slope of the tangent line at any point on the graph of a function, $f(x)$. We will explore what all of that means in the next unit after you have had an opportunity to practice taking derivatives.

Through the combined used of the following rules, you should now be able to take the derivative of any polynomial function.

If $f(x) = x^n$, then $f'(x) = nx^{n-1}$ for all $n$, $n$ being a real number.

If $f(x) = c$, then $f'(x) = 0$, where $c$ is a constant.

If $f(x) = c \cdot u(x)$, then $f'(x) = c \cdot u'(x)$.

If $f(x) = cx^n$, then $f'(x) = cnx^{n-1}$.

If $f(x) = u(x) + v(x)$, then $f'(x) = u'(x) + v'(x)$.

If $f(x) = u(x) - v(x)$, then $f'(x) = u'(x) - v'(x)$.

# EXERCISES

Find $f'(x)$ for each of the following:

1.   $f(x) = 3x^{15}$

2.   $f(x) = -7x^2$

3.   $f(x) = 21x^7$

4.   $f(x) = x$

5.   $f(x) = -11x^9$

6.   $f(x) = 10$

7.   $f(x) = -8$

8.   $f(x) = 4x + 3$

9.   $f(x) = -3x + 17$

10.   $f(x) = \dfrac{2}{5}x$

11.   $f(x) = 3x^2 + 2x - 6$

12.   $f(x) = -8x^3 - 7x^2 + 11x - 5$

13.   $f(x) = 5x^3 - 4x^2 + 3x + 20$

14.   $f(x) = 34x^2 + x^5$

15.   $f(x) = \frac{1}{3}x^3 + \frac{1}{2}x^2 - 2x + 13$

16.   $f(x) = ax^2 + bx + c$

17.   $f(x) = mx + b$

18.   $f(x) = ax^3 + bx^2 + cx + d$

---

If additional practice is needed:

Budnick, page 451, problems 1–4
Piascik, page 92, problems 2–3

# UNIT 10

# More About the Derivative

In this unit we will continue our introduction to a derivative by illustrating the relationship between the slope of the curve and the derivative. And at the end of the unit, alternative notation for denoting a derivative will be described.

To illustrate the relationship between the derivative and the slope of the tangent line to a curve, let's return to a familiar function, the second-degree polynomial.

## EXAMPLE 1

Graph: $y = f(x) = x^2 - 6x + 5$.

Solution:
1. This is a quadratic with $a = 1$, $b = -6$, $c = 5$.

2. The parabolic curve opens up.

3. The $y$-intercept is at 5.

4. The $x$-intercepts are at 1 and 5 because

$$0 = x^2 - 6x + 5$$

$$0 = (x - 1)(x - 5)$$

$$x - 1 = 0 \quad x - 5 = 0$$

$$x = 1 \qquad x = 5$$

5.  The vertex is located at $(3, -4)$ because

    the axis of symmetry is $x = 3$

    and $y = f(3) = (3)^2 - 6(3) + 5 = 9 - 18 + 5 = -4$

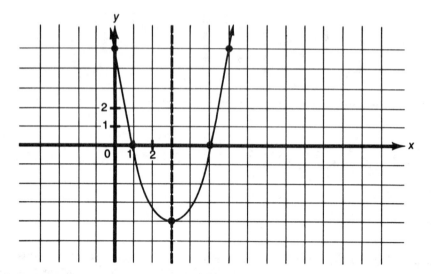

Answer:

___

In Unit 9, we stated that from a geometric interpretation the derivative is a formula, expressed in terms of $x$, for finding the slope of the tangent line at any point on the graph of a function, $f(x)$.

## EXAMPLE 2

Find the derivative for $f(x) = x^2 - 6x + 5$.

___

Solution:    $f(x) = x^2 - 6x + 5$

$f'(x) = 2x - 6(1) + 0$ ◄───────────┐

$\quad\quad = 2x - 6$

The sooner you are able to skip that first step, the better.

___

| Definition:    $f'(a)$ is the slope of the tangent line to the curve at the specific point where $x = a$. |
|---|

## EXAMPLE 3

Compute and interpret:    $f'(5)$, $f'(3)$, $f'(0)$, and $f'(2)$ from Examples 1 and 2.

___

Solution:   From Example 1:   $f(x) = x^2 - 6x + 5$

From Example 2:   $f'(x) = 2x - 6$

$$f'(5) = 2(5) - 6 = 4$$

From the graph, when $x = 5$, $y = 0$, which is the point (5, 0).

The slope of the tangent line to the curve at the point (5, 0) is 4.

Remember that tangent lines are represented by *very short* line segments. On the next grid I have drawn just such a line segment at the point (5, 0) with a slope of 4. I have tried to aid your understanding by showing, off to the side, what a line looks like with a slope of 4. The lines are meant to be parallel.

$$f'(3) = 2(3) - 6 = 0$$

From the graph, when $x = 3$, $y = -4$, which is the point (3, -4).

The slope of the tangent line to the curve at the point (3, -4) is 0.

A line with 0 slope is horizontal and the tangent is shown as such.

$$f'(0) = 2(0) - 6 = 0 - 6 = -6$$

From the graph, when $x = 0$, $y = 5$, which is the point (0, 5).

The slope of the tangent line to the curve at the point (0, 5) is -6.

The tangent line is shown as a very short line segment with slope of -6. Again I have tried to aid you by showing, off to the side, what a line looks like with slope of -6.

$$f'(2) = 2(2) - 6 = 4 - 6 = -2$$

In this instance, the $y$ value for the point needs to be determined.

If $x = 2$, what is $y$?   $f(2) = (2)^2 - 6(2) + 5$

$$= 4 - 12 + 5$$

$$= -3, \text{ which is the point } (2, -3).$$

The slope of the tangent line to the curve at the point (2, -3) is -2.

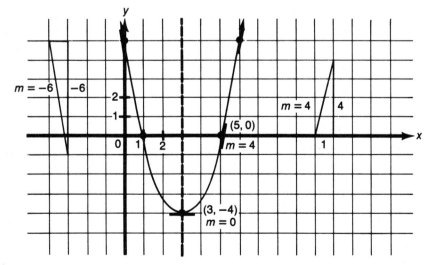

Answer:

There was nothing special about the points selected for the previous example. Indeed, any value can be used for $x$ as the next problem is intended to show.

## Problem 1

Using the information provided in Examples 1 and 2, find and interpret $f'(10)$.

Solution:

Answer:   $f'(10) = 14$. The slope of the tangent line to the curve at $x = 10$ is 14. Or better still, the slope of the tangent line to the curve at the point $(10, 45)$ is 14.

Sometimes the derivative is referred to as the **instantaneous rate of change** of $y$ with respect to $x$. It is an estimate or approximation of the change in $y$ for a unit change in $x$. Notice the similarity of the wording to that of the slope of a line.

In the previous problem you found the slope of the tangent to be 14. Another interpretation of the number 14, would be that as $x$ increases by 1 from the point $(10, 45)$, the $y$ value will increase by approximately 14. It is an approximation because 14 represents the slope of the tangent line, whereas the function itself is a curve.

## EXAMPLE 4

Find and interpret $f'(x)$ for $f(x) = 5x - 2$.

Solution:    $f(x) = 5x - 2$

$f'(x) = 5$

The derivative is a formula for finding the slope of the tangent line to the curve at any point. In this example, the formula equals the constant 5, regardless of the values of $x$. But that is consistent with what we already know. $f(x) = 5x - 2$ is a linear function with slope of 5 and the slope is constant everywhere on the line. To say that $f'(x) = 5$ is the same thing as $m = 5$, but in different symbols.

Answer:   $f'(x) = 5$ means that the slope of the line is 5.

Consider another similar situation.

## EXAMPLE 5

Find and interpret $f'(x)$ for $f(x) = 3$.

Solution:    $f(x) = 3$
$f'(x) = 0$

Answer:   $f'(x) = 0$ means that the slope of the line is 0.

Again this is consistent with what we know. The graph of $f(x) = 3$ is a horizontal line with $y$-intercept of 3 and slope of 0. The graph of $f(x) = c$ always represents a horizontal line and the slope of a horizontal line is 0. That is the reasoning behind the rule that the derivative of a constant is zero.

___

And now for some unfamiliar examples.

# EXAMPLE 6

Find and interpret $f'(0)$ for $f(x) = 7x^5 + 3x^4 - 11x^3 + x^2 + 25x + 1$.

___

Solution:   $y = f(x) = 7x^5 + 3x^4 - 11x^3 + x^2 + 25x + 1$

$y = f(0) = 0 + 0 - 0 + 0 + 0 + 1$

$\qquad = 1$ which is the point $(0, 1)$ on the curve.

$f'(x) = 35x^4 + 12x^3 - 33x^2 + 2x + 25$

$f'(0) = 0 + 0 - 0 + 0 + 25$

$\qquad = 25$

Answer:   The slope of the tangent line to the curve at the point $(0, 1)$ is 25.

Comment:   This is a fifth-degree polynomial function. I have no more idea of what the graph of this curve looks like than you do. What we have determined is that at the point $(0, 1)$, the curve is increasing at the approximate rate of 25; this means if $x$ increases by 1 (*from* 0), the $y$ value will increase (*from* 1) by approximately 25. Or the next point would be at about $(0 + 1, 1 + 25) = (1, 26)$. If need be, we could calculate the next point by substituting 1 into the function to determine $y$:

$$y = f(1) = 7(1)^5 + 3(1)^4 - 11(1)^3 + (1)^2 + 25(1) + 1$$

$$= 7 + 3 - 11 + 1 + 25 + 1$$

$$= 26$$

In this example, our approximation turned out to be the exact answer.

___

The next example will use the derivative as an aid in curve sketching.

# EXAMPLE 7

Find $f'(0)$, $f'(1)$, $f'(2)$, $f'(3)$, and $f'(4)$ where $y = f(x) = x^3 - 6x^2 + 9x$ and use the results to sketch the graph of $f(x)$.

___

Solution:     $y = f(x) = x^3 - 6x^2 + 9x$

$y = f(0) = 0 - 0 + 0$

$\qquad = 0$ which means $(0, 0)$ is a point on the curve.

$f'(x) = 3x^2 - 12x + 9$

$f'(0) = 0 - 0 + 9$

$\qquad = 9$

The slope of the tangent line to the curve at the point $(0, 0)$ is 9.

$y = f(x) = x^3 - 6x^2 + 9x$

$y = f(1) = 1 - 6 + 9$

$\qquad = 4$ which means $(1, 4)$ is a point on the curve.

$f'(x) = 3x^2 - 12x + 9$

$f'(1) = 3 - 12 + 9$

$\qquad = 0$

The slope of the tangent line to the curve at the point $(1, 4)$ is 0.

$y = f(x) = x^3 - 6x^2 + 9x$

$y = f(2) = (2)^3 - 6(2)^2 + 9(2)$

$\qquad = 8 - 24 + 18$

$\qquad = 2$ which means $(2, 2)$ is a point on the curve.

$f'(x) = 3x^2 - 12x + 9$

$f'(2) = 3(2)^2 - 12(2) + 9$

$\qquad = 12 - 24 + 9$

$\qquad = -3$

The slope of the tangent line to the curve at the point $(2, 2)$ is $-3$.

$y = f(x) = x^3 - 6x^2 + 9x$

$y = f(3) = (3)^3 - 6(3)^2 + 9(3)$

$\qquad = 27 - 54 + 27$

$\qquad = 0$ which means $(3, 0)$ is a point on the curve.

$f'(x) = 3x^2 - 12x + 9$

$f'(3) = 3(3)^2 - 12(3) + 9$

$\qquad = 27 - 36 + 9$

$\qquad = 0$

The slope of the tangent line to the curve at the point $(3, 0)$ is 0.

$$y = f(x) = x^3 - 6x^2 + 9x$$

$$y = f(4) = (4)^3 - 6(4)^2 + 9(4)$$

$$= 64 - 96 + 36$$

$$= 4 \text{ which means } (4, 4) \text{ is a point on the curve.}$$

$$f'(x) = 3x^2 - 12x + 9$$

$$f'(4) = 3(4)^2 - 12(4) + 9$$

$$= 48 - 48 + 9$$

$$= 9$$

The slope of the tangent line to the curve at the point (4, 4) is 9.

The above information is plotted on the next grid. Are you able to see in your own mind the shape of the curve?

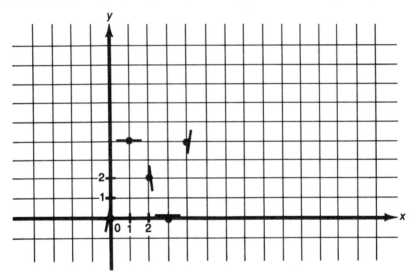

Now combine the above information with our previous knowledge. The function $y = f(x) = x^3 - 6x^2 + 9x$ is a cubic with $a = 1$. The curve is opening up:

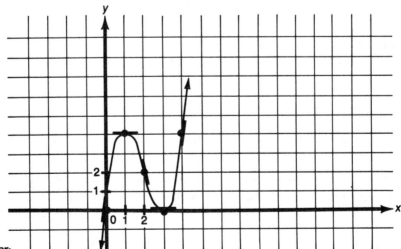

Answer:

Return to Problem 2 of Unit 8, and compare the results. What important information has been added?

Two important items have been determined. One, the basic location of the graph now has been established. Two, the high point of the left peak is at (1, 4). Do you see why?

It is time to reverse the procedure slightly.

## EXAMPLE 8

If $f(x) = x^2 - 3x - 4$, find the value of $x$ such that the slope is 11.

Solution:    $f(x) = x^2 - 3x - 4$

$f'(x) = 2x - 3$

The derivative is a formula for finding the slope of the tangent line to the curve at a point. This problem is asking for the value of $x$ when the slope is 11. Stated another way, when $f'(x) = 11$, what is $x$? The answer is the solution to the following equation:

$$11 = 2x - 3$$

$$14 = 2x$$

$$x = 7$$

Answer:    When $x = 7$, the slope of the tangent line to the curve is 11. Or shortened, when $x = 7$ the slope of the curve is 11. You might wish to verify the answer for yourself.

## Problem 2

If $f(x) = x^2 - 3x - 4$, find the value of $x$ such that the slope is 0.

Solution:

Answer:   $x = 3/2$

Do you recognize what you have found? Can you relate the value of $x$ to anything that is already known about this parabola? If not, I suggest you graph the function.

# OTHER NOTATION FOR DERIVATIVES

Thus far $f'(x)$ has been used exclusively to denote the derivative. For a variety of reasons, which we will not discuss here, there are numerous other notations that are used as well. Some have more advantages than others. The important thing is that you be able to recognize each, and understand the notation.

Consider the function:   $y = f(x) = 3x^2 - 7x + 2$

We know

1.  $f'(x)$ denotes the derivative and $f'(x) = 6x - 7$.

Other notations that also denote the derivative are:

2.  $y'$ read "$y$ prime" and $y' = 6x - 7$.

3.  $f'$ read "$f$ prime" and $f' = 6x - 7$.

In addition there are some new symbols that denote a derivative:

4.  $\dfrac{dy}{dx}$   read "$dy\ dx$" and $\dfrac{dy}{dx} = 6x - 7$.

$\dfrac{dy}{dx}$ is a recognized symbol in itself. In other words, it is *not* a fraction whereby the $d$'s may be cancelled.

$\dfrac{dy}{dx}$ stresses the rate of change concept, much like the notation $\dfrac{\Delta y}{\Delta x}$ did for the slope of a line.

One of the major advantages of this notation is that by definition the dependent variable appears in the numerator, the independent variable appears in the denominator, and any other variables appearing in the function are to be considered constants.

5.  $\dfrac{df}{dx}$   read "$df\ dx$" uses the name of the function rather than the dependent variable.

And then there are a few variations, the necessity for which you will not see until the following unit.

6.  $\dfrac{d(3x^2 - 7x + 2)}{dx}$   read "the derivative of $3x^2 - 7x + 2$ with respect to $x$" and again

$\dfrac{d(3x^2 - 7x + 2)}{dx} = 6x - 7$.

7.  $D_x(3x^2 - 7x + 2)$   read the same as number 6 and $D_x(3x^2 - 7x + 2) = 6x - 7$.

So all total there are seven common notations, all used to represent the derivative. Although I have favorites, I will try to use all seven so that you may get used to them.

## EXAMPLE 9

If $y = f(x) = 5x^3$, find $f'(x)$, $y'$, $f'$, $\dfrac{dy}{dx}$, $\dfrac{df}{dx}$, $\dfrac{d(5x^3)}{dx}$ and $D_x(5x^3)$.

Solution:   All of the above notations denote the derivative.

Answer:   The derivative of $5x^3$ is $15x^2$.

# VOCABULARY LESSON

Derivative is a noun. Differentiate is a verb. And differentiable is an adjective. Therefore, the two sentences mean exactly the same thing:

    a.   Find the derivative of $y = f(x) = 3x^2$.

    b.   Differentiate the function $y = f(x) = 3x^2$.

And to say that the function, $y = f(x) = 3x^2$, is differentiable means that the derivative exists.

  The next two examples pull together the concepts of this unit with that of writing linear functions.

## EXAMPLE 10

Write the equation of the tangent line to the curve, $y = f(x) = x^2$, at the point $(-3, 9)$.

Solution:   To write the equation of a line, the point-slope formula is used.

In order to use the formula, we need two items, the slope and a given point on the line. The point was given as $(-3, 9)$. What is the slope?

$$y = f(x) = x^2 \text{ is the function.}$$

$$f'(x) = 2x \text{ is the formula for the slope of the tangent line.}$$

$$f'(-3) = 2(-3) = -6 \text{ is the slope of the tangent line at the point.}$$

Use point-slope formula.      $y - y_1 = m(x - x_1)$

Substitute.                   $y - 9 = -6(x - (-3))$

Remove parentheses.       $y - 9 = -6(x + 3)$

                              $y - 9 = -6x - 18$

Solve for $y$.                     $y = -6x - 9$

Answer:   The equation of the tangent line to the curve at the point $(-3, 9)$ is $y = -6x - 9$.

## EXAMPLE 11

Write the equation of the tangent line to the curve, $y = f(x) = x^2$, where $x = 5$.

Solution:   To write the equation of a line, the point-slope formula is used.

In order to use the formula, we need two items, the slope and a given point on the line. All that is given is that $x = 5$. What is the slope? What is the $y$-coordinate?

$$y = f(x) = x^2 \text{ is the function.}$$

$$y = f(5) = (5)^2 = 25 \text{ means that } (5, 25) \text{ is the point on the curve.}$$

$$f'(x) = 2x \text{ is the formula for the slope of the tangent line.}$$

$$f'(5) = 2(5) = 10 \text{ is the slope of the tangent line at the point.}$$

Use point-slope formula.      $y - y_1 = m(x - x_1)$

Substitute.                   $y - 25 = 10(x - 5)$

|  |  |  |
|---|---|---|
| Remove parentheses. | | $y - 25 = 10(x - 5)$ |
|  |  | $y - 25 = 10x - 50$ |
| Solve for $y$. | | $y = 10x - 25$ |

Answer:   The equation of the tangent line to the curve at the point $(5, 25)$ is $y = 10x - 25$.

---

You should now be able to evaluate and explain expressions such as $f'(a)$ where $x = a$. In addition you should be able to write the equation of the tangent line to the curve at the point where $x = a$.

Further, you should be able to recognize the seven common notations used to denote a derivative: $f'(x)$, $f'$, $y'$, $\dfrac{dy}{dx}$, $\dfrac{df}{dx}$, $\dfrac{df(x)}{dx}$, and $D_x$.

## EXERCISES

1.  Find $\dfrac{dy}{dx}$ if $y = 11 + 2x - x^3$.

2.  Find $D_x(7x + 2 - 3x^5)$.

3.  If $f(x) = 3 - 7x^2 + 21x + 2x^5$, find $f'(1)$.

4.  Find and interpret $f'$ for $f(x) = 11x - 2$.

5.  Find $\dfrac{d(1 - 3x^2)}{dx}$.

6.  If $y = x^6/3 - 2x$, find $\dfrac{dy}{dx}$. Hint: $x^6/3 = \dfrac{x^6}{3} = \dfrac{1}{3}x^6$

7.  Differentiate: $f(x) = x^4/4 - x^3/3 + 5x^2$

8.  Let $f(x) = 4x^2 - 2x + 9$.

    a.   Find $f'(x)$.

    b.   Determine the slope at $x = 0$.

    c.   Find $f'(2)$.

    d.   Determine the value of $x$ such that the slope is 0.

9.  Find the equation of the line tangent to the graph of $y = x^9 - 5x^8 + x + 12$ at the point $(-1, 5)$.

10.  Given the equation $y = f(x) = x^4 - 3x^3 + 2x^2 - 6$, find:

    a.   $f'(x)$

    b.   The equation of the line tangent to the graph at $x = 2$.

---

If additional practice is needed:

Barnett and Ziegler, pages 148–150, problems 9–12, 21–24; pages 157–159, problems 1–8, 15–20, 29–32, 46–48
Budnick, page 444, problems 15–22
Hoffmann, pages 86–87, problems 1–4, 26, 30
Piascik, pages 92–93, problems 4–10; pages 110–112, problems 4–6

# UNIT 11

## Instantaneous Rates of Change

In previous units the geometric interpretation of a derivative as a formula that expresses the slope of the tangent to the curve as a function of $x$ was used. This unit will show how the derivative can be interpreted as a rate of change.

As stated in Unit 10, the derivative also is referred to as the instantaneous rate of change of $y$ with respect to $x$. The derivative is used to estimate the change in $y$ for a unit change in $x$. It is an approximation because the slope of the tangent line is used to provide the estimation of the change rather than the function itself.

Here is a sampling of various application problems illustrating the derivative as a rate of change.

### EXAMPLE 1

Mr. Smith is president of the Board of Trustees for the city's Repertory Theatre. The Board's current effort is a major fund-raising project being conducted by direct-mail solicitations. Next week, a letter requesting donations to support the Repertory Theatre will be sent to a group of people whose names appear on a mailing list compiled by the Board. Mr. Smith has formulated that if $x$ letters (in hundreds) are sent, the amount of money donated will be

$$f(x) = .5x^2 + 14x + 50$$

where $x$ = the number of letters sent (in hundreds)

and $f(x)$ = money donated in dollars (in hundreds).

a. Find and interpret $f(10)$.

b. Express the rate of change of money donated with respect to the number of letters sent.

c. Find the rate of change of money donated at $x = 10$.

d. Interpret the answer to part c.

e. Compute the actual change in the amount of money donated that will result if 100 additional letters are mailed.

Solution:   a.   $f(x) = .5x^2 + 14x + 50$

$f(10) = .5(10)^2 + 14(10) + 50$

$= .5(100) + 140 + 50$

$= 50 + 140 + 50$

$= 240$

Remember both variables are measured in hundreds.

Answer:   If 1,000 letters of solicitation are sent, it is estimated that $24,000 will be donated.

b.   The rate of change of money donated, $f(x)$, with respect to the number, $x$, of letters sent is the derivative.

$$\text{Rate of change} = f'(x) = .5(2x) + 14$$

$$= x + 14$$

For any value of $x$, this derivative is an approximation to the *additional* amount of money in hundreds of dollars that will be donated by mailing out an *additional* 100 letters.

Note:   Recall the derivative is stated as the change in $y$ for a unit change in $x$. In this example a unit change in $x$ represents a change of 100 letters.

c.   Rate of change $= f'(x) = x + 14$

$$f'(10) = 10 + 14$$

$$= 24$$

d.   Currently 1,000 letters are being sent. If an additional 100 letters are sent, it is estimated that an additional $2,400 would be donated.

e.   The actual change is the difference between the amount donated if 1,100 letters are sent and the amount donated if 1,000 letters are sent.

$$f(x) = .5x^2 + 14x + 50$$

$$f(11) = .5(11)^2 + 14(11) + 50$$

$$= .5(121) + 154 + 50$$

$$= 264.5$$

$$f(10) = 240$$

$$\text{Actual change} = f(11) - f(10)$$

$$= 264.5 - 240$$

$$= 24.5$$

24.5 represents $2,450

The reason for the difference between the actual change and the estimated change is that the rate of change of the money donated varied with the number of letters sent. The instantaneous rate of change in part c can be thought of as the change in money donated that would occur if the rate of change remained constant.

In geometric terms, the difference between these two quantities is the difference between using the tangent line to estimate the next point one unit away versus using the actual curve to determine the point.

## Problem 1

Continue with Mr. Smith's fund-raising project.

a.  Find $f(16)$.

b.  Find the rate of change of money donated at $x = 16$. Are the donations increasing or decreasing at this point?

c.  Interpret $f'(16)$.

d.  Compute the actual change in the amount of money donated if 100 additional letters are sent.

Solution:

Answers:    a. 402; b. 30, increasing; c. If the number of letters sent is increased from 1,600 to 1,700, it is estimated that an additional $3,000 will be collected; d. 30.5.

Here is another example.

## EXAMPLE 2

It is estimated that $t$ months from now, the number of out-of-state vehicles on North Carolina ferries in thousands will be $N(t) = -t^2 + 13t$.

a.  Derive an expression for the rate at which the number of out-of-state vehicles will be changing with respect to time $t$ months from now.

b.  At what rate will the number of vehicles be changing 4 months from now?

c.  Will the number of vehicles be increasing or decreasing?

d.  By how much will the number of vehicles actually change during the fifth month?

Solution:    a.   The rate of change of the number of out-of-state vehicles with respect to time is the derivative of the function $N(t)$.

$$\text{Rate of change} = N'(t) = -2t + 13.$$

b.   The rate of change of the number of out-of-state vehicles 4 months from now will be

$$N'(4) = -2(4) + 13 = -8 + 13 = 5 \text{ or } 5,000 \text{ cars per month.}$$

It is an estimate of the change in $N(t)$ for a unit change in $t$. That is, as $t$ increases by 1 (from 4 to 5), the value of $N(t)$ will increase by approximately 5.

c.   The number of vehicles will be increasing.

d.   The actual change in the number of vehicles during the 5th month is the difference between the number at the end of 5 months and the number at the end of 4 months. Or,

$$\text{Change in number of vehicles} = N(5) - N(4)$$
$$= [-(5)^2 + 13(5)] - [-(4)^2 + 13(4)]$$
$$= [-25 + 65] - [-16 + 52]$$
$$= -25 + 65 + 16 - 52$$
$$= 4 \text{ or } 4,000 \text{ cars}$$

Why the difference between the actual change and the monthly rate of change? The situation is illustrated below. In part b the tangent line is being used to approximate a unit change of one on the curve. The instantaneous rate of change can be thought of as the change in the number of vehicles that would occur during the fifth month if the rate of change remained constant.

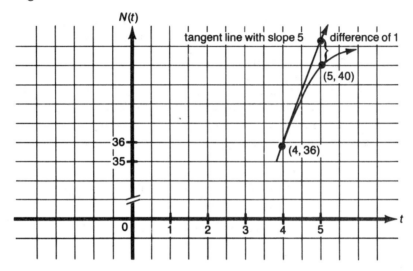

# INSTANTANEOUS VELOCITY

A formula giving the **instantaneous velocity** of an object at any point in time is given by the instantaneous rate of change of distance with respect to time. Or simply, velocity equals the derivative of the distance function expressed in terms of time.

For our example we will return to the rubber ball being thrown in Exercise 4 of Unit 7. It would be helpful if you review your answers to that question before continuing.

## EXAMPLE 3

Unit 7, Exercise 4—Revisited

If a rubber ball is thrown vertically upward from the top of a building, its height (in feet) $t$ seconds later is given by $H(t) = -16t^2 + 128t + 320$.

a.  Find the formula giving the ball's instantaneous velocity.

b.  Find the ball's instantaneous velocity at $t = 5$.

c.   What was the initial velocity of the ball when it was thrown upward?

d.   What is the velocity of the rubber ball at the instant it hits the ground?

---

Solution:   a.   Height $= H(t) = -16t^2 + 128t + 320$

Velocity $= H'(t) = -32t + 128$

b.   To determine the instantaneous velocity at $t = 5$, the derivative, $H'(t)$, must be evaluated at $t = 5$.

$$\text{Velocity} = H'(t) = -32t + 128$$

$$H'(5) = -32(5) + 128$$

$$= -160 + 128$$

$$= -32 \text{ feet per second}$$

The minus sign indicates the direction of the velocity (down).

An instantaneous rate of 32 feet per second at the end of 5 seconds means that if the rate were to remain constant for the next second, the ball would **fall** an additional 32 feet. If the ball is accelerating or decelerating (that is, if the rate does not remain constant), then the instantaneous rate is an approximation of what actually happens during the next second.

c.   To determine the initial velocity, which would be at $t = 0$, the derivative is evaluated at time $t = 0$.

$$\text{Velocity} = H'(t) = -32t + 128$$

$$H'(0) = 0 + 128$$

$$= 128 \text{ feet per second}$$

The rubber ball was thrown vertically upward with an initial velocity of 128 feet per second.

d.   In order to determine the velocity of the ball when it hits the ground, we must know when it will hit the ground. The ball will hit the ground when $H(t) = 0$, or

$$0 = -16t^2 + 128t + 320.$$

From Unit 7, Exercise 4,
the solution was found to be 10.

$$\text{Velocity} = H'(t) = -32t + 128$$

$$H'(10) = -32(10) + 128$$

$$= -320 + 128$$

$$= -192 \text{ feet per second}$$

A negative sign implies a downward direction.

The rubber ball will have a velocity of $-192$ feet per second when it hits the ground.

---

## Problem 2

Continue with Example 3.

a. Determine the ball's instantaneous velocity at $t = 2$.

b. Is the ball's height increasing or decreasing at this point in time?

c. Determine the ball's instantaneous velocity at $t = 4$.

d. How do you explain your answer?

Solution:

Answers: a. 64 feet per second; b. increasing; c. 0 feet per second; d. The ball has reached its maximum height and is about to start falling back to the ground.

# MARGINAL ANALYSIS

In economics, the term **marginal analysis** is used to denote the approximation procedure of using the derivative of a function to estimate the change in the function produced by a one unit increase in the size of the variable.

The next several examples will introduce the concept of marginal cost and marginal revenue.

## EXAMPLE 4

Suppose the total cost of producing $x$ units of some commodity is given by the cost function

$$C(x) = 20 + 2x + .01x^2 \quad x \geq 0$$

where $x$ = number of units,

and $C(x)$ = total cost in dollars to produce $x$ units.

a. Determine the cost of producing 100 units.

b. Determine the cost of producing 101 units.

c. Determine the cost of producing the 101st unit.

Solution: a. $C(100) = 20 + 2(100) + .01(100)^2$

$= 20 + 200 + .01(10{,}000)$

$= 20 + 200 + 100$

$= 320$

Answer: The cost of producing 100 units is $320.

b.  $C(101) = 20 + 2(101) + .01(101)^2$

$$= 20 + 202 + .01(10,201)$$

$$= 20 + 202 + 102.01$$

$$= 324.01$$

Answer:   The cost of producing 101 units is $324.01.

c.  The cost of producing the 101st unit is found by subtracting:

$$C(101) - C(100) = 324.01 - 320$$

$$= 4.01$$

Answer:   The cost of producing one additional unit after 100, or the cost of producing the 101st unit is $4.01.

---

We just computed the additional cost of producing one more unit, in this example, when $x = 100$. In practice, this quantity is usually approximated by the derivative $C'(x)$, which is called the **marginal cost**. Marginal cost is the name given to the rate of change of the total production cost with respect to the number of units produced. It is a good approximation to the cost of producing one additional unit.

> Definition:   Marginal Cost = $MC = C'(x)$,
>
> where $C(x)$ = total production cost of producing $x$ units.

# EXAMPLE 5

Suppose the total cost of producing $x$ units of some commodity is given by the cost function

$$C(x) = 20 + 2x + .01x^2 \quad x \geq 0$$

where $x$ = number of units,

and $C(x)$ = total cost in dollars to produce $x$ units.

a.  Find the equation for marginal cost.

b.  Find the marginal cost at $x = 100$.

c.  Interpret the answer to part b.

---

Solution:  a.  By definition, marginal cost equals the derivative of cost; therefore
$MC = C'(x) = 2 + .02x$

b.  $MC = C'(x) \quad = 2 + .02x$

$MC = C'(100) = 2 + .02(100)$

$$= 2 + 2$$

$$= 4$$

c.  At a production level of $x = 100$, one more unit costs approximately $4 to produce. Or stated another way, the marginal cost of the 101st unit is $4.

Did you notice how very close the approximation was to the actual cost of producing the 101st unit? The answers varied by only $.01.

---

## Problem 3

Use the information provided in Example 5 to answer the following:

a.  Find the marginal cost at $x = 200$.

b.  Interpret your answer to part a.

---

Solution:

Answers:    a. 6; b. At production level $x = 200$, one more unit costs approximately $6 to produce. Or, the marginal cost of the 201st unit is $6.

---

By a similar line of reasoning, given a revenue function $R(x)$, its derivative, $R'(x)$, is called the **marginal revenue**. Marginal revenue is the name given to the rate of change of total revenue with respect to the number of units sold. It is a good approximation to the revenue derived from selling one additional unit.

> Definition:    Marginal Revenue $= MR = R'(x)$,
>
> where $R(x) =$ total revenue derived from selling $x$ units.

---

## EXAMPLE 6

Given the sales revenue function

$$R(x) = 8x \quad x \geq 0$$

where $x =$ number of units

and $R(x) =$ total revenue in dollars for selling $x$ units.

a.  Find the equation for marginal revenue.

b.  Find the marginal revenue at $x = 100$.

c.  Interpret the answer to part b.

---

Solution:    a.    By definition, marginal revenue equals the derivative of the revenue function,

$$MR = R'(x).$$

Therefore                          $MR = R'(x) = 8.$

In this example, the marginal revenue formula is a constant, meaning that revenue increases by $8 for each additional unit sold.

b.    $MR = R'(100) = 8$

c.   At production level $x = 100$, approximately \$8 of additional sales revenue is derived from selling one more unit. Or stated another way, the marginal revenue derived from selling the 101st unit would be \$8.

You should now have a workable understanding of the derivative as a rate of change and be able to use it to estimate the change in $y$ for a unit change in $x$.

Furthermore, given an expression for distance as a function of time, you should be able to determine the instantaneous velocity of the object at any given point in time. And you should be able to determine marginal cost and marginal revenue equations, as well as being able to interpret their meanings.

## EXERCISES

1.   An ant colony was treated with an insecticide and the number of survivors $N(t)$, in hundreds, after $t$ hours was found to be given approximately by $N(t) = -t^2 - 5t - 750$.

   a.   Find $N'(t)$.

   b.   Find the rate of change of the colony at $t = 10$.

2.   The total cost in dollars, $C(x)$, for manufacturing $x$ units is given by

$$C(x) = 10 - 2.5x^2 + x^3 \quad x \geq 2$$

   a.   Find the marginal cost equation.

   b.   Find the marginal cost at $x = 10$ level of production.

   c.   Interpret answer to part b.

3.   Total cost in dollars of manufacturing $q$ units is given by

$$C(q) = 25q^2 + q + 100 \quad q > 0.$$

   a.   Use the marginal cost equation to estimate the cost of manufacturing the 101st unit.

   b.   Compute the actual cost of manufacturing the 101st unit.

4.   In April, 1986 the number of manufactured housing sales for one state was found to be 1,757. At that time it was projected that $t$ months from then ($t = 0$ for April, 1986), the number of manufactured housing sales for the state would be $N = \frac{1}{3}t^3 - 6t^2 + t + 1,757$.

   a.   Determine the formula for the instantaneous rate of change for the number of manufactured housing sales with respect to time.

   b.   At what rate was the number of housing sales changing with respect to time 6 months from then? Were sales increasing or decreasing at that point in time?

   c.   At what rate was the number of housing sales changing with respect to time 12 months from then? Were sales increasing or decreasing at that point in time?

5.   It is estimated that the daily output at a certain plant is $Q(x) = .1x^3 + 3x^2 + 1$, where $x$ is the number of workers employed at the plant. Currently there are 10 workers employed at the plant.

   a.   Estimate the effect that one additional worker will have on the daily output.

   b.   Compute the actual change in the daily output that will result if one additional worker is hired.

6.   Unit 7, Exercise 5—Revisited

Jim's profit at his restaurant is related to the number of customers as defined by

$$P(x) = -x^2 + 140x - 4{,}000$$

where $P(x)$ = profit (or loss) in dollars per day
and      $x$ = number of customers served per day

Marginal profit is defined in the same manner as marginal cost and marginal revenue.

a.   Find the equation for marginal profit.

b.   Find the marginal profit at $x = 50$.

c.   Is profit increasing or decreasing at this point?

7.   A ball is tossed upward from the top of a tower. The function which describes the height above the ground of the ball is

$$y = H(t) = -16t^2 + 64t + 80$$

where $H(t)$ = the height in feet above the ground
and      $t$ = time measured in seconds from when the ball was tossed.

a.   What is the instantaneous velocity at $t = 1$?

b.   What is the height of the ball after 4 seconds?

c.   What is the velocity of the ball after 4 seconds?

A negative sign implies a downward direction.

d.   What was the initial velocity with which the ball was tossed?

e.   When will the ball reach its maximum height?

Hint:   Graph the function.

f.   How high will the ball rise?

g.   When will the ball hit the ground?

h.   What is the velocity of the ball at the instant it hits the ground?

---

If additional practice is needed:

Barnett and Ziegler, pages 149–150, problems 19, 33–36
Budnick, page 452, problems 47–50; page 455, problems 11–12
Hoffmann, page 94, problems 1, 3–4, 6, 9–11, 16–17, 21–24
Piascik, page 92, problems 11–12; pages 111–113, problems 11–19

# UNIT 12

## More Differentiation Rules

This unit will continue our discussion of finding derivatives. Specifically, you will learn three more rules that will enable you to take the derivative of additional groups of functions besides polynomial functions.

Recall from Unit 9 the following rules:

Power:       If $f(x) = x^n$, then $f'(x) = nx^{n-1}$

Constant:    If $f(x) = c$, then $f'(x) = 0$

Sum:         If $f(x) = u(x) + v(x)$, then $f'(x) = u'(x) + v'(x)$

Difference:  If $f(x) = u(x) - v(x)$, then $f'(x) = u'(x) - v'(x)$

## PRODUCT RULE

The Product Rule allows us to take the derivative of a product of two functions without necessarily multiplying them out first.

---

**Product Rule**

If $f(x) = F(x) \cdot S(x)$, then $f'(x) = F(x) \cdot S'(x) + S(x) \cdot F'(x)$

In words, the derivative of a product of two functions is the first function times the derivative of the second function plus the second function times the derivative of the first function.

---

The above statement can be shortened somewhat to the following: the derivative of a product of two functions is the first times the derivative of the second plus the second times the derivative of the first. I used $F(x)$ and $S(x)$ to denote the first and second functions so as to make it easier to follow.

In all of the examples for this unit, an extra step will be added using the $\dfrac{d\,f(x)}{dx}$ notation to indicate how the rule is applied before actually taking the derivative. You may wish to omit the step yourself after you become more familiar with the rules, but at the beginning it is best to leave it in.

# EXAMPLE 1

If $f(x) = x^3(x - 5)$, find the derivative.

---

Solution:  One approach would be to remove the parentheses and then take the derivative of the polynomial function term by term. But I want to illustrate the use of the Product Rule.

The original function:     $f(x) = x^3(x - 5)$

Product Rule          $f'(x) = F(x) \cdot S'(x) + S(x) \cdot F'(x)$

Extra step to show        $f'(x) = x^3 \dfrac{d(x - 5)}{dx} + (x - 5) \dfrac{d(x^3)}{dx}$
what I intend to do

Take the derivatives.        $= x^3(1) \quad + (x - 5)(3x^2)$

Remove parentheses.       $= x^3 + 3x^3 - 15x^2$

Combine like terms.       $= 4x^3 - 15x^2$

Answer:   $f'(x) = 4x^3 - 15x^2$

---

# Problem 1

If $f(x) = x^3(x - 5)$, remove the parentheses and take the derivative term by term to verify that the answer is the same.

---

Solution:

---

Which method was easier? I prefer doing it term by term, but in this unit we will practice using the Product Rule. After that, the choice will be yours.

# EXAMPLE 2

Use the Product Rule to find the derivative of $f(x) = (x^4 + 3)(3x^3 + 1)$.

---

Solution:   Start with function.     $f(x) = (x^4 + 3)(3x^3 + 1)$

Product Rule          $f'(x) = F(x) \cdot S'(x) + S(x) \cdot F'(x)$

Extra step to show        $f'(x) = (x^4 + 3) \dfrac{d(3x^3 + 1)}{dx} + (3x^3 + 1) \dfrac{d(x^4 + 3)}{dx}$
what is intended

Take the derivatives       $= (x^4 + 3)(9x^2) + (3x^3 + 1)(4x^3)$

Remove parentheses.       $= 9x^6 + 27x^2 + 12x^6 + 4x^3$

Combine like terms.                $= 21x^6 + 4x^3 + 27x^2$

$$\text{Answer:} \quad f'x = 21x^6 + 4x^3 + 27x^2$$

Here's a word of warning before you try one. Be sure to keep each factor in parentheses until the very end; otherwise factors easily become mistaken for separate terms.

## Problem 2

Use the Product Rule to find $y'$ if $y = (x^2 + 1)(3x - 2)$.

Solution:   Start with function.

Product Rule        $f'(x) = F(x) \cdot S'(x) + S(x) \cdot F'(x)$

Extra step to show
   what is intended

Take the derivatives.

Remove parentheses.

Combine like terms.

$$\text{Answer:} \quad y' = 9x^2 - 4x + 3$$

In case you did not get the correct answer for that problem, we will do another example using the Product Rule before continuing.

## EXAMPLE 3

Use the Product Rule to find $\dfrac{dy}{dx}$ if $y = (x - 3)(x^2 + 7x + 10)$.

Solution:   $y = (x - 3)(x^2 + 7x + 10)$

$$\frac{dy}{dx} = F(x) \cdot S'(x) + S(x) \cdot F'(x)$$

$$\frac{dy}{dx} = (x - 3)\frac{d(x^2 + 7x + 10)}{dx} + (x^2 + 7x + 10)\frac{d(x - 3)}{dx}$$

$$= (x - 3)(2x + 7) + (x^2 + 7x + 10)(1)$$

$$= 2x^2 + 7x - 6x - 21 + x^2 + 7x + 10$$

$$= 3x^2 + 8x - 11$$

$$\text{Answer:} \quad \frac{dy}{dx} = 3x^2 + 8x - 11$$

# QUOTIENT RULE

Thus far all that has happened since the initial introduction of the Power Rule is that the functions have become progressively more complicated. But all of the functions have been polynomial functions. And the derivative remained a formula for finding the slope of the tangent line to the curve at any point.

The only part that is about to change with the next rule is that we are no longer necessarily dealing with polynomial functions. The function might be but it probably is not a polynomial function. The next rule is for finding the derivative of a quotient of two functions.

---

**Quotient Rule**

If $f(x) = \dfrac{N(x)}{D(x)}$, then $f'(x) = \dfrac{D(x) \cdot N'(x) - N(x) \cdot D'(x)}{[D(x)]^2}$

In words, the derivative of a quotient of two functions is the denominator times the derivative of the numerator minus the numerator times the derivative of the denominator, all over the denominator squared.

---

It is not really as bad as it looks. Believe it or not it is easier to use than the Product Rule once you get used to it. Again I have tried to make it easier to follow by letting $N(x)$ denote the function in the numerator and $D(x)$ denote the function in the denominator.

We will do a variety of examples.

---

# EXAMPLE 4

If $f(x) = \dfrac{x}{x^2 + 5}$, find $f'(x)$.

---

Solution:   Start with function.  $f(x) = \dfrac{x}{x^2 + 5}$

Quotient Rule   $f'(x) = \dfrac{D(x) \cdot N'(x) - N(x) \cdot D'(x)}{[D(x)]^2}$

Extra step to show what is intended   $f'(x) = \dfrac{(x^2 + 5)\dfrac{d(x)}{dx} - x\dfrac{d(x^2 + 5)}{dx}}{(x^2 + 5)^2}$

Take the derivatives.   $= \dfrac{(x^2 + 5)(1) - x(2x)}{(x^2 + 5)^2}$

Remove parentheses, but leave denominator as is.   $= \dfrac{x^2 + 5 - 2x^2}{(x^2 + 5)^2}$

Combine like terms.                 $= \dfrac{-x^2 + 5}{(x^2 + 5)^2}$

Answer:   $f'(x) = \dfrac{-x^2 + 5}{(x^2 + 5)^2}$

Comment:   The preferred form for the answer is to leave the denominator in factored form; i.e., not multiplied.

---

# EXAMPLE 5

If $y = \dfrac{3}{x}$, find $\dfrac{dy}{dx}$.

Solution:   Start with function.        $y = \dfrac{3}{x}$

Quotient Rule          $f'(x) = \dfrac{D(x) \cdot N'(x) - N(x) \cdot D'(x)}{[D(x)]^2}$

Extra step          $\dfrac{dy}{dx} = \dfrac{(x)\dfrac{d(3)}{dx} - 3\dfrac{d(x)}{dx}}{x^2}$

Take the derivatives.          $= \dfrac{(x)0 - 3(1)}{x^2}$

Remove parentheses.          $= \dfrac{0 - 3}{x^2}$

Combine like terms.          $= \dfrac{-3}{x^2}$

Answer:   $\dfrac{dy}{dx} = \dfrac{-3}{x^2}$

Here is a rather short one for you to try and then I'll do some more.

---

# Problem 3

If $f(x) = \dfrac{5}{x^2}$, find $f'(x)$. Be sure to simplify the final answer.

Solution:   Start with function.

Quotient Rule          $f'(x) = \dfrac{D(x) \cdot N'(x) - N(x) \cdot D'(x)}{[D(x)]^2}$

Extra step to show
  what is to be done

Take the derivatives.

Remove parentheses.

Combine like terms.

Answer:   $f'(x) = \dfrac{-10}{x^3}$

---

Recall that fractions should always be simplified; that means reduce fractions to lowest terms.

To simplify an algebraic fraction:

   a.   factor the numerator and denominator

   b.   cancel all common factors

## EXAMPLE 6

If $f(x) = \dfrac{x+2}{x^3}$, find $f'(x)$.

---

Solution:   Start with function.     $f(x) = \dfrac{x+2}{x^3}$

Quotient Rule     $f'(x) = \dfrac{D(x) \cdot N'(x) - N(x) \cdot D'(x)}{[D(x)]^2}$

Extra step     $f'(x) = \dfrac{(x^3)\dfrac{d(x+2)}{dx} - (x+2)\dfrac{d(x^3)}{dx}}{x^6}$

Take the derivatives.     $= \dfrac{(x^3)(1) - (x+2)(3x^2)}{x^6}$

Remove parentheses.     $= \dfrac{x^3 - 3x^3 - 6x^2}{x^6}$

Combine like terms.     $= \dfrac{-2x^3 - 6x^2}{x^6}$

Simplify.

   a.   Factor     $= \dfrac{-2x^2(x+3)}{x^6}$

   b.   Cancel     $= \dfrac{-2\cancel{x^2}(x+3)}{\cancel{x^6}^{\,4}}$

               $= \dfrac{-2(x+3)}{x^4}$

Answer:   $f'(x) = \dfrac{-2(x+3)}{x^4}$

There are two words of warning concerning the Quotient Rule: one, at the beginning of the problem put both the numerator and denominator in parentheses; two, do not attempt to simplify by cancelling common factors until the very **last** step. A common error is to cancel a factor in the denominator with a term, rather than a factor, in the numerator.

## EXAMPLE 7

If $y = \dfrac{x+2}{x-5}$, find $\dfrac{dy}{dx}$.

Solution:   Start with function.                 $y = \dfrac{x+2}{x-5}$

Quotient Rule                 $f'(x) = \dfrac{D(x) \cdot N'(x) - N(x) \cdot D'(x)}{[D(x)]^2}$

Extra step                 $\dfrac{dy}{dx} = \dfrac{(x-5)\dfrac{d(x+2)}{dx} - (x+2)\dfrac{d(x-5)}{dx}}{(x-5)^2}$

Take the derivatives.                 $= \dfrac{(x-5)(1) - (x+2)(1)}{(x-5)^2}$

Remove parentheses,                 $= \dfrac{x-5-x-2}{(x-5)^2}$
except for denominator.

Combine like terms.                 $= \dfrac{-7}{(x-5)^2}$

Answer:   $\dfrac{dy}{dx} = \dfrac{-7}{(x-5)^2}$

Now it's your turn to try one that is more complicated.

## Problem 4

If $y = \dfrac{4x}{1+x^2}$, find $\dfrac{dy}{dx}$.

Solution:   Start with function.

Quotient Rule                 $f'(x) = \dfrac{D(x) \cdot N'(x) - N(x) \cdot D'(x)}{[D(x)]^2}$

Extra step

Take the derivatives.

Remove parentheses.


Combine like terms.


Simplify:


   a.   Factor

   b.   Cancel


Answer: $\dfrac{dy}{dx} = \dfrac{4 - 4x^2}{(1 + x^2)^2}$

---

# CHAIN RULE/POWER OF A FUNCTION

The last rule of the unit is for taking the derivative of a function raised to some power, $n$. It goes by two different names, either the Chain Rule, or the Power of a Function, which is more descriptive.

---

**Chain Rule/Power of a Function**

If $f(x) = [u(x)]^n$, then $f'(x) = n[u(x)]^{n-1}u'(x)$, where $n$ is a real number.

In words, the derivative of a function raised to a power is equal to the power times the function raised to the power minus one, and all of that is multiplied times the derivative of the function.

---

An easy way to think of this rule is that it usually takes the form of having a function in parentheses raised to some power.

In symbols, the Chain Rule would look like this:

$$\text{If } f(x) = (\quad)^n, \text{ then } f'(x) = n(\quad)^{n-1}\frac{d(\quad)}{dx}$$

Stating the above in words, the derivative of a quantity raised to a power is equal to the power times the quantity raised to the power minus 1 times the derivative of what is in the parentheses.

After the last two rules, this one is really short and fast.

---

# EXAMPLE 8

Find the derivative of $f(x) = (7x^2 - 2x + 13)^5$.

---

Solution:   If we had nothing better to do with our time, we could multiply the expression out and take the derivative term by term. Obviously the Chain Rule is meant to be the better way.

Start with the function.     $f(x) = (7x^2 - 2x + 13)^5$

Use the Chain Rule.     $f'(x) = n[u(x)]^{n-1}u'(x)$

Extra step to show     $f'(x) = 5(7x^2 - 2x + 13)^4 \dfrac{d(7x^2 - 2x + 13)}{dx}$
what is intended

Take the derivative.     $= 5(7x^2 - 2x + 13)^4(14x - 2)$

And that is the answer.

Nothing more need be done unless you prefer to rewrite it as:

$$f'(x) = 5(14x - 2)(7x^2 - 2x + 13)^4$$
$$\text{or}$$
$$f'(x) = 10(7x^2 - 1)(7x^2 - 2x + 13)^4$$

# EXAMPLE 9

Find the derivative of $f(x) = (3x + 1)^{10}$.

Solution:   Start with the function.     $f(x) = (3x + 1)^{10}$

Use Chain Rule.     $f'(x) = n[u(x)]^{n-1}u'(x)$

Extra step     $f'(x) = 10(3x + 1)^9 \dfrac{d(3x + 1)}{dx}$

Take the derivative.     $= 10(3x + 1)^9(3)$

Answer:   $f'(x) = 30(3x + 1)^9$ would be the preferred way to write it.

Since these are so short, try two.

# Problem 5

If $f(x) = (x^2 - 3x + 5)^{12}$, find $f'(x)$.

Solution:   Start with the function.

Use Chain Rule.     $f'(x) = n[u(x)]^{n-1}u'(x)$

Take the derivative.

Answer:   $f'(x) = 12(x^2 - 3x + 5)^{11}(2x - 3)$

## Problem 6

Find $D_x(3x - 5)^{14}$.

Solution:   Start with the function.

   Use Chain Rule.          $f'(x) = n[u(x)]^{n-1}u'(x)$

Answer:   $D_x(3x - 5)^{14} = 42(3x - 5)^{13}$

You should now be able to take the derivative of the product of two functions, of the quotient of two functions, and of a function raised to a power.

A summary of the rules thus far are:

Power:              If $f(x) = x^n$, then $f'(x) = nx^{n-1}$

Constant:           If $f(x) = c$, then $f'(x) = 0$

Constant · Function:    If $f(x) = c \cdot u(x)$, then $f(x) = c \cdot u'(x)$

Special Case:        If $f(x) = c \cdot x^n$, then $f'(x) = cnx^{n-1}$

Sum:               If $f(x) = u(x) + v(x)$, then $f'(x) = u'(x) + v'(x)$

Difference:          If $f(x) = u(x) - v(x)$, then $f'(x) = u'(x) - v'(x)$

Product Rule:        If $f(x) = F(x) \cdot S(x)$, then $f'(x) = F(x)S'(x) + S(x)F'(x)$

Quotient Rule:       If $f(x) = \dfrac{N(x)}{D(x)}$, then $f'(x) = \dfrac{D(x)N'(x) - N(x)D'(x)}{[D(x)]^2}$

Chain Rule:          If $f(x) = [u(x)]^n$, then $f'(x) = n[u(x)]^{n-1}u'(x)$

Before beginning the next unit, use the appropriate rules to differentiate the various functions listed below. Final answers should be simplified.

# EXERCISES

1.   Use the Product Rule to find $f'(x)$ for $f(x) = (2x + 1)(x^2 - 3)$.

2.   Use the Product Rule to find $\dfrac{dy}{dx}$ for $y = (x^2 + 1)(2 - x^3)$.

3.   Differentiate $y = f(x) = 3(x^7 + 2x^5 - x^3 + 6x - 1)$.

4.   Find the derivative of $f(u) = (u^3 + 5)^7$.

5.   If $f(x) = \dfrac{-1}{x^2}$, find $f'(x)$ and simplify.

6.   Find $D_x\left(\dfrac{3}{x + 4}\right)$.

7.  Find $y'$ for $y = \dfrac{x^2 + 2}{x - 3}$.

8.  Given $N = \dfrac{100t}{t - 9}$, find the rate of change of $N$ with respect to $t$.

9.  Given $f(x) = (6x^2 + 12x + 1)^5$

    a.  Find $f(-2)$ and interpret.

    b.  Find $f'(-2)$ and interpret.

10. Consider the revenue function defined by

    $$R(x) = (x + 1)(2x^2 + x + 3) \quad x \geq 0$$

    where $x = $ the number of units sold

    and $R(x) = $ total revenue in dollars for selling $x$ units.

    a.  Find the equation for marginal revenue.

    b.  Find the marginal revenue at $x = 10$. Is revenue increasing or decreasing at this point?

11. Suppose that the number of pounds of jelly beans people are willing to buy per day in a small candy store at a price of $p$ is given by

    $$D(p) = 6 + p - p^2 \quad 0 \leq p \leq 3$$

    where $p = $ price per pound of jelly beans in dollars

    and $D(p) = $ number of pounds sold per day at price $p$.

    a.  Find $D'(p)$, the rate of change of demand with respect to price.

    b.  Find $D(1.5)$ and $D'(1.5)$ and interpret.

12. It is estimated that $t$ years from now, the population of people aged 65 and over in Japan will be

    $$P(t) = 15 - \frac{6}{t + 1} \text{ million.}$$

    a.  Derive a formula for the rate at which the population will be changing with respect to time $t$ years from now.

    b.  At what rate will the population be growing 1 year from now?

    c.  By how much will the population actually increase during the second year?

    d.  At what rate will the population be growing 9 years from now?

    e.  What will happen to the rate of population growth in the long run?

---

If additional practice is needed:

Barnett and Ziegler, pages 167–168, problems 1–32, 49–55; pages 175–176, problems 1–8
Budnick, pages 451–452, problems 15–24, 33–43
Hoffmann, page 86, problems 11–28
Piascik, pages 120–121, problems 1–16; page 126, problems 1–8

# UNIT 13

## The Power Rule Revisited

In this unit you will learn how to take the derivative of functions with negative or fractional exponents.

## NEGATIVE EXPONENTS

Thus far all of our experience taking derivatives has been with exponents that were whole numbers. That is about to change. First we will consider negative exponents. As a review, the definition of a negative exponent is repeated below along with some sample problems.

$$\text{Definition:} \quad \textbf{Negative Exponent} \quad b^{-n} = \frac{1}{b^n} \text{ and } \frac{1}{b^{-n}} = b^n$$

Therefore a negative exponent has the effect of moving a factor from top to bottom (or vice versa).

Consider Examples 1–8, and be certain you understand the simplification.

EXAMPLE 1     $2^{-1} = \dfrac{1}{2}$

EXAMPLE 2     $x^{-1} = \dfrac{1}{x}$

EXAMPLE 3     $q^{-3} = \dfrac{1}{q^3}$

EXAMPLE 4     $2x^{-1} = \dfrac{2}{x}$

EXAMPLE 5     $3t^{-2} = \dfrac{3}{t^2}$

EXAMPLE 6     $\dfrac{1}{x^{-2}} = x^2$

EXAMPLE 7     $\dfrac{5}{x^{-3}} = 5x^3$

EXAMPLE 8     $\dfrac{2}{p^{-1}} = 2p$

# FUNCTIONS WITH NEGATIVE EXPONENTS

We are now ready to apply the Power Rule to functions containing $x$ raised to a negative power. Keep in mind that derivatives always should be simplified, meaning that final answers should be written using positive integral exponents only—no fractional, negative, or zero exponents.

## Taking Derivatives of Functions with Negative Exponents

The basic procedure is as follows:

1.  Rewrite the function as a power of $x$, if not already written that way.

2.  Use the Power Rule to take the derivative.

3.  Simplify by removing parentheses and whatever else may be done.

4.  If any negative exponents exist, use the definition of a negative exponent to rewrite the expression with positive exponents.

The Power Rule is repeated here for reference.

> **Power Rule**
> If $f(x) = x^n$, then $f'(x) = nx^{n-1}$ for all $n$, $n$ being a real number.

# EXAMPLE 9

If $f(x) = x^{-3}$, find $f'(x)$.

Solution:   Given.                      $f(x) = x^{-3}$

Power Rule.          If $f(x) = x^n$,   then   $f'(x) = nx^{n-1}$

Use Power Rule.                              $f'(x) = -3x^{-3-1}$

Simplify.                                          $= -3x^{-4}$

Rewrite with positive exponent.          $= \dfrac{-3}{x^4}$

## EXAMPLE 10

If $f(x) = 2x^{-5}$, find $f'(x)$.

Solution:  Given.                                   $f(x) = 2x^{-5}$

Use Power Rule.                      $f'(x) = 2(-5x^{-5-1})$

Simplify.                                      $= -10x^{-6}$

Rewrite with positive exponent.            $= \dfrac{-10}{x^6}$

## EXAMPLE 11

Find $f'(x)$ if $f(x) = \dfrac{1}{x^2}$.

Solution:  We could use the Quotient Rule, but there is a faster method.

Given.                                   $f(x) = \dfrac{1}{x^2}$

Rewrite as a power of $x$.                    $= x^{-2}$

Use Power Rule.                        $f'(x) = -2x^{-2-1}$

Simplify.                                      $= -2x^{-3}$

Rewrite with positive exponent.              $= \dfrac{-2}{x^3}$

For comparison purposes, rework the above problem.

## Problem 1

Use the Quotient Rule to find the derivative of $f(x) = \dfrac{1}{x^2}$.

Solution:

Which method was shorter? Which method did you like better? I prefer rewriting and doing the problem with negative exponents, but the choice is yours.

I'll do one more before having you try one.

## EXAMPLE 12

Find $f'(x)$ if $f(x) = \dfrac{2}{x} - \dfrac{7}{x^4}$.

Solution:   Given.

$$f(x) = \frac{2}{x} - \frac{7}{x^4}$$

Rewrite as powers of $x$.

$$= 2x^{-1} - 7x^{-4}$$

Use Power Rule.

$$f'(x) = 2(-1x^{-1-1}) - 7(-4x^{-4-1})$$

Simplify.

$$= -2x^{-2} + 28x^{-5}$$

Rewrite with positive exponents.

$$= \frac{-2}{x^2} + \frac{28}{x^5}$$

## Problem 2

Use the Power Rule to find the derivative of $f(x) = \dfrac{1}{x}$.

Solution:   Given.

Rewrite as a power of $x$.

Use Power Rule.

Simplify.

Rewrite with positive exponent.

Answer:  $f'(x) = \dfrac{-1}{x^2}$

Try another with no hints this time.

## Problem 3

Use the Power Rule to find the derivative of $f(x) = \dfrac{3}{x^2}$.

Solution:

Answer:  $f'(x) = \dfrac{-6}{x^3}$

# FRACTIONAL EXPONENTS

Now it is time to turn our attention to fractional exponents. Fractional exponents are more troublesome to work with for a lot of reasons, but I'll try to make it as straightforward as possible. Don't get discouraged.

$$\text{Definition:} \quad \textbf{Fractional Exponents} \ b^{\frac{n}{d}} = \left(\sqrt[d]{b}\right)^n = \sqrt[d]{b^n}$$

$$\text{Note:} \quad b^{\frac{n}{d}} \quad \begin{matrix} \longleftarrow \text{denotes the power} \\ \longleftarrow \text{denotes the root} \end{matrix}$$

Note that two forms of the definition are given.

$$b^{\frac{n}{d}} = \left(\sqrt[d]{b}\right)^n = \sqrt[d]{b^n}$$

numerical    algebraic

The first form is useful in numerical calculations, provided that the $d$th root of $b$ is a known integer. It is then convenient to take the root before raising to the $n$th power in order to work with smaller numbers.

The second form is the more common way of rewriting algebraic expressions with fractional exponents.

As the two forms suggest, we can do either the root or the power first.

We will do some numerical problems first and use the numerical form of the definition. For a more detailed explanation, see Unit 11, Fractional Exponents, from FORGOTTEN ALGEBRA.

Numerical form:

$$b^{\frac{n}{d}} = \left(\sqrt[d]{b}\right)^n \quad \begin{matrix}\nearrow \text{power} \\ \searrow \text{root}\end{matrix}$$

---

EXAMPLE 13 $\qquad 8^{\frac{1}{3}} = \left(\sqrt[3]{8}\right)^1 = (2)^1 = 2$

EXAMPLE 14 $\qquad 27^{\frac{2}{3}} = \left(\sqrt[3]{27}\right)^2 = (3)^2 = 9$

EXAMPLE 15 $\qquad (-8)^{\frac{4}{3}} = \left(\sqrt[3]{-8}\right)^4 = (-2)^4 = 16$

EXAMPLE 16 $\qquad 9^{-\frac{1}{2}} = \left(\sqrt{9}\right)^{-1} = (3)^{-1} = \dfrac{1}{3}$

or, alternatively,

$$= \dfrac{1}{9^{\frac{1}{2}}} = \dfrac{1}{\sqrt{9}} = \dfrac{1}{3}$$

---

Now consider a few algebraic problems and use the algebraic form of the definition to simplify the expressions.

Algebraic form:

$$b^{\frac{n}{d}} = \sqrt[d]{b^n} \quad \begin{matrix}\nearrow \text{power} \\ \searrow \text{root}\end{matrix}$$

EXAMPLE 17    $x^{\frac{1}{2}} = \sqrt{x}$

EXAMPLE 18    $p^{\frac{2}{3}} = \sqrt[3]{p^2}$

Three problems should be enough for you to try because we want to get on to the real task of taking derivatives with fractional exponents. Simplify each of the following.

Problem 4    $32^{\frac{2}{5}}$

Problem 5    $4^{-\frac{1}{2}}$

Problem 6    $x^{\frac{2}{5}}$

Answers:    $4, \dfrac{1}{2}, \sqrt[5]{x^2}$

# FUNCTIONS WITH FRACTIONAL EXPONENTS

We are now ready to use the Power Rule to take derivatives of functions with fractional and/or negative exponents.

## Taking Derivatives of Functions with Fractional Exponents

The basic procedure remains as stated earlier with a fifth step added:

1.  Rewrite the function as a power of $x$, if not already written that way.

2.  Use the Power Rule to take the derivative.

3.  Simplify by removing parentheses and whatever else may be done.

4.  If any negative exponents exist, use the definition of a negative exponent to rewrite the expression with positive exponents.

5.  If any fractional exponents exist, use the definition of a fractional exponent to rewrite the expression with radical signs.

The Power Rule is restated here one more time.

---

**Power Rule**

If $f(x) = x^n$, then $f'(x) = nx^{n-1}$ for all $n$, $n$ being a real number.

---

To repeat in words, if $x$ is raised to some exponent, the derivative is equal to the exponent times $x$ raised to the exponent minus 1. In the following examples, the exponent will be a fraction, but the rule remains unchanged. Merely the calculations become more troublesome.

# EXAMPLE 19

Find $f'(x)$ if $f(x) = x^{\frac{1}{2}}$.

Solution: Given. $$f(x) = x^{\frac{1}{2}}$$

Power Rule. If $f(x) = x^n$, then $f'(x) = n\,x^{n-1}$

Use Power Rule. $$f'(x) = \frac{1}{2}x^{\frac{1}{2}-1}$$

Simplify. $$= \frac{1}{2}x^{-\frac{1}{2}}$$

Rewrite with positive exponents. $$= \frac{1}{2}\cdot\frac{1}{x^{\frac{1}{2}}}$$

Rewrite without fractional exponents. $$= \frac{1}{2\sqrt{x}}$$

# EXAMPLE 20

If $f(x) = 5x^{\frac{1}{3}}$, find $f'(x)$.

Solution: Given $$f(x) = 5x^{\frac{1}{3}}$$

Use Power Rule. $$f'(x) = 5\left(\frac{1}{3}x^{\frac{1}{3}-1}\right)$$

Simplify. $$= 5\cdot\frac{1}{3}x^{-\frac{2}{3}}$$

Rewrite with positive exponents. $$= \frac{5}{3}\cdot\frac{1}{x^{\frac{2}{3}}}$$

Rewrite without fractional exponents. $$= \frac{5}{3\sqrt[3]{x^2}}$$

# FUNCTIONS WITH RADICALS

## Taking Derivatives of Functions with Radicals

The basic procedure remains as stated; however Step 1 might be reworded as follows:

1. If the function contains a radical sign, rewrite the function as a power of $x$.

The remaining steps are the same as the following examples will illustrate.

# EXAMPLE 21

If $f(x) = 7\sqrt{x}$, find $f'(x)$.

Solution:

| | |
|---|---|
| | $f(x) = 7\sqrt{x}$ |
| Rewrite as a power of $x$. | $= 7x^{\frac{1}{2}}$ |
| Use Power Rule. | $f'(x) = 7\left(\dfrac{1}{2}x^{\frac{1}{2}-1}\right)$ |
| Simplify. | $= 7 \cdot \dfrac{1}{2} \cdot x^{-\frac{1}{2}}$ |
| Rewrite with positive exponents. | $= \dfrac{7}{2} \cdot \dfrac{1}{x^{\frac{1}{2}}}$ |
| Rewrite without fractional exponents. | $= \dfrac{7}{2\sqrt{x}}$ |

# EXAMPLE 22

If $f(x) = 3\sqrt[3]{x^2}$, find $f'(x)$.

Solution:

| | |
|---|---|
| | $f(x) = 3\sqrt[3]{x^2}$ |
| Rewrite as a power of $x$. | $= 3x^{\frac{2}{3}}$ |
| Use Power Rule. | $f'(x) = 3\left(\dfrac{2}{3}x^{\frac{2}{3}-1}\right)$ |
| Simplify. | $= 3 \cdot \dfrac{2}{3} x^{-\frac{1}{3}}$ |
| Rewrite with positive exponents. | $= 2 \cdot \dfrac{1}{x^{\frac{1}{3}}}$ |
| Rewrite without fractional exponents. | $= \dfrac{2}{\sqrt[3]{x}}$ |

The next two problems are for you to do.

# Problem 4

If $f(x) = x^{\frac{1}{4}}$, find $f'(x)$.

Solution:  

| | |
|---|---|
| Given. | $f(x) =$ |
| Use Power Rule. | $f'(x) =$ |
| Simplify. | $=$ |
| Rewrite with positive exponents. | $=$ |

Rewrite without fractional exponents.          =

Answer: $\dfrac{1}{4\sqrt[4]{x^3}}$

## Problem 5

If $f(x) = 6\sqrt{x}$, find $f'(x)$.

Solution:                                      $f(x) =$

Rewrite as a power of $x$.                     $=$

Use Power Rule.                                $f'(x) =$

Simplify.                                       $=$

Rewrite with positive exponents.               $=$

Rewrite without fractional exponents.          $=$

Answer: $\dfrac{3}{\sqrt{x}}$

The next examples illustrate how to handle negative fractional exponents.

## EXAMPLE 23

If $f(x) = \dfrac{1}{\sqrt{x}}$, find $f'(x)$.

Solution:                                      $f(x) = \dfrac{1}{\sqrt{x}}$

Rewrite as a power of $x$.                     $= \dfrac{1}{x^{\frac{1}{2}}}$

                                               $= x^{-\frac{1}{2}}$

Use Power Rule.                                $= -\dfrac{1}{2} x^{-\frac{1}{2}-1}$

Simplify.                                       $= -\dfrac{1}{2} x^{-\frac{3}{2}}$

Rewrite with positive exponents.               $= \dfrac{-1}{2} \cdot \dfrac{1}{x^{\frac{3}{2}}}$

Rewrite without fractional exponents.          $= \dfrac{-1}{2\sqrt{x^3}}$ or $\dfrac{-1}{2x\sqrt{x}}$

# EXAMPLE 24

If $f(x) = \dfrac{3}{\sqrt[5]{x^2}}$, find $f'(x)$.

Solution:

$$f(x) = \frac{3}{\sqrt[5]{x^2}}$$

Rewrite as a power of $x$.

$$= 3x^{-\frac{2}{5}}$$

Use Power Rule.

$$f'(x) = 3\left(-\frac{2}{5}x^{-\frac{2}{5}-1}\right)$$

Simplify.

$$= 3 \cdot \frac{-2}{5}x^{-\frac{7}{5}}$$

Rewrite with positive exponents.

$$= \frac{-6}{5} \cdot \frac{1}{x^{\frac{7}{5}}}$$

Rewrite without fractional exponents.

$$= \frac{-6}{5\sqrt[5]{x^7}} \quad \text{or} \quad \frac{-6}{5x\sqrt[5]{x^2}}$$

Before concluding this unit, there is one additional type of problem to be considered—functions, not just $x$, under radical signs. Now that you are able to use the Power Rule with fractional exponents, the same procedure carries over to functions requiring the use of the Chain Rule.

Recall the Chain Rule.

If $f(x) = [u(x)]^n$, then $f'(x) = n[u(x)]^{n-1}u'(x)$ for all $n$.

# EXAMPLE 25

If $f(x) = \sqrt{x^2 + 5x - 3}$, find $f'(x)$.

Solution:

$$f(x) = \sqrt{x^2 + 5x - 3}$$

Rewrite as a power of $x$.

$$= (x^2 + 5x - 3)^{\frac{1}{2}}$$

Use Chain Rule.

$$f'(x) = \frac{1}{2}(x^2 + 5x - 3)^{\frac{1}{2}-1}\frac{d}{dx}(x^2 + 5x - 3)$$

Simplify.

$$= \frac{1}{2}(x^2 + 5x - 3)^{-\frac{1}{2}}(2x + 5)$$

Rewrite with positive exponents.

$$= \frac{1}{2} \cdot \frac{1}{(x^2 + 5x - 3)^{\frac{1}{2}}} \cdot (2x + 5)$$

Rewrite without fractional exponents.

$$= \frac{2x + 5}{2\sqrt{x^2 + 5x - 3}}$$

Let's make the last example easier.

## EXAMPLE 26

If $f(x) = \sqrt{3x}$, find $f'(x)$.

Solution:

|  |  |
|---|---|
|  | $f(x) = \sqrt{3x}$ |
| Rewrite as a power of $x$. | $= (3x)^{\frac{1}{2}}$ |
| Use Chain Rule. | $f'(x) = \frac{1}{2}(3x)^{\frac{1}{2}-1} \dfrac{d(3x)}{dx}$ |
| Simplify. | $= \frac{1}{2}(3x)^{-\frac{1}{2}} 3$ |
| Rewrite with positive exponents. | $= \frac{1}{2} \cdot \dfrac{1}{(3x)^{\frac{1}{2}}} \cdot 3$ |
| Rewrite without fractional exponents. | $= \dfrac{3}{2\sqrt{3x}}$ |

You should now be able to take the derivative of functions with negative and/or fractional exponents. Implied within that statement is that you should also be able to take the derivative of radical functions. In all cases, the final answer should be simplified, meaning written without the use of negative, fractional, or zero exponents.

Briefly, the basic procedure for taking derivatives of such functions is:

1.  Rewrite the function as a power of $x$.
2.  Use either the Power Rule or Chain Rule to take the derivative.
3.  Simplify by removing parentheses and such.
4.  Rewrite any negative exponents with positive exponents.
5.  Rewrite any fractional exponents with radical signs.

## EXERCISES

Differentiate each of the following and simplify:

1.  $f(x) = 3x^{-4}$
2.  $f(x) = 2x^{\frac{3}{2}}$
3.  $f(x) = x^{-\frac{1}{3}}$
4.  $f(x) = \dfrac{5}{x^2} - \dfrac{1}{x}$
5.  $y = 2\sqrt{x}$
6.  $y = \sqrt{2x}$
7.  $y = \dfrac{2}{\sqrt{x}}$

8.  $g(x) = \dfrac{1}{3x^2}$

9.  $y = \sqrt{1 - 3x^2}$

10.  $D(p) = \sqrt[3]{p^2 + 1}$

11.  If $f(x) = x^{\frac{1}{2}}$, find $f'\left(\dfrac{1}{4}\right)$.

12.  If $g(u) = 2u - 4u^{-1}$, find $g'(2)$.

---

If additional practice is needed:

Barnett and Ziegler, pages 157–160, problems 9–12, 21–28, 33–44, 67–70
Budnick, pages 451, problems 5–14
Hoffmann, page 86, problems 5–10; page 96, problems 12–15; page 115, problems 17–35
Piascik, pages 110–111, problems 1–3, 7–10; pages 126–127, problems 9–14

# UNIT 14

## Maximum and Minimum Points

The major objectives of the next few units are to develop methods for determining where a function achieves maximum or minimum values and what they are. In preparation for those methods, this unit consists solely of terminology and definitions. It is brief, but it will provide the foundation for the techniques that follow in the next three units.

## HIGHER ORDER DERIVATIVES

Thus far $f'(x)$ has been discussed as the derivative of the function $f(x)$. Actually it is the first derivative of the function. There are second derivatives, third derivatives, and so on.

---

Definition:   The **second derivative** of a function $f(x)$ is the derivative of $f'(x)$ and is denoted by $f''(x)$.

---

Finding a second derivative is a straightforward matter. You differentiate the first derivative.

====

## EXAMPLE 1

If $f(x) = 5x^4 - 7x^3 + x^2 + 34x - 1$, find $f'(x)$ and $f''(x)$.

---

Solution:    $f(x) = 5x^4 - 7x^3 + x^2 + 34x - 1$

$\qquad f'(x) = 5(4x^3) - 7(3x^2) + (2x) + 34$

$\qquad\qquad = 20x^3 - 21x^2 + 2x + 34$

$\qquad f''(x) = 20(3x^2) - 21(2x) + 2$

$\qquad\qquad = 60x^2 - 42x + 2$

---

I am confident that if I were to ask, without your even seeing an example, you could correctly find the third derivative of the above function.

Only first and second derivatives will be of interest to us.

# OTHER NOTATION FOR SECOND DERIVATIVES

As with first derivatives, there are just as many different notations for second derivatives. Again, the important thing is that you are able to recognize and understand each notation.

Consider the function:   $y = f(x) = x^3 - 7x^2 + x - 6$.

We know

1.   $f'(x)$ denotes the first derivative and $f'(x) = 3x^2 - 14x + 1$.

$f''(x)$ denotes the second derivative and $f''(x) = 6x - 14$.

Other notations that denote second derivatives are:

2.   $y''$ read "y double prime" and $y'' = 6x - 14$.

3.   $f''$ read "f double prime" and $f'' = 6x - 14$.

4.   $\dfrac{d^2y}{dx^2}$ read simply as "the second derivative of $y$ with respect to $x$."

Do notice the location of the 2s; it is not a typographical error. In the numerator, the 2 is between the $d$ and $y$. In the denominator, the 2 is after the $d$ and $x$.

## Problem 1

If $f(x) = x^5 - 3x^2 + 27$, find $f'(x)$ and $f''(x)$.

Solution:    $f(x) =$

$f'(x) =$

$f''(x) =$

Answers:   $f'(x) = 5x^4 - 6x,\ f''(x) = 20x^3 - 6$

It will be three more units before we use second derivatives, but the concept of a first derivative was needed for a definition in this unit.

# CRITICAL POINTS

As before, this will be an intuitive approach. We will use another picture to explain the words. This picture has everything possible happening in it. Some of it is beyond the scope of this book, but I wanted the definitions to be complete in the event that you continue on with additional studies in calculus.

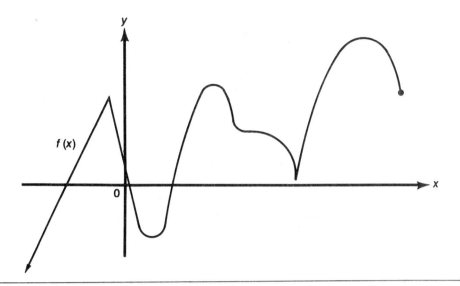

---

Definition: A **critical point** of a function is a point where the first derivative is zero or undefined.

---

What does that mean in terms of our picture?

To say that the first derivative is zero, really means that the slope of the tangent line is zero, and that in turn means the tangent line is horizontal and the curve "levels off."

To say that the first derivative is undefined has two possible meanings: one, the slope of the tangent line is undefined and therefore the tangent line is vertical. Or two, the tangent line fails to exist and the curve is "pointed."

Before reading the answer on the next page, how many critical points are there in the drawing?

# CLASSIFICATION OF CRITICAL POINTS

All critical points can be classified into one of three categories.

**Relative Maximum:** The words *relative maximum* are used to describe a peak, a point that is higher than any nearby point on the graph. Local maximum is used frequently in place of relative maximum.

**Relative Minimum:** The words *relative minimum* are used to describe a low point, a point that is lower than any nearby point on the graph. Again local minimum and relative minimum are used interchangeably.

**Stationary Inflection Point:** The words *stationary inflection point* are used to describe critical points that are neither high points nor low points.

A word of caution: **first the point must be a critical point.** Only a critical point is classified further as to whether it is a relative maximum, a relative minimum, or a stationary inflection point.

To return to the question posed earlier, there are *six* critical points in the above drawing. Tangent lines have been added to help you "see" the critical points. And coordinates have been added so that we can distinguish among the various points.

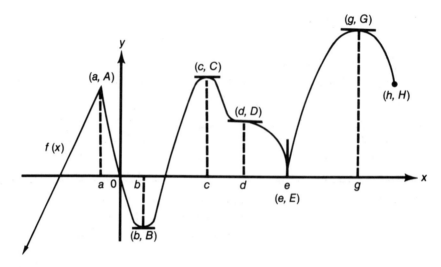

Recall a critical point exists where either the tangent line is horizontal, vertical, or nonexistent. Be sure you understand why each of the points indicated is a critical point.

Critical points:  $(a, A)$; $(b, B)$; $(c, C)$; $(d, D)$; $(e, E)$; and $(g, G)$.

Using the drawing as our guide, each of the critical points can be classified further as to whether it is a relative maximum, a relative minimum, or a stationary inflection point. Relative maxima (maxima is the plural of maximum) are peaks or high points. Relative minima (minima is the plural of minimum) are low points. Stationary inflection points are what is left over, neither high points nor low points. Try classifying the critical points yourself, before reading the answer.

1.  $(a, A)$    relative maximum

2.  $(b, B)$    relative minimum

3.  $(c, C)$    relative maximum

4.  $(d, D)$    stationary inflection point

5.  $(e, E)$    relative minimum

6.  $(g, G)$    relative maximum

Stated in words:

1.  A relative maximum value for $y$ is $A$ and it occurs when $x = a$.

2.  A relative minimum value for $y$ is $B$ and it occurs when $x = b$.

3.  A relative maximum value for $y$ is $C$ and it occurs when $x = c$.

4.  Although $(d, D)$ is a critical point, it is neither a relative maximum nor a relative minimum. It is a stationary inflection point.

5.  A relative minimum value for $y$ is $E$ and it occurs when $x = e$.

6.  A relative maximum value for $y$ is $G$ and it occurs when $x = g$.

Although $(h, H)$ is a low point it is *not* a relative minimum. Why? Because it is not a critical point, and therefore, by definition, cannot be a relative minimum.

# ABSOLUTE OR GLOBAL

**Absolute Maximum or Global Maximum:** The words, absolute maximum or global maximum, are used to denote the largest value of the function, if it exists. The absolute maximum would be located at the highest point on the graph. It need not be a critical point.

**Absolute Minimum or Global Minimum:** The words, absolute minimum or global minimum, are used to denote the smallest value of the function, if it exists. The absolute minimum would be located at the lowest point on the graph. It need not be a critical point.

Referring back to the drawing, $(g, G)$ is the highest point on the graph. Thus $G$ is the absolute maximum value and it occurs when $x = g$. There is no absolute minimum. Even though $B$ is the smallest (lowest) critical point, the arrow on the left side of the graph indicates the curve continues downward thus having values of $y$ that would be smaller (lower) than $B$. Had the arrow not been there and an endpoint instead, we would have had an absolute minimum.

# INCREASING AND DECREASING

There is one last concept before leaving our picture. Remember we always read graphs from left to right, exactly the same as you read a sentence.

---

Definition: If the curve is rising as you move from left to right, the function is said to be **increasing**.

---

Definition: If the curve is falling as you move from left to right the function is said to be **decreasing**.

---

Starting at the arrow on the left side of the curve and moving to the right, let me verbally describe the graph using the words increasing and decreasing.

The curve is increasing until you get to $x = a$,

at $x = a$, there is a sharp corner,

then it is decreasing until you get to $x = b$,

at $x = b$, the curve levels off,

then it increases until $x = c$,

at $x = c$, the curve levels off again,

then it decreases until $x = d$,

at $x = d$, the curve levels off again,

then it decreases until $x = e$,

at $x = e$, there is a sharp point,

then it increases until $x = g$,

at $x = g$, the curve levels off again,

then it decreases until $x = h$.

and at $x = h$, the curve ends.

In symbols, the above information on increasing and decreasing would look like this:

If $x < a$, $f(x)$ is increasing.

If $a < x < b$, $f(x)$ is decreasing.

If $b < x < c$, $f(x)$ is increasing.

If $c < x < d$, $f(x)$ is decreasing.

If $d < x < e$, $f(x)$ is decreasing.

If $e < x < g$, $f(x)$ is increasing.

If $g < x \leq h$, $f(x)$ is decreasing.

Notice that intervals of $x$ are used to describe where the function is increasing and decreasing. The $y$ values are not used, only the $x$ values. Also notice that the values where the critical points are located are *not* included as part of the interval.

Critical points were defined in terms of derivatives. Can we do the same thing for increasing and decreasing? Yes. Consider the section of the graph from $b < x < c$. Sketch in three or four tangent lines along that section. What do they all have in common? Their slopes are all positive. Do the same thing for the section from $c < x < d$. This time all the slopes are negative. That leads us to the following theorems.

---

Theorem:    If $f'(x)$ is positive for each value of $x$ in some interval, then $f(x)$ is increasing on that interval.

---

Theorem:    If $f'(x)$ is negative for each value of $x$ in some interval, then $f(x)$ is decreasing on that interval.

---

Summarizing what we have said so far, if the derivative at some point is positive, the curve is increasing. If the derivative is negative, the curve is decreasing. And if the derivative is zero, the curve levels off, and it is a critical point.

Critical points are points where the first derivative is 0 or undefined. Graphically that implies critical points occur where the tangent line is horizontal, vertical, or nonexistent.

Critical points can be classified as either

1.  Relative maximum ⟍――――― a high point,
    Local maximum ――⟋

2.  Relative minimum ⟍――――― a low point, or
    Local minimum ――⟋

3.  Stationary Inflection Point—neither a high nor a low point.

A global or absolute maximum is the largest value (highest point on the curve), if it exists. Similarly, a global or absolute minimum is the smallest value (lowest point on the curve), if it exists. They need not be critical points.

For the remainder of this book, the only critical points we will consider are those where the derivative is zero; i.e., the tangent line is horizontal. Keep in mind that others do exist, but they are beyond the scope of this presentation.

Answer the following questions before continuing.

## EXERCISES

Use the drawing provided to answer the following questions:

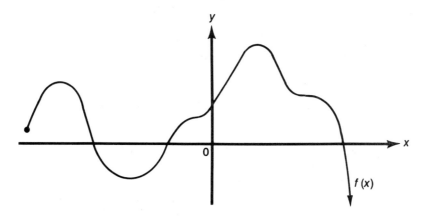

1. How many critical points are there?

2. Indicate where they are located.

3. Sketch the tangent line at each critical point.

4. Classify each critical point.

5. Does the function have an absolute maximum?

6. Does the function have a global minimum?

7. Over what intervals is the function increasing?

# UNIT 15

## Classifying Critical Points— One Procedure

In this unit you will learn one method for determining where a function achieves maximum or minimum values.

In the previous unit, you learned to classify critical points as either relative maxima, relative minima, or stationary inflection points by whether or not they were high points, low points, or neither based on the drawing. But what happens when the drawing is not available? How can we determine where the relative maximum and relative minimum occur, or if they occur at all? There is a theorem to help us.

---

**Theorem:** If a function has a relative maximum or minimum, it will occur at a critical point.

---

In an effort to locate the relative maximum or minimum points, the first step then is to locate all the critical points. Recall a critical point is where the first derivative is zero or undefined. From this unit on, we will consider only those that are zero.

## LOCATING CRITICAL POINTS

Critical points occur where the first derivative equals zero, $f'(x) = 0$.

To locate critical points:

1. Take the first derivative.

2. Set it equal to zero.

3. Solve for $x$. ($x^*$ is used frequently to denote a critical point.)

4. Find the $y$-coordinate, $f(x^*)$.

    Reminder: To find the $y$-coordinate, you must substitute $x^*$ into the original function, $y = f(x)$.

178

# EXAMPLE 1

Find all critical points for $y = f(x) = x^2 - 4x + 5$.

Solution:   Given.                                      $y = f(x) = x^2 - 4x + 5$

Critical points occur where $f'(x) = 0$.

Take the first derivative.                  $f'(x) = 2x - 4$

Set it equal to zero.                       $0 = 2x - 4$

Solve for $x$.                              $4 = 2x$

$x^* = 2$

Find the $y$-coordinate.            $y = f(2) = (2)^2 - 4(2) + 5$

Reminder:   Substitute into the                 $= 4 - 8 + 5$
original function, $y = f(x)$.                   $= 1$

Answer:   $f(x)$ has a critical point at $(2, 1)$.

# EXAMPLE 2

Find all critical points of $y = f(x) = x^3 + 3x^2 + 1$.

Solution:   Given.                              $y = f(x) = x^3 + 3x^2 + 1$

Critical points are where $f'(x) = 0$.

Take the first derivative.                  $f'(x) = 3x^2 + 6x$

Set it equal to zero.                       $0 = 3x^2 + 6x$

Solve for $x$.                              $0 = 3x(x + 2)$

$3x = 0 \quad x + 2 = 0$

$x^* = 0 \qquad x^* = -2$

Find the $y$-coordinates.           If $x^* = 0$,

$y = f(0) = (0)^3 + 3(0)^2 + 1 = 1$

and $(0, 1)$ is a critical point.

If $x^* = -2$,

$f(-2) = (-2)^3 + 3(-2)^2 + 1$

$= -8 + 12 + 1$

$= 5$

and $(-2, 5)$ is a critical point.

Answer:   $f(x)$ has two critical points, one at $(0, 1)$ and another located at $(-2, 5)$.

The next one is for you, but it is easy.

## Problem 1

Find all critical points for the function $y = f(x) = x^{16}$.

Solution:    Given.

Critical points occur where $f'(x) = 0$.

Take the first derivative.

Set it equal to zero.

Solve for $x$.

Find the $y$-coordinate.

Answer:    $f(x)$ has a critical point at $(0, 0)$.

We are halfway to our objective. You now should be able to locate a function's critical points; but how do we classify them without the aid of the drawing? At the start, I stated you would learn one method in this unit. That is not really true. You will learn two, but they are related.

# CLASSIFYING CRITICAL POINTS

## Using What You Already Know About Graphs

One way to classify critical points is to use your knowledge of the graph of the function. In other words, sketch or visualize your own drawing. Considering all the time we spent on first-, second-, and third-degree polynomial functions, this part should be easy.

We will return to our first example.

## EXAMPLE 3

Find and classify all critical points for $y = f(x) = x^2 - 4x + 5$.

Solution:    From Example 1, we found that $f(x)$ has a critical point at $(2, 1)$.

To classify it, ask yourself what do you know about its graph.

We don't need everything, just enough for a very quick sketch.

The point $(2, 1)$ is plotted on the grid below, along with its horizontal tangent. The horizontal tangent is most important; that shows it is a critical point and reminds us that the curve levels off at that point.

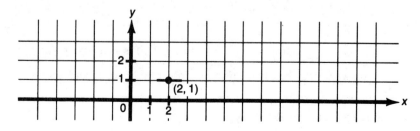

This is a quadratic with $a = 1$.

The parabolic curve opens up like such ⌣ and we have enough information to complete our quick sketch.

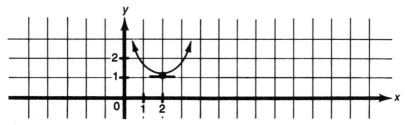

Therefore the critical point $(2, 1)$ is a relative minimum. In fact, it is an absolute minimum because we know the vertex is the lowest point (or smallest value) on the curve.

Stated another way, the minimum value for $y$ is 1 and it occurs when $x = 2$.

## EXAMPLE 4

Find and classify all critical points for $y = f(x) = x^2 - 12x + 1$.

Solution:   Given.                              $y = f(x) = x^2 - 12x + 1$

Critical points are where $f'(x) = 0$.

Take the first derivative          $f'(x) = 2x - 12$

Set it equal to zero.                  $0 = 2x - 12$

Solve for $x$.                             $12 = 2x$

$$x^* = 6$$

Find the $y$-coordinate.        $y = f(6) = 36 - 72 + 1 = -35$

$f(x)$ has a critical point at $(6, -35)$.

To classify the critical point, use your knowledge of the graph.

Plot the critical point $(6, -35)$ along with its horizontal tangent to show that the curve levels off there.

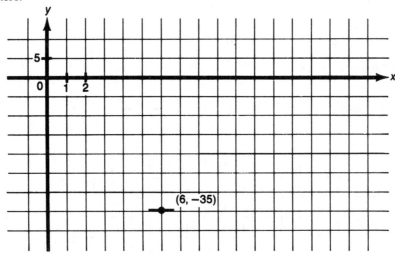

This is a quadratic with $a = 1$ and the parabola opens up: .

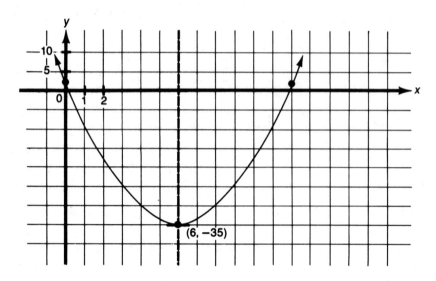

Therefore the critical point at $(6, -35)$ is a relative minimum. Again, you could be more precise in that it is an absolute.

Stated another way, the minimum value for $y$ is $-35$ and it occurs when $x = 6$.

Ready to try one?

## Problem 2

Find and classify all critical points for $y = f(x) = 100 - 10x - x^2$.

Solution:   Given.

Critical points occur where $f'(x) = 0$.

Take the first derivative.

Set it equal to zero.

Solve for $x$.

Find the $y$-coordinate.

Use your knowledge of the graph to classify the critical point.

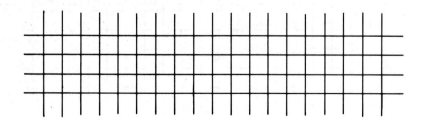

Answer: The critical point at $(-5, 125)$ is an absolute maximum. Or, the maximum value for $y$ is 125 and it occurs when $x = 5$.

---

## EXAMPLE 5

Find all critical points for $y = f(x) = 5x - 1$.

---

Solution: Given. $\qquad\qquad\qquad\qquad y = f(x) = 5x - 1$

Critical points occur where $f'(x) = 0$.

Take the first derivative. $\qquad\qquad f'(x) = 5$

Set it equal to zero. $\qquad\qquad\qquad 0 = 5$

Solve for $x$. $\qquad\qquad\qquad\qquad$ This equation has no solution.

Answer: $f(x)$ has no critical points.

---

Can you explain why there are no critical points? What does the graph of $f(x)$ look like? Does a straight line have any critical points? That is like asking if a straight line has any high or low points. The answer is no.

We have done examples with first- and second-degree functions, it is time to consider third-degree. To save time, we will return to Example 2.

---

## EXAMPLE 6

Find and classify all critical points for $y = f(x) = x^3 + 3x^2 + 1$.

---

Solution: From Example 2, we found that $f(x)$ had critical points at $(-2, 5)$ and $(0, 1)$.

To classify the critical points, use your knowledge of what the graph of $f(x)$ looks like.

First, plot both critical points. Be sure to include the horizontal tangents to remind yourself that the curve levels off at each of the points.

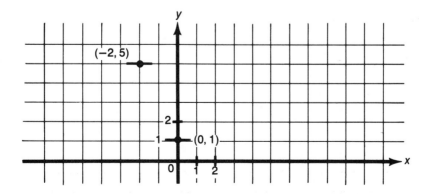

The function is a cubic with $a = 1$; therefore it opens up as ⌒⌣. Since we have already plotted our two critical points, we can sketch the curve in without much effort.

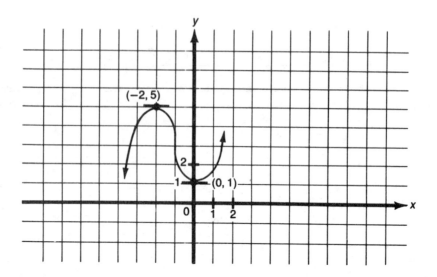

Notice from our sketch that the cubic has two critical points. The one on the left is a relative maximum and the one of the right is a relative minimum. There are no absolutes; because the curve continues downward on the left and upward on the right.

Answer:   $(-2, 5)$ is a relative maximum and $(0, 1)$ is a relative minimum.

# EXAMPLE 7

Find and classify all critical points for $y = f(x) = \dfrac{1}{3}x^3 - x^2 - 3x + 1$.

Solution:   Given.

$$y = f(x) = \frac{1}{3}x^3 - x^2 - 3x + 1$$

Take the first derivative.

$$f'(x) = \frac{1}{3}(3x^2) - 2x - 3$$

$$= x^2 - 2x - 3$$

Set it equal to zero.

$$0 = x^2 - 2x - 3$$

Solve for $x$.

$$0 = (x - 3)(x + 1)$$

$$x^* = 3 \quad x^* = -1$$

Find the $y$-coordinates.   If $x^* = 3$,

$$f(3) = 9 - 9 - 9 + 1 = -8$$

and $(3, -8)$ is a critical point.

If $x^* = -1$,

$$f(-1) = \frac{1}{3}(-1)^3 - (-1)^2 - 3(-1) + 1$$

$$= \frac{1}{3}(-1) - (1) + 3 + 1$$

$$= -\frac{1}{3} - 1 + 3 + 1$$

and $\left(-1, 2\frac{2}{3}\right)$ is a critical point.

Plot both critical points along with their horizontal tangents. Using your knowledge of third-degree polynomial functions, you know that the function is a cubic with $a = 1$; therefore it is shaped as ⌢⌣. Finish the sketch.

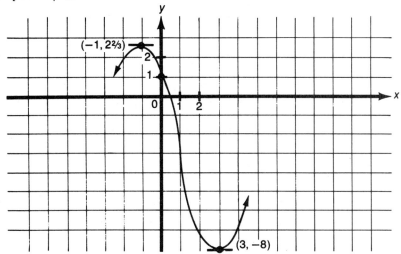

Answer:   The critical point $\left(-1, 2\frac{2}{3}\right)$ is a relative maximum.

The critical point $(3, -8)$ is a relative minimum.

I will do one more before having you try one.

## EXAMPLE 8

Find and classify all critical points for $y = f(x) = -x^3 - 3x^2 + 1$.

Solution:   Given.

$$y = f(x) = -x^3 - 3x^2 + 1$$

Take the first derivative.

$$f'(x) = -3x^2 - 6x$$

Set it equal to zero.

$$0 = -3x^2 - 6x$$

Solve for $x$.

$$0 = -3x(x + 2)$$

$$x^* = 0 \quad x^* = -2$$

Find the $y$-coordinates.

If $x^* = 0$,

$$y = f(0) = 0 - 0 + 1 = 1$$

and $(0, 1)$ is a critical point.

If $x^* = -2$

$$y = f(-2) = 8 - 12 + 1 = -3$$

and $(-2, -3)$ is a critical point.

The critical points are plotted along with their horizontal tangents. The function is a cubic with $a = -1$; therefore it is shaped as $\curvearrowright$.

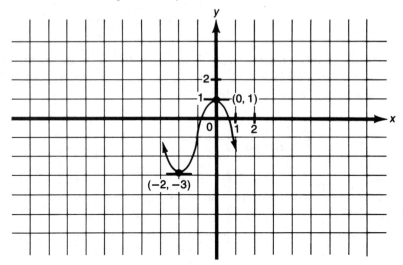

Answer:    The critical point $(-2, -3)$ is a relative minimum.

The critical point $(0, 1)$ is a relative maximum.

Here is your problem.

## Problem 3

Find and classify all critical points for $y = f(x) = 7x^3 - 21x + 2$.

Solution:    Given.

Take the first derivative.

Set it equal to zero.

Solve for $x$.

Find the $y$-coordinates.

To classify the critical points, use your knowledge of the graph. Plot both critical points along with their horizontal tangents. Finish the sketch.

Answer:    The critical point $(-1, 16)$ is a relative maximum.

The critical point $(1, -12)$ is a relative minimum.

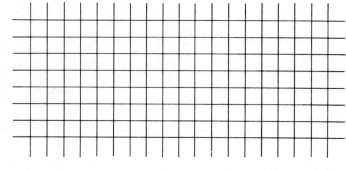

The next one is a bit unusual and I had best explain it.

## EXAMPLE 9

Find and classify all critical points for $y = f(x) = 4x^3 + 1$.

Solution:   Given.

$$y = f(x) = 4x^3 + 1$$

Take the first derivative.

$$f'(x) = 12x^2$$

Set it equal to zero.

$$0 = 12x^2$$

Solve for $x$.

$$0 = x^2$$

$$x^* = 0$$

Find the $y$-coordinate.      If $x^* = 0$,

$$y = f(0) = 0 + 1 = 1$$

and $(0, 1)$ is a critical point.

Plot the critical point along with its horizontal tangent. The function is a cubic with $a = 4$; therefore it is shaped as ⌒⌣. But this cubic function has only one critical point! How can it be ⌒⌣ shaped?

The best way that I can explain it to you is for you to imagine you are holding onto both ends of the curve and you gently pull until both the peak and valley disappear, but still leaving one leveling-off point in the middle, like the sketch below.

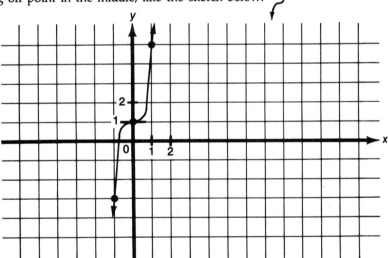

The above situation can be generalized to conclude that if a cubic has only one critical point it will be a stationary inflection point.

Answer:   The critical point $(0, 1)$ is a stationary inflection point.

The vast majority of functions that you will encounter in calculus as well as in various economics courses tend to be first-, second-, or third-degree polynomial functions. As you have seen, these functions are predictable and their maximum and minimum values rather easily determined. But, of course, there are other functions and we do need a method for classifying all critical points.

One method for classifying critical points is the **original-function test**.

# THE ORIGINAL-FUNCTION TEST FOR CLASSIFYING CRITICAL POINTS

Let $f(x)$ be a function with a critical point at $x^* = a$ and determine the values of $f(x)$ to the left and right of the critical point.

1.  If $f(x) > f(x^*)$ to the left of $x^* = a$ and if $f(x) > f(x^*)$ to the right of $x^* = a$, then the critical point at $x^* = a$ is a relative minimum.

2.  If $f(x) < f(x^*)$ to the left of $x^* = a$ and if $f(x) < f(x^*)$ to the right of $x^* = a$, then the critical point at $x^* = a$ is a relative maximum.

3.  If neither statements 1 nor 2 are satisfied, then the critical point at $x^* = a$ is a stationary inflection point.

Although the above test is a bit wordy, it is rather straightforward to use. Let me first explain the reasoning before doing any examples.

A relative minimum by definition is a low point. The original-function test examines what happens with the graph on either side of the critical point. If, as in statement 1 above, the points on the curve on either side of the critical point are both higher, then the critical point must be a relative minimum.

A relative maximum is defined as a high point. So if, as in statement 2 above, the points on the curve on either side of the critical point are both lower, then the critical point must be a relative maximum.

If neither of the two situations exists, the critical point must be a stationary inflection point.

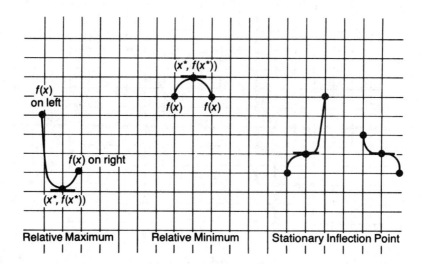

To use the original-function test, select an $x$ value on either side of the critical point $x^* = a$. It should be a value quite close. Evaluate $f(x)$, the original function, at both values and compare the $y$-coordinates according to the test.

The examples will be of functions whose graphs are unfamiliar to us. We will start by returning to Problem 1 of this unit.

## EXAMPLE 10

Find and classify all critical points for $y = f(x) = x^{16}$.

Solution: From Problem 1, $y = f(x) = x^{16}$ was found to have a critical point at $(0, 0)$.

To classify the critical point, use the original-function test. Select a value for $x$ on either side of the critical point and substitute the values into the original function, then compare.

On the left:        Let $x = -1$, then $y = f(-1) = (-1)^{16} = 1$

At the critical point,      $x = 0$, and $y = f(0)$                $= 0$

On the right:       Let $x = 1$, then $y = f(1) = (1)^{16}$     $= 1$

Do a quick sketch. The point on the left and the point on the right are both higher than critical point.

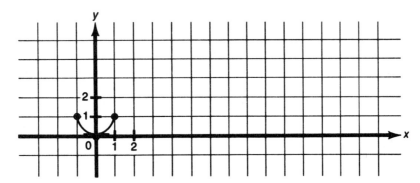

Answer:    The critical point $(0, 0)$ is a relative minimum.

## EXAMPLE 11

Find and classify all critical points for $y = f(x) = x^5$.

Solution: Given.                   $y = f(x) = x^5$

          Take the first derivative.       $f'(x) = 5x^4$

          Set it equal to zero.           $0 = 5x^4$

          Solve for $x$.                  $x^* = 0$

          Find the $y$-coordinate.        If $x^* = 0$,

                                  $y = f(0) = 0$

                              and $(0, 0)$ is a critical point.

To classify the critical point, use the original-function test.

For the point $(0, 0)$:

On the left:     Let $x = -1$, then $y = f(-1) = (-1)^5 = -1$

At the point:        $x = 0$,    and  $y = f(0)$    =              0
On the right:  Let $x = 1$,    then $y = f(1)$    $= (1\quad)^5 =$    1

Do a quick sketch. Notice that the point on the left is lower than the critical point and the point on the right is higher than the critical point.

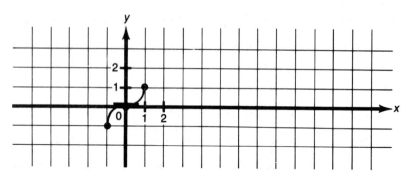

Answer:   The critical point (0, 0) is a stationary inflection point.

The last two examples were selected with the critical point occurring at the origin, (0, 0), so as to make our calculations easier. Obviously that will not always be the case as the last example illustrates.

## EXAMPLE 12

Find and classify all critical points for $y = f(x) = x^5 - 5x^4 + 1$.

Solution:   Given.                                     $y = f(x) = x^5 - 5x^4 + 1$

Take the first derivative.        $f'(x) = 5x^4 - 20x^3$

Set it equal to zero.             $0 = 5x^4 - 20x^3$

Solve for $x$.                    $0 = 5x^3(x - 4)$

                                  $x^* = 0 \quad x^* = 4$

Find the $y$-coordinates.        If $x^* = 0$,

                              $y = f(0) = 0 - 0 + 1 = 1$

                                  and (0, 1) is a critical point.

                              If $x^* = 4$,

                              $y = f(4) = 1{,}024 - 1{,}280 + 1 = -255$

                                  and (4, $-255$) is a critical point.

To classify the critical points, use the original-function test.

For the point (0, 1):

On the left:   Let $x = -1$, then $y = f(-1) = -1 - 5 + 1 = -5$

At the point:        $x =$    0, and  $y = f(0)$      =      $0 - 0 + 1 = 1$

On the right: Let $x =$    1, then $y = f(1)$     =      $1 - 5 + 1 = -3$

This is the second situation described in the test. With the points on both sides of the critical point being lower than the critical point, it is a relative maximum. A sketch is provided.

For the point $(4, -255)$:
On the left:   Let $x = 3$, then $y = f(3) = 243 - 405 + 1 = -161$

At the point:      $x = 4$, and   $y = f(4)$                      $= -255$

On the right:  Let $x = 5$, then $y = f(5) = 3,125 - 3,125 + 1 = 1$

This is the first situation described in the test. The point on the left of the critical point is higher and the point on the right is higher; the result is that the critical point is a relative minimum. The points are shown on the sketch.

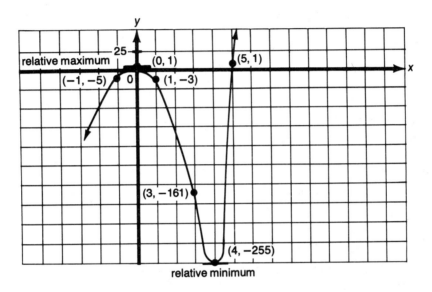

Answer:   The critical point $(0, 1)$ is a relative maximum.

The critical point $(4, -255)$ is a relative minimum.

---

You should be able now to locate critical points and classify them.

Critical points are located by taking the first derivative, setting it equal to zero, solving for $x$, and then determining the corresponding $y$-coordinate.

To classify critical points, if the function is a first-, second-, or third-degree polynomial function, you should use your knowledge of the graph of the function.

1. First-degree polynomial functions have no critical points.

2. Second-degree polynomial functions have one critical point, either an absolute maximum or an absolute minimum depending on which way the parabola is opening.

3. Two possibilities exist with third-degree polynomial functions. If there are two critical points, one is a relative maximum and one is a relative minimum, again depending on which way the curve is opening. If there is only one critical point, it is a stationary inflection point.

To classify critical points for functions other than first-, second-, or third-degree polynomial functions, one method is the original-function test. Using the original function, points are examined on either side of the critical point and briefly:

1. If both points are higher than the critical point, the point is a relative minimum.

2. If both points are lower than the critical point, the point is a relative maximum.

3. If neither situation exists, the critical point is a stationary inflection point.

## EXERCISES

Find and classify all critical points for each of the following:

1. $f(x) = 2x^2 - 4x$

2. $f(x) = -x^2 - 1$

3. $f(x) = 3x - 2$

4. $f(x) = 6x^2 + x - 1$

5. $f(x) = x^3 + 3x^2$

6. $f(x) = -x^3 + 7$

7. $f(x) = x^3 - 3x^2 - 9x + 5$

8. $f(x) = x^6 + 2$

9. $f(x) = 5x^7 - 35x^6 + 63x^5$

   This one is a bit more difficult than the others, but that also makes it more interesting.

10. $f(x) = \dfrac{1}{3}x^3 - 3x^2 + 5x + 2$

    When you have finished this problem, return to Example 5 of Unit 8 and see what you have found.

If additional practice is needed:

Barnett and Ziegler, pages 205–206, problems 1–26
Budnick, page 492, problems 1–18
Hoffmann, page 131, problems 1–16
Piascik, page 165, problems 1–14

# UNIT 16

## Classifying Critical Points— A Second Procedure

In this unit you will learn a second method for determining where a function achieves maximum or minimum values.

In the previous unit, first you learned how to locate critical points, followed by how to classify them based on your knowledge of what the graph looked like. If you were unfamiliar with the particular function, the original-function test was to be used.

The technique for locating critical points remains the same, but we will introduce a second procedure for classifying the points. It is called the **first-derivative test**.

The basis for the first-derivative test is a theorem from Unit 14, which is repeated here in a shortened version.

> Theorem:   If $f'(x) > 0$ on an interval, then $f(x)$ is increasing.
>
> If $f'(x) < 0$ on an interval, then $f(x)$ is decreasing.

Recall we said that if the derivative is positive on an interval, the curve is increasing. If the derivative is negative, the curve is decreasing. And if the derivative is zero, the curve is leveling off, and you have a critical point.

The first-derivative test examines what is happening on either side of the critical point, much like the original-function test. Except with this method, we use the first derivative to determine whether the curve is increasing or decreasing.

# THE FIRST-DERIVATIVE TEST FOR CLASSIFYING CRITICAL POINTS

Let $y = f(x)$ be a function with a critical point at $x^* = a$ and determine the values of $f'(x)$ to the left and right of the critical point.

1.  If $f'(x) < 0$ to the left of $x^* = a$ and

    if $f'(x) > 0$ to the right of $x^* = a$, then

    the critical point at $x^* = a$ is a relative minimum.

2.  If $f'(x) > 0$ to the left of $x^* = a$ and

    if $f'(x) < 0$ to the right of $x^* = a$, then

    the critical point at $x^* = a$ is a relative maximum.

3.  If neither statements 1 nor 2 are satisfied,

    the critical point at $x^* = a$ is a stationary inflection point.

Before doing any examples, let me explain and draw a picture of each situation.

1.  $f'(x) < 0$ to the left of $x^* = a$ means the curve is decreasing to the left of the critical point. At the critical point the curve levels off. $f'(x) > 0$ to the right of $x^* = a$ means the curve is increasing to the right of the critical point. The sketch below reflects this situation and should illustrate why the critical point is a relative minimum.

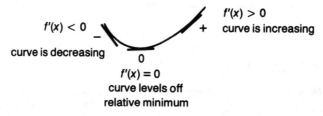

2.  By similar reasoning, the second situation has the curve increasing to the left, leveling off at the critical point, and decreasing to the right. A sketch should help to convince you that the critical point is a relative maximum.

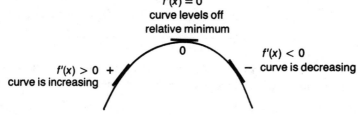

3.  Two possibilities remain for the third situation.

    a.  If $f'(x) > 0$ both to the left and right, then the curve is increasing, leveling off, and increasing again.

    b.  If $f'(x) < 0$ both to the left and right, then the curve is decreasing, leveling off, and decreasing again.

In either case, the critical point is a stationary inflection point.

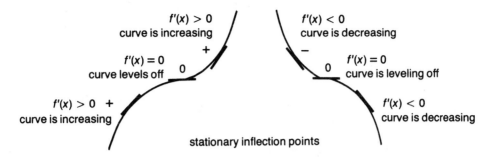

$f'(x) > 0$
curve is increasing

$f'(x) < 0$
curve is decreasing

$f'(x) = 0$
curve levels off

$f'(x) = 0$
curve is leveling off

$f'(x) > 0$ +
curve is increasing

$f'(x) < 0$
curve is decreasing

stationary inflection points

To use the first-derivative test, select a value for $x$ on either side of the critical point $x^* = a$. As with the original-function test, the values should be quite close to $a$. This time, however, we use $f'(x)$ to calculate the two values and compare their signs. The actual numerical answer is unimportant. It is only the sign, positive or negative, that is of concern in the test.

Enough words; we will work through some examples.

## EXAMPLE 1

Find all critical points for $y = f(x) = x^4 - 4x$.

Use the first-derivative test to classify the critical points.

Solution:   $y = f(x) = x^4 - 4x$

$f'(x) = 4x^3 - 4$

Find all critical points. Critical points occur where $f'(x) = 0$.

$0 = 4x^3 - 4$

$4x^3 = 4$

$x^3 = 1$

$x^* = 1$

If $x^* = 1$,

$y = f(1) = 1 - 4 = -3$ and $(1, -3)$ is the only critical point.

Use the first-derivative test with $f'(x) = 4x^3 - 4$.

For the point $(1, -3)$                                      the curve is:

On the left:    Let $x = 0$, then $f'(0) = 0 - 4 = -$      decreasing

At the point:   $(1, -3)$ the curve levels off             levels off

On the right:   Let $x = 2$, the $f'(2) = 32 - 4 = +$     increasing

This is the first situation described in the test. The first derivative is negative to the left of the critical point and positive to the right. Can you visualize the graph? It is decreasing, leveling off, then increasing. What must the critical point be?

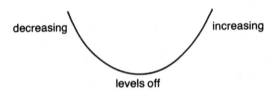

decreasing                                             increasing

levels off

Answer:    The critical point $(1, -3)$ is a relative minimum.

Be sure to notice the following two items regarding the use of the first-derivative test:

1.  The values selected on either side of the critical point are substituted into the first derivative. That is where the test gets its name. Similarly with the original-function test, we substituted the numbers into the original function.

2.  The actual numerical value of the derivative is unimportant; only the sign, positive or negative, need be determined.

# EXAMPLE 2

Sketch the graph of the $y = f(x) = x^4 - 4x$.

Solution:    $f(x)$ is a fourth-degree polynomial function.

The $y$-intercept is at 0.

The $x$-intercepts are 0 and $\sqrt[3]{4}$ because

$$0 = x^4 - 4x$$

$$0 = x(x^3 - 4)$$

$$x = 0 \qquad x^3 - 4 = 0$$

$$x^3 = 4$$

$$x = \sqrt[3]{4}$$

From Example 1, the function has a relative minimum at $(1, -3)$, which implies the curve is decreasing on the left and increasing on the right.

Using the fact that the graph of a polynomial function is a smooth, continuous curve without any breaks or sharp corners, a sketch of the curve would appear as shown.

Answer:

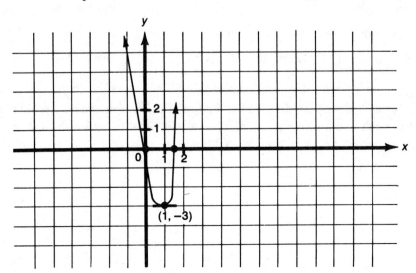

How do I know the curve is continuing upward on the right where I have the arrow? Because in order for the curve to turn downward again, there would need to be another critical point. But we found there was only one and it is shown. The same line of reasoning applies to the left-hand arrow.

---

# EXAMPLE 3

Use the graph from Example 2 to answer the following questions:

a.  Identify any absolute maximum and absolute minimum values of $f(x)$.

b.  State the interval(s) over which $f(x)$ is increasing.

c.  State the interval(s) over which $f(x)$ is decreasing.

---

Solution:  a.  From the drawing, the absolute minimum value of $f(x)$ is $-3$ and occurs when $x = 1$. $f(x)$ does not have an absolute maximum.

b.  The function $f(x)$ is increasing over the interval $1 < x$.

c.  The function $f(x)$ is decreasing over the interval $x < 1$.

Remember we read graphs from left to right. So although the arrow at the left is pointing upward, think of starting at the arrow and moving to the right. The curve is falling, hence the function is said to be decreasing.

---

Consider another fourth-degree polynomial function.

# EXAMPLE 4

Find all critical points for $y = f(x) = x^4 - 8x^3 + 18x^2 - 27$.

Use the first-derivative test to classify the critical points.

---

Solution:  $y = f(x) = x^4 - 8x^3 + 18x^2 - 27$

$$f'(x) = 4x^3 - 24x^2 + 36x$$

$$= 4x(x^2 - 6x + 9)$$

$$= 4x(x - 3)(x - 3)$$

$$= 4x(x - 3)^2$$

Find all critical points. Critical points occur where $f'(x) = 0$.

$$0 = 4x^3 - 24x^2 + 36x$$

$$0 = 4x(x - 3)^2$$

$$4x = 0 \qquad x - 3 = 0$$

$$x^* = 0 \qquad\quad x^* = 3$$

If $x^* = 0$,

$y = f(0) = 0 - 0 + 0 - 27 = -27$

and $(0, -27)$ is a critical point.

If $x^* = 3$,

$y = f(3) = 81 - 8(27) + 18(9) - 27 = 81 - 216 + 162 - 27 = 0$

and $(3, 0)$ is another critical point.

Use the first-derivative test with $f'(x) = 4x(x - 3)^2$.

Remember we need only determine the sign, not the actual value, so I will shortcut some of the work.

For the point $(0, -27)$ with $f'(x) = 4x^3 - 24x + 36x$         the curve is:

On the left:     Let $x = -1$, then $f'(-1) = -4 - 24 - 36 = -$     decreasing

At the point:     $(0, -27)$ the curve levels off               levels off

On the right:    Let $x = 1$,     then $f'(1)$   $= 4 - 24 + 36 = +$    increasing

Try to visualize the curve. It is decreasing, leveling off, then increasing. The critical point $(0, -27)$ is a relative minimum.

decreasing                  increasing

levels off

Remember we need only really determine the sign, not the actual value, of the derivative. So I will show you how to shortcut some of the calculations. In order to do so, you must use the first derivative in factored form rather than multiplied out. Then only the sign of the factor is considered. For example, if $x = 2$, then $(x - 3)$ would be indicated as a minus factor, $(-)$, because $2 - 3$ is a negative number.

For the point $(3, 0)$ with $f'(x) = 4x(x - 3)^2$          the curve is:

On the left:     Let $x = 2$, then $f'(2) = (+)(+)(-)^2 = +$      increasing

At the point:     $(3, 0)$ the curve levels off                 levels off

On the right:    Let $x = 4$, then $f'(4) = (+)(+)(+)^2 = +$      increasing

At this critical point, the curve is increasing, leveling off, then increasing again. Therefore, the critical point $(3, 0)$ is a stationary inflection point.

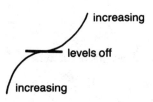

increasing

levels off

increasing

---

I should add that this is my least favorite of the three procedures because it is rather long. Nevertheless, it is important enough that we consider it and you become proficient with its use. You might even like it. After the next unit, the choice will be yours as to which of the three procedures you use to classify critical points.

# EXAMPLE 5

Sketch the graph of $y = f(x) = x^4 - 8x^3 + 18x^2 - 27$.

a.   Identify any absolute maximum and absolute minimum values of $f(x)$.

b.   State the interval(s) over which $f(x)$ is increasing.

c.   State the interval(s) over which $f(x)$ is decreasing.

Solution:   1.   The function is a fourth-degree polynomial function.

2.   The $y$-intercept is at $-27$.

3.   There are at most four $x$-intercepts.

4.   From Example 4, the function has a relative minimum at $(0, -27)$, which implies the curve is decreasing on the left and increasing on the right.

5.   Also from Example 4, the function has a stationary inflection point at $(3, 0)$ with the curve increasing on the right.

6.   And because the graph of a polynomial function is known to be a smooth, continuous curve without any breaks or sharp corners, we can connect the critical points with a nice smooth curve as shown.

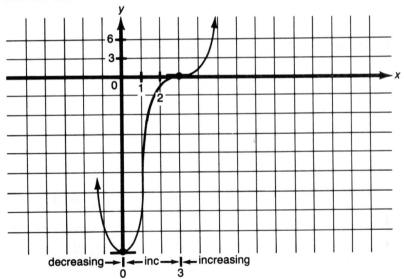

Answer:   a.   From the drawing, the absolute minimum value of $f(x)$ is $-27$ and occurs when $x = 0$. $f(x)$ does not have an absolute maximum.

b.   The function $f(x)$ is increasing over the intervals $0 < x < 3$ and $3 < x$.

c.   The function $f(x)$ is decreasing over the interval $x < 0$.

Here are three related problems for you to try before we continue with some functions that are more complicated.

## Problem 1

Use the first-derivative test to determine any relative maxima and minima for $y = f(x) = -x^5 + 1$.

Solution:    $y = f(x) =$

$\qquad f'(x) =$

Find all critical points. Critical points occur where $f'(x) = 0$.

Use the first-derivative test.

For the point

On the left:

At the point:

On the right:

Visualize the graph.

Answer:    The critical point $(0, 1)$ is a stationary inflection point.

## Problem 2

Sketch the graph of the $y = f(x) = -x^5 + 1$.

Solution:

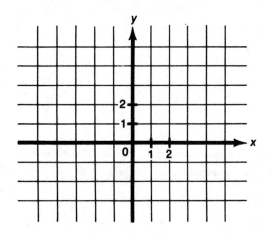 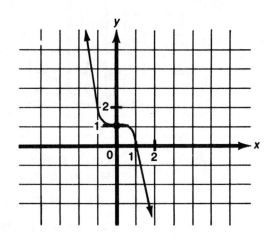

Answer:

## Problem 3

Use the information from Problems 1 and 2, to answer the following questions:

a. Identify any absolute maximum and absolute minimum values of $f(x)$.

b. State the interval(s) over which $f(x)$ is increasing.

c. State the interval(s) over which $f(x)$ is decreasing.

Solution:

Answers:  a.  $f(x)$ has neither an absolute maximum nor a minimum.

b.  $f(x)$ is never increasing.

c.  $f(x)$ is decreasing over the intervals $x < 0$ and $0 < x$. Note: At $x = 0$, the curve is neither increasing nor decreasing. At $x = 0$, the slope of the tangent line is 0, hence a critical point.

The procedure for using the first-derivative test is always the same. The only thing that remains for us to do is to consider functions that are more complicated. The examples that follow are no longer polynomial functions and we will not attempt to sketch their graphs.

## EXAMPLE 6

Find all critical points for $y = f(x) = \dfrac{2x^2 + 18 + x}{x}$ for $x > 0$.

Use the first-derivative test to classify the critical points.

Solution:  $y = f(x) = \dfrac{2x^2 + 18 + x}{x}$

$$f'(x) = \frac{x\dfrac{d(2x^2 + 18 + x)}{dx} - (2x^2 + 18 + x)\dfrac{d(x)}{dx}}{x^2}$$

$$= \frac{x(4x + 0 + 1) - (2x^2 + 18 + x)(1)}{x^2}$$

$$= \frac{4x^2 + x - 2x^2 - 18 - x}{x^2}$$

$$= \frac{2x^2 - 18}{x^2}$$

Find all critical points. Critical points occur where $f'(x) = 0$.

$$0 = \frac{2x^2 - 18}{x^2}$$

$0 = 2x^2 - 18$        (Multiplied entire equation by $x^2$.)

$0 = x^2 - 9$        (Divided equation by 2.)

$0 = (x - 3)(x + 3)$

$x - 3 = 0$        $x + 3 = 0$

$x^* = 3$        $x^* = -3$, but not in the domain.

If $x^* = 3$,

$$y = f(3) = \frac{2(9) + 3 + 18}{3} = 13$$

and $(3, 13)$ is the only critical point.

Use the first-derivative test with $f'(x) = \frac{2x^2 - 18}{x^2}$.

| For the point $(3, 13)$ | | the curve is: |
|---|---|---|
| On the left: | Let $x = 2$, then $f'(2) = (8 - 18)/4 = -$ | decreasing |
| At the point: | $(3, 13)$ the curve levels off | levels off |
| On the right: | Let $x = 4$, then $f'(4) = (32 - 18)/4 = +$ | increasing |

At this critical point, the curve is decreasing, leveling off, then increasing. Therefore, the critical point $(3, 13)$ is a relative minimum.

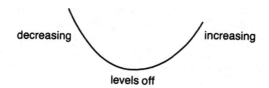

decreasing                                        increasing

levels off

# EXAMPLE 7

Find all critical points for $y = f(x) = \dfrac{x}{x + 1}$. Use the first-derivative test.

Solution:   $y = f(x) = \dfrac{x}{x + 1}$

$$f'(x) = \frac{(x + 1)\dfrac{d(x)}{dx} - x\dfrac{d(x + 1)}{dx}}{(x + 1)^2}$$

$$= \frac{(x + 1)(1) - x(1)}{(x + 1)^2}$$

$$= \frac{x + 1 - x}{(x + 1)^2}$$

$$= \frac{1}{(x + 1)^2}$$

Find all critical points.

$$0 = \frac{1}{(x + 1)^2}$$

$$0 = 1 \qquad \text{(Multiplied equation by } (x + 1)^2)$$

But $0 = 1$ is a false statement.

The original equation has no solution.

Answer:   The function has no critical points.

---

That last example was too easy. We need to end the unit with a more challenging problem.

## EXAMPLE 8

Find all critical points for $y = f(x) = \dfrac{x^2}{x - 1}$.

Use the first-derivative test to classify the critical points.

---

Solution:   $y = f(x) = \dfrac{x^2}{x - 1}$

$$f'(x) = \frac{(x - 1) \dfrac{d(x^2)}{dx} - x^2 \dfrac{d(x - 1)}{dx}}{(x - 1)^2}$$

$$= \frac{(x - 1)(2x) - x^2(1)}{(x - 1)^2}$$

$$= \frac{2x^2 - 2x - x^2}{(x - 1)^2}$$

$$= \frac{x^2 - 2x}{(x - 1)^2}$$

$$= \frac{x(x - 2)}{(x - 1)^2}$$

Find all critical points.

$$0 = \frac{x(x - 2)}{(x - 1)^2} \qquad \text{(Clear of fractions by multiplying)}$$

$$0 = x(x - 2)$$

$$x = 0 \qquad x - 2 = 0$$

$$x^* = 0 \qquad x^* = 2$$

If $x^* = 0$,

$y = f(0) = \dfrac{0}{0-1} = 0$ and $(0, 0)$ is a critical point.

If $x^* = 2$,

$y = f(2) = \dfrac{(2)^2}{2-1} = \dfrac{4}{1}$ and $(2, 4)$ is a critical point.

Use the first-derivative test with $f'(x) = \dfrac{x(x-2)}{(x-1)^2}$.

For the point $(0, 0)$                                                the curve is:

On the left:     Let $x = -1$, then $f'(-1) = \dfrac{(-)(-)}{(-)^2} = \dfrac{+}{+} = +$      increasing

At the point:     $(0, 0)$ the curve levels off                                 levels off

On the right:     What happens if we let $x = 1$? If $x = 1$, the denominator of the function is zero and undefined. Therefore we need to select another value, but it must be close to 0, such as .5.

                    Let $x = .5$, then $f'(.5) = \dfrac{(+)(-)}{(-)^2} = \dfrac{-}{+} = -$      decreasing

At this critical point, the curve is increasing, leveling off, then decreasing. Therefore, the critical point $(0, 0)$ is a relative maximum.

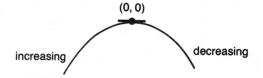

For the point $(2, 4)$                                                the curve is:

On the left:     We have the same situation as above. If we let $x = 1$ both the function and derivative are undefined. Therefore we need to select some other value that is closer to 2 on the left than 1. One choice might be 1.5.

                    Let $x = 1.5$, then $f'(1.5) = \dfrac{(+)(-)}{(+)^2} = \dfrac{-}{+} = -$      decreasing

At the point:     $(2, 4)$ the curve levels off                                 levels off

On the right:     Let $x = 3$, then $f'(3) = \dfrac{(+)(+)}{(+)^2} = \dfrac{+}{+} = +$      increasing

At this critical point, the curve is decreasing, leveling off, then increasing. Therefore, the critical point $(2, 4)$ is a relative minimum.

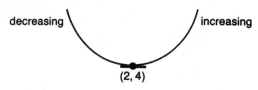

For your information, here is the graph of the function $f(x)$.

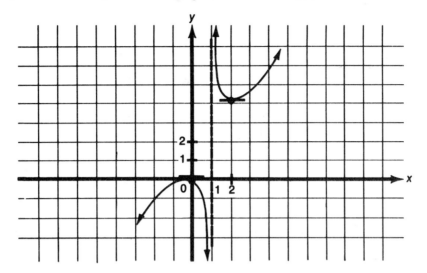

You now should be able to use the first-derivative test to classify critical points. And further, for polynomial functions you should be able to state the intervals over which the function is increasing and decreasing.

To classify critical points for functions using the first-derivative test, the signs of the first derivative are examined on either side of the critical point and briefly:

1. If the derivative is negative on the left and positive on the right, the point is a relative minimum.

2. If the derivative is positive on the left and negative on the right, the point is a relative maximum.

3. If neither situation exists, the critical point is a stationary inflection point.

## EXERCISES

Find all relative maxima and minima using the first-derivative test whenever it applies.

1. $y = f(x) = x^4 - 4x^3 + 10$

2. Sketch the graph of $y = f(x) = x^4 - 4x^3 + 10$.

   a. Identify any absolute maximum and absolute minimum values.

   b. State the interval(s) over which $f(x)$ is increasing.

   c. State the interval(s) over which $f(x)$ is decreasing.

3. $y = f(x) = \dfrac{1}{4}x^4 + x$

4. $y = f(x) = x^5 - 5x$

5. Sketch the graph of $y = f(x) = x^5 - 5x$.

   a. Identify any absolute maximum and absolute minimum values.

   b. State the interval(s) over which $f(x)$ is increasing.

   c. State the interval(s) over which $f(x)$ is decreasing.

6.  $y = f(x) = \dfrac{1}{x - 2}$

7.  $y = f(x) = \dfrac{x}{x^2 + 1}$

8.  Ken is the General Operations Manager for an exclusive resort club with golf, tennis, marina, and food and beverage facilities on the island of Cat Cay in the Bahamas. Since taking the job as manager in 1980, Ken has determined that the number of guests per year staying at the club can be estimated by the following function:

$$f(x) = 2x^3 - 21x^2 + 60x + 10$$

where $x$ is the number of years since 1980.

For example, $f(1) = 51$ means that in 1981 there were 51 guests at the club.

a.  Graph $f(x)$. Be sure to include critical points.

b.  What is the restricted domain for $f(x)$?

c.  How many guests were there in 1987?

d.  At what year between 1980 and the present did the club have the greatest number of guests?

e.  Which year after the first had the fewest number of guests?

f.  What was the fewest number of guests after the first year?

g.  Describe the number of guests between 1983 and 1984.

h.  How would you describe the future prospects for the club?

---

If additional practice is needed:

Barnett and Ziegler, pages 205–206, problems 1–36
Budnick, page 492, problems 1–18
Hoffmann, page 149, problems 1–30
Piascik, page 177, problems 1–3

# UNIT 17

## Classifying Critical Points— A Third Procedure

In this unit you will learn yet a third method, the second-derivative test, for determining where a function achieves maximum or minimum values.

Before introducing the test, we need some terms defined. There are numerous definitions for concavity; this is the one I happen to like.

---

Definition:    A curve is **concave up** over an interval if every chord in that interval lies above the curve.

---

Definition:    A curve is **concave down** over an interval if every chord in that interval lies below the curve.

---

Think of a chord as being a line segment connecting any two points on the curve. If, no matter which two points you select, the line segment connecting them is above the curve, the curve is said to be concave up.

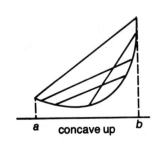 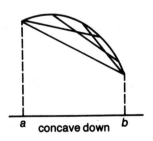 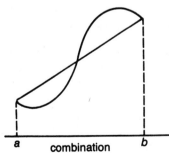

| a    concave up    b | a    concave down    b | a    combination    b |

> Definition:   The point at which the curve changes from concave up to down or vice versa is called an **inflection point**.

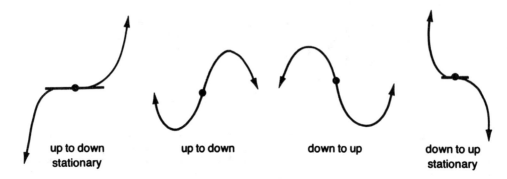

| up to down | up to down | down to up | down to up |
| stationary |            |            | stationary |

All of the above are examples of inflection points. However, two of them, the first one and the last one, are stationary inflection points which means they are critical points as well.

# CONCEPT OF DERIVATIVES

Recall that the first derivative at a point gives the slope of the tangent line at that point or the instantaneous rate of change of $y$ with respect to $x$.

The second derivative (the derivative of the first derivative) gives the rate of change of the slope of the tangent.

Consider a relative minimum point:

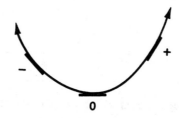

—at points to the left of $x^* = a$, the tangent lines have negative slopes,

—at the minimum point, the slope is zero, and

—at points to the right of $x^* = a$, the tangent lines have positive slopes.

Thus as you move from left to right, the slope of the tangent line is increasing from negative values, to zero, to positive values. Hence, the rate of change of the slope (the second derivative) should be positive, and the curve is concave up.

By a similar argument, if the second derivative is negative on an interval, then the first derivative is decreasing, and the curve is concave down. Stated mathematically, we have the following theorem.

> Theorem:   If $f''(x) > 0$ over an interval,
>
>                  then $f(x)$ is concave up over the interval.

> Theorem:   If $f''(x) < 0$ over an interval,
>
> then $f(x)$ is concave down over the interval.

And now we have the basis for a new way of testing for critical points; it is called the **second-derivative test.**

# THE SECOND-DERIVATIVE TEST FOR CLASSIFYING CRITICAL POINTS

Let $f(x)$ be a function with a critical point at $x^* = a$ and determine the sign of $f''(x^*)$.

1.   If $f''(x^*) > 0$, then $f(x)$ is concave up, ∖___◢ , and the critical point is a relative minimum.

2.   If $f''(x^*) < 0$, then $f(x)$ is concave down, ◤‾‾◥ , and the critical point is a relative maximum.

3.   If $f''(x^*) = 0$, the test fails.

To say that the test fails means this procedure gives no information that would allow you to classify the critical point. You must rework the problem using either of the other two methods.

## EXAMPLE 1

Use the second-derivative test to determine all relative maxima and minima for $f(x) = x^4 - 32x + 48$.

Solution:   $f(x) = x^4 - 32x + 48$

$f'(x) = 4x^3 - 32$

$f''(x) = 12x^2$

Find all critical points. Critical points occur where $f'(x) = 0$.

$0 = 4x^3 - 32$

$4x^3 = 32$

$x^3 = 8$

$x^* = 2$

If $x^* = 2$,

$f(2) = (2)^4 - 32(2) + 48 = 16 - 64 + 48 = 0$

and $(2, 0)$ is a critical point.

Use the second-derivative test with $f''(x) = 12x^2$.

At the point $(2, 0)$,

$f''(2) = 12(2)^2 = +$; hence the curve is concave up, ∖___◢ , and the critical point is a relative minimum.

Be sure to notice the following two items regarding the use of the second-derivative test.

1.  The $x$-coordinate of the critical point is used to evaluate the second derivative.

2.  The actual numerical value of the second derivative is unimportant; only the sign, positive or negative, need be determined.

## EXAMPLE 2

Sketch the graph of $y = f(x) = x^4 - 32x + 48$.

Solution:   $f(x)$ is a fourth-degree polynomial function.

The $y$-intercept is at 48.

From Example 1, the function has a relative minimum at (2, 0).

The graph will be a smooth, continuous curve without any breaks or sharp corners.

Answer:

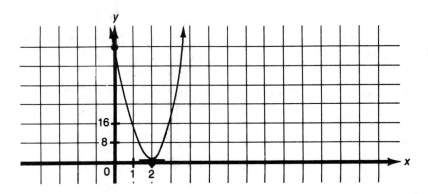

## EXAMPLE 3

Use the second-derivative test to determine all relative maxima and minima for $y = f(x) = 3x^4 - 16x^3 + 18x^2$.

Solution:   $y = f(x) = 3x^4 - 16x^3 + 18x^2$

$\quad\quad\quad\quad f'(x) = 12x^3 - 48x^2 + 36x$

$\quad\quad\quad\quad f''(x) = 36x^2 - 96x + 36$

Find all critical points. Critical points occur where $f'(x) = 0$.

$\quad\quad\quad\quad 0 = 12x^3 - 48x^2 + 36x$

$\quad\quad\quad\quad 0 = x^3 - 4x^2 + 3x$

$\quad\quad\quad\quad 0 = x(x^2 - 4x + 3)$

$\quad\quad\quad\quad 0 = x(x - 1)(x - 3)$

$\quad\quad\quad\quad x^* = 0 \quad x^* = 1 \quad x^* = 3$

If $x^* = 0$,

$y = f(0) = 0 - 0 + 0 = 0$, and $(0, 0)$ is a critical point.

If $x^* = 1$,

$y = f(1) = 3 - 16 + 18 = 5$, and $(1, 5)$ is another critical point.

If $x^* = 3$,

$y = f(3) = 3(81) - 16(27) + 18(9) = 243 - 432 + 162 = -27$, and $(3, -27)$ is still another critical point.

Use the second-derivative test with $f''(x) = 36x^2 - 96x + 36$.

At the point $(0, 0)$,

$f''(0) = 0 - 0 + 36 = +$; hence the curve is concave up, ⌣ , and the critical point is a relative minimum.

At the point $(1, 5)$,

$f''(1) = 36 - 96 + 36 = -$; hence the curve is concave down, ⌢ , and the critical point is a relative maximum.

At the point $(3, -27)$,

$f''(3) = 36(9) - 96(3) + 36 = 324 - 288 + 36 = +$; hence the curve is concave up, and the critical point is a relative minimum.

---

Since this is the first example we have had with three critical points, let's stop and sketch it as well.

# EXAMPLE 4

Sketch the graph of $y = f(x) = 3x^4 - 16x^3 + 18x^2$.

---

Solution:   The function is a fourth-degree polynomial function.

The $y$-intercept is at 0.

There are three critical points:

      $(0, 0)$ is a relative minimum,

      $(1, 5)$ is a relative maximum,

      $(3, -27)$ is a relative minimum.

Because the graph of a polynomial function is a continuous curve, connect the critical points with a smooth curve as shown.

Answer:

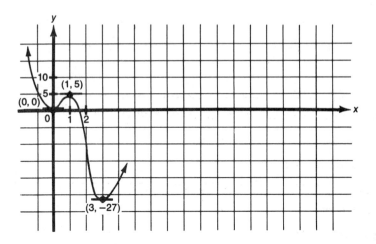

Time for a review problem on increasing and decreasing intervals.

## Problem 1

Use the graph from Example 4 to answer the following questions:

a.  Identify any absolute maximum and absolute minimum values of $f(x)$.

b.  State the interval(s) over which $f(x)$ is increasing.

c.  State the interval(s) over which $f(x)$ is decreasing.

Solution:

Answers:    a.  The absolute minimum value is $f(x) = -27$ and it occurs when $x = 3$. The function has no absolute maximum.

b.  The function is increasing over the intervals $0 < x < 1$ and $3 < x$.

c.  The function is decreasing over the intervals $x < 0$ and $1 < x < 3$.

Since you need the practice more than I do, I am going to have you work the first part of the next example, that of finding the critical points. Then I will classify them afterward.

## Problem 2

Find all critical points for $y = f(x) = 3x^5 - 5x^3$. Hint: There are three.

Solution:

Answer:    $(1, -2), (-1, 2), (0, 0)$

# EXAMPLE 5

Classify all critical points for $y = f(x) = 3x^5 - 5x^3$.

Solution:    $y = f(x) = 3x^5 - 5x^3$

$$f'(x) = 15x^4 - 15x^2$$

$$f''(x) = 60x^3 - 30x$$

Use the second-derivative test with $f''(x) = 60x^3 - 30x$.

At the point $(1, -2)$,

$f''(1) = 60 - 30 = +$; hence the curve is concave up, ⌣ , and the critical point is a relative minimum.

At the point $(-1, 2)$,

$f''(-1) = 60(-1) - 30(-1) = -60 + 30 = -$; hence the curve is concave down, ⌢ , and the critical point is a relative maximum.

At the point $(0, 0)$,

$f''(0) = 0 - 0 = 0$ and the test fails.

To classify the critical point, we must rework the problem using either the original-function test or the first-derivative test.

I will use the first-derivative test with $f'(x) = 15x^4 - 15x^2$.

To simplify our work, factor the first derivative.

$$f'(x) = 15x^4 - 15x^2$$

$$= 15x^2(x^2 - 1)$$

$$= 15x^2(x + 1)(x - 1)$$

Recall that to use the first-derivative test, we select a value on either side of the critical point and determine the sign of the first-derivative. Typically we would use $-1$ on the left and $1$ on the right. But in this example those two values are too far away; they are critical points themselves. Therefore we need to use some values closer to 0; I will use .5 and $-.5$. Remember we need only the signs of the derivative, not the numerical values.

At the point $(0, 0)$        with $f'(x) = 15x^2(x + 1)(x - 1)$

On the left:   Let $x = -.5$, $f'(-.5) = 15(-.5)^2(\underline{-.5 + 1})(\underline{-.5 - 1})$

$$= (+)(+)(+)(-)$$

$$= -, \text{ curve is decreasing}$$

At the point: (0, 0) the curve is leveling off.

On the right: Let $x = .5$, $f'(.5) \qquad = 15(.5)^2(.5 + 1)(.5 - 1)$

$$= (+)(+)(+)(-)$$

$$= -, \text{ curve is decreasing}$$

According to the first-derivative test, the curve is decreasing, leveling off, then decreasing. The critical point (0, 0) is a stationary inflection point.

## EXAMPLE 6

Sketch the graph of $y = f(x) = 3x^5 - 5x^3$.

a.   Identify any absolute maximum and absolute minimum values of $f(x)$.

b.   State the interval(s) over which $f(x)$ is increasing.

c.   State the interval(s) over which $f(x)$ is decreasing.

Solution:   The function is a fifth-degree polynomial function.

The $y$-intercept is at 0.

The $x$-intercepts are 0, $\sqrt{5/3}$, and $-\sqrt{5/3}$ because

$$0 = 3x^5 - 5x^3$$

$$0 = x^3(3x^2 - 5)$$

$$x^3 = 0 \qquad 3x^2 - 5 = 0$$

$$x = 0 \qquad\qquad x^2 = 5/3$$

$$x = \pm\sqrt{\frac{5}{3}}$$

From Example 5, $(-1, 2)$ is a relative maximum,

(0, 0) is a stationary inflection point, and

$(1, -2)$ is a relative minimum.

Answers:

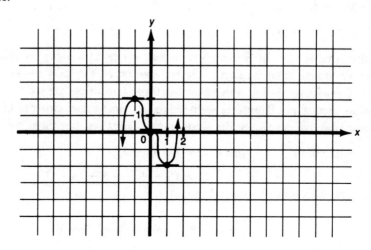

   a.   The function has neither an absolute maximum nor an absolute minimum value.

   b.   The function is increasing over the intervals $x < -1$ and $1 < x$.

   c.   The function is decreasing over the intervals $-1 < x < 0$ and $0 < x < 1$.

# A COMPARISON OF THE THREE PROCEDURES

Let us conclude this entire section with a brief summary of the three methods for classifying critical points.

The original-function test:

Use $f(x)$ and determine a point on either side of the critical point. Two complete numerical calculations are required.

The first-derivative test:

Use $f'(x)$ and determine the sign of a point on either side of the critical point. Two calculations are required, but only the sign has to be computed, not the numerical value.

The second-derivative test:

Use $f''(x)$ and determine the sign of the critical point. Only one calculation is required and that of the sign only. The disadvantage is that you must find the second derivative, which is sometimes more work than it is worth.

Clearly there are advantages and disadvantages to each method. So that you will have a better appreciation of each procedure, you will be directed to work one of the exercise problems using each test. After this unit, you may select the method of your choosing.

You should now be able to classify critical points using the second-derivative test.

To classify critical points for functions using the second-derivative test, the sign of the second derivative is determined for the critical point and briefly:

1.   If the second derivative is positive, the curve is concave up, and the point is a relative minimum.

2.   If the second derivative is negative, the curve is concave down, and the point is a relative maximum.

3.   If the second derivative is zero, the test fails, and the point must be reworked using either of the other two tests.

## EXERCISES

Find all relative maxima and minima using the second-derivative test whenever it applies. If the second-derivative test fails, use the first-derivative test.

1.   $y = f(x) = x^4 - 18x^2$

2.   Sketch the graph of $y = f(x) = x^4 - 18x^2$.

   a.   Identify any absolute maximum and absolute minimum values.

   b.   State the interval(s) over which $f(x)$ is increasing.

   c.   State the interval(s) over which $f(x)$ is decreasing.

3.  $y = f(x) = 2 - x^5$

4.  $y = f(x) = 10x^6 + 24x^5 + 15x^4 + 2$

5.  Sketch the graph of $y = f(x) = 10x^6 + 24x^5 + 15x^4 + 2$.

6.  Find all critical points for $y = f(x) = \dfrac{x^2 + 1}{x}$.

7.  Use the original-function test to classify the critical points for the function in question 6.

8.  Use the first-derivative test to classify the critical points for the function in question 6.

9.  Use the second-derivative test to classify the critical points for the function in question 6.

10.  Compare the three methods used in questions 7, 8, and 9. Which test do you prefer?

---

If additional practice is needed:

Barnett and Ziegler, pages 219–220, problems 1–26, 35–42
Budnick, page 501, problems 1–10
Hoffmann, 160, problems 1–14
Piascik, page 186, problems 1–2, 4

# UNIT 18

## Absolute Maxima and Minima

In this unit you will learn how to locate the absolute maximum and absolute minimum for a continuous function on a closed interval.

Recall that the words absolute maximum, or global maximum, are used to denote the largest value of the function, if it exists. The absolute maximum would be located at the highest point on the graph. Similarly, absolute minimum, or global minimum, refers to the smallest value of the function, if it exists, and would be located at the lowest point on the graph. Absolute maxima and minima need not be at a critical point.

We need the graph of a function to use as an example. I am confident you can do the next problem with ease.

## Problem 1

Graph: $y = f(x) = x^3 + 3x^2 + 1$.

Solution:

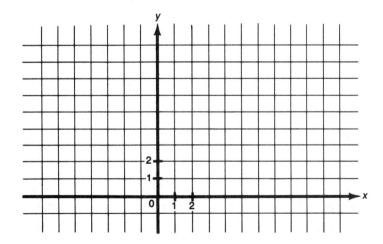

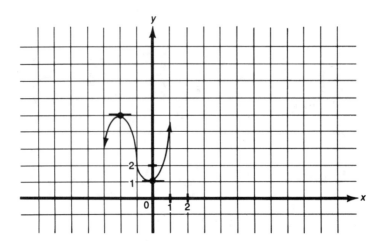

Answer:

---

As we already know and can observe from our graph, a cubic has neither an absolute maximum nor an absolute minimum. On the right the curve is continuing upward and on the left the curve is continuing downward. But if we restrict our attention to just a portion of the curve, say from $-1 \le x \le 2$, all of that changes.

A solid line in the next drawing indicates the portion of the curve from $-1 \le x \le 2$ with the remainder being shown with a dashed line. From the drawing it should be obvious that the solid portion of the curve has an absolute maximum and an absolute minimum.

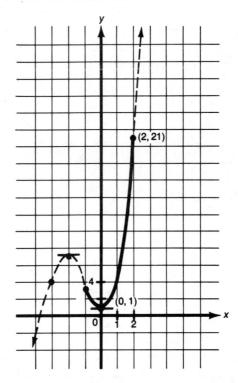

The absolute maximum occurs at an endpoint and the absolute minimum occurs at a critical point.

From previous units we have used the graph to determine the absolute maxima and minima for various examples. In this unit we will introduce a procedure for locating absolute maxima and minima without having to draw a graph, provided the function is continuous and defined over a closed interval. Such continuous functions, meaning there are no breaks in the curve, always will have an absolute maximum and an absolute minimum.

> Theorem:  If $y = f(x)$ is a continuous function over the interval $a \leq x \leq b$, its absolute maximum and absolute minimum will occur at either a critical point or at an endpoint.

Two criteria must be met before the theorem applies. First, the function must be defined over a closed interval, meaning both the lower and upper limits of the interval must be included. And second, the function must be continuous over the entire interval, meaning there are no breaks in the curve. In other words, the function must be defined for all $x$ in the interval. Stated still another way, the function has a restricted domain.

It is actually less work to find absolute maxima and minima than classifying critical points. It is a three-step procedure.

# PROCEDURE FOR LOCATING ABSOLUTE MAXIMUM AND MINIMUM ON A CLOSED INTERVAL

1. Find the $x$-coordinates of all critical points *within* the interval.

2. Determine the $y$-coordinate, using $f(x)$, for each of the critical points within the interval, and for the lower and upper limits of the interval.

3. The absolute maximum is the largest of these $y$-coordinates.
   The absolute minimum is the smallest of these $y$-coordinates.

Notice it is a rather short procedure. There is no test to perform. No graph is necessary. The values must exist. The absolute maximum will be the largest and the absolute minimum will be the smallest of the $y$-coordinates found in Step 2.

## EXAMPLE 1

Find the absolute maximum and absolute minimum for $y = f(x) = x^3 + 3x^2 + 1$ for $-1 \leq x \leq 2$.

Solution:  1. Find the $x$-coordinates of critical points within the interval.

$$y = f(x) = x^3 + 3x^2 + 1 \qquad -1 \leq x \leq 2$$

$$f'(x) = 3x^2 + 6x$$

$$0 = 3x^2 + 6x$$

$$= 3x(x + 2)$$

$$x = 0 \qquad x = -2$$

(Note:   $-2$ is not within the interval and therefore not used in Step 2.)

2. Determine the $y$-coordinate for each of the critical points within the interval and for the lower and upper limits of the interval.

critical:   $y = \quad f(0) = 0 + 0 + 1$                                                  $= 1$

lower:   $y = f(-1) = (-1)^3 + 3(-1)^2 + 1 = -1 + 3 + 1 = 3$

upper:   $y = \quad f(2) = (2)^3 + 3(2)^2 + 1 = 8 + 12 + 1 \quad = 21$

3.   The absolute maximum is the largest of the $y$-coordinates and the absolute minimum is the smallest.

The absolute maximum is 21 and occurs when $x = 2$.

The absolute minimum is 1 and occurs when $x = 0$.

Note that our answer is consistent with that of Problem 1 worked several pages earlier.

# EXAMPLE 2

Given:   $y = f(x) = x^2 - 12x - 64 \qquad -4 \leq x \leq 4.$

Find the absolute maximum and minimum for the function over the interval.

Solution:   Before starting, what do we know and what do we really want to find? We know that the function is a quadratic, opening up, with an absolute minimum at its vertex and no absolute maximum. But in this example we are considering only the portion of the curve from $-4 \leq x \leq 4$. We want to locate its absolute maximum and absolute minimum values just over the given interval.

1.   Find the $x$-coordinates of critical points within the interval.

$$y = f(x) = x^2 - 12x - 64 \qquad -4 \leq x \leq 4$$

$$f'(x) = 2x - 12$$

$$0 = 2x - 12$$

$$2x = 12$$

$$x \neq 6$$

(6 is not within the interval and therefore not used in Step 2.)

2.   Determine the $y$-coordinate for each of the critical points within the interval and for the lower and upper limits of the interval.

$$y = f(-4) = (-4)^2 - 12(-4) - 64 = 0$$

$$y = f(4) = (4)^2 - 12(4) - 64 = -96$$

3.   The absolute maximum is the largest of the $y$-coordinates and the absolute minimum is the smallest.

The absolute maximum is 0 and occurs when $x = -4$.

The absolute minimum is $-96$ and occurs when $x = 4$.

For reference, the graph of the function is shown below. The section of the curve under consideration is shown with a solid line, the remainder of the parabola is shown with a dashed line.

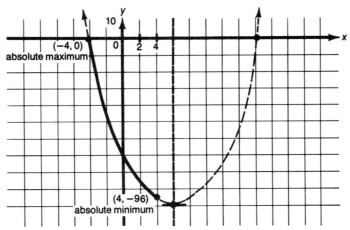

Try one yourself.

## Problem 2

Given:  $y = f(x) = x^2 - 8x + 6$     $0 \le x \le 6$.

Find the absolute maximum and minimum for the function over the given interval.

---

Solution:  1.  Find the $x$-coordinates of critical points within the interval.

2.  Determine the $y$-coordinate for each of the critical points within the interval and for the lower and upper limits of the interval.

critical:

lower:

upper:

3.  The absolute maximum is the largest of the $y$-coordinates and the absolute minimum is the smallest.

Answers:  The absolute maximum is 6 and occurs when $x = 0$.

The absolute minimum is $-10$ and occurs when $x = 4$.

---

I know you got the right answer, but, just to be on the safe side, we will work one more example for practice.

## EXAMPLE 3

Given:  $y = f(x) = -x^5 + 5x^4 + 1$     $1 \le x \le 3$.

Find the absolute maximum and minimum for the function over the given interval.

---

Solution:  1.  Find the $x$-coordinates of critical points within the interval.

$$y = f(x) = -x^5 + 5x^4 + 1 \qquad 1 \leq x \leq 3$$

$$f'(x) = -5x^4 - 20x^3$$

$$0 = -5x^3(x + 4)$$

$$x \neq 0 \qquad x \neq -4$$

(Neither of these values are within the closed interval and therefore not used in Step 2.)

2.  Determine the $y$-coordinate for each of the critical points within the interval and for the lower and upper limits of the interval.

$$y = f(1) = -1 + 5 + 1 \qquad\qquad\qquad = 5$$

$$y = f(3) = -(3)^5 + 5(3)^4 + 1 = -243 + 405 + 1 = 163$$

3.  The absolute maximum is the largest of the $y$-coordinates and the absolute minimum is the smallest.

The absolute maximum is 163 and occurs when $x = 3$.

The absolute minimum is 1 and occurs when $x = 1$.

---

You should now be able to locate both the absolute maximum and absolute minimum for a continuous function defined on a closed interval. It is a three-step procedure and does not require the graph to be drawn.

Briefly the procedure is to find the $x$-coordinates of all critical points within the interval, then determine the corresponding $y$-coordinates for the critical points as well as for the endpoints of the interval. The absolute maximum will be the largest and the absolute minimum will be the smallest value of these $y$-coordinates.

## EXERCISES

Determine the location and values of the absolute maximum and absolute minimum for $y = f(x)$ over the given interval:

1.  $y = f(x) = x^2 - 12x + 1$                        $1 \leq x \leq 10$

2.  $y = f(x) = 5x - 1$                               $0 \leq x \leq 5$

3.  $y = f(x) = \dfrac{1}{3}x^3 - x^2 - 3x + 1$          $0 \leq x \leq 3$

4.  $y = f(x) = -x^3 - 3x^2 + 1$               $-2 \leq x \leq 2$

5.  $y = f(x) = x^4 - 8x^3 + 18x^2 - 27$       $1 \leq x \leq 2$

6.  $y = f(x) = 3x^4 - 16x^3 + 18x^2$           $0 \leq x \leq 2$

---

If additional practice is needed:

Barnett and Ziegler, pages 235–236, problems 1–6, 11–14
Budnick, pages 495–496, problems 1–10
Hoffmann, page 173, problems 1–9
Piascik, page 178, problems 4–5

# UNIT 19

## Optimization Problems

In this unit we will consider a variety of application problems, each of which has the objective of maximizing (or minimizing) a particular quantity, such as profit, output, or even an area. The emphasis will be not only on the mathematical answer, but also the interpretation of the answer in terms of the problem.

If you are tempted to skip this unit, please don't. Trust me. It is meant to be an interesting and fun unit. We are now able to put together the technical information and the skills you have been mastering to solve practical problems. Hopefully you will start to appreciate the power of calculus and the usefulness of what we have been doing. Although the problems here are of my choosing, I am optimistic that you will be able to relate the concepts to problems that are of interest to you.

Word problems or application problems sometimes bring terror to students. Having to write the necessary equations to solve a problem is ever so much harder than simply solving a given equation. I will not lie. There is no easy answer. What I will do is to set down a general technique that I use that is applicable to the majority of optimization problems. After that it takes practice and a bit of self confidence on your part.

## GENERAL TECHNIQUE FOR OPTIMIZATION PROBLEMS

1.   Decide what quantity is to be maximized or minimized and write the equation for this quantity—in words first, if necessary.

2.   Use the constraints of the problem to write the equation in terms of only one independent variable and simplify.

3.   Find the first derivative, set it equal to zero, and solve. Test to determine if critical points are maxima or minima. Stated another way, find all critical points and classify.

4.   Answer the question posed in the original problem.

Some of the above should sound very familiar. Step 3 is basically what we spent the last four units doing; by now it should be second nature to you. From here on the unit will be a variety of application problems, all of which will be worked using this four step general technique.

Reminder, application problems often require restricted domains.

# EXAMPLE 1

$y = f(x) = -\dfrac{2}{3}x^3 + \dfrac{19}{2}x^2 + 10x$ with $x \geq 0$ is the profit in dollars when $x$ units are produced. How many units should be produced to make profit as large as possible?

Solution: 1. What is to be maximized in the problem? Profit

   Write the equation for profit.

   The equation for profit was given as:-

$$y = f(x) = -\frac{2}{3}x^3 + \frac{19}{2}x^2 + 10x \qquad x \geq 0$$

2. Write function in terms of a single independent variable.

   The function was given in terms of a single variable.

$$y = f(x) = -\frac{2}{3}x^3 + \frac{19}{2}x^2 + 10x \qquad x \geq 0$$

3. Find critical points and classify.

$$y = f(x) = -\frac{2}{3}x^3 + \frac{19}{2}x^2 + 10x \qquad x \geq 0$$

$$f'(x) = -\frac{2}{3}(3x^2) + \frac{19}{2}(2x) + 10$$

$$= -2x^2 + 19x + 10$$

$$0 = 2x^2 - 19x - 10$$

$$0 = (2x + 1)(x - 10)$$

$$\cancel{x^* = -\frac{1}{2}} \qquad x^* = 10$$

$\left( -\dfrac{1}{2}$ is not within the domain and should not be considered further. $\right)$

   If $x^* = 10$,

$$f(10) = -\frac{2}{3}(10)^3 + \frac{19}{2}(10)^2 + 10(10)$$

$$= -\frac{2}{3}(1,000) + \frac{19}{2}(100) + 100$$

$$= -666.67 + 950 + 100$$

$$= 383.33$$

Test: The function is a cubic with $a = -\dfrac{2}{3}$ and opening ∽.

The $y$-intercept is at 0.

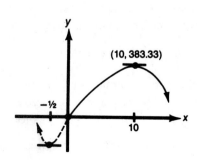

The point $(10, 383.33)$ is a relative maximum.

The point $(10, 383.33)$ is an absolute maximum on the restricted domain.

4.  Answer question posed in original problem.

Answer:  Maximum profit of $383.33 will be achieved when 10 units are produced.

---

Some preliminary explanations will be helpful before our next example.

Recall that demand functions describe the relationship between the number of units sold (the demand for a product) and its selling price.

Let the demand function for a particular product be

$$q = f(p) = 120{,}000 - 3p$$

where $q$ = the quantity sold at price $p$

and     $p$ = price in dollars.

Suppose we price our product at $1, how many units would we sell? And what would be our total revenue? Recall that total revenue equals (selling price) (quantity sold).

Let     $p = \$1$,

then     $q = f(1) = 120{,}000 - 3(1) = 120{,}000 - 3 = 119{,}997$ units.

Revenue = price $\cdot$ quantity = $(1)(119{,}997) = \$119{,}997$

Suppose we increase our price to $10, what happens?

Let     $p = \$10$,

then     $q = f(10) = 120{,}000 - 30 = 119{,}970$ units,

Revenue = price $\cdot$ quantity = $(10)(119{,}970) = \$1{,}199{,}700$.

We certainly increased our revenue with a small increase in price.

Business must be great. If we increase the price to $100, what happens?

Let     $p = \$100$,

then     $q = f(100) = 120{,}000 - 300 = 119{,}700$ units

Revenue = price $\cdot$ quantity = $(100)(119{,}700) = \$11{,}970{,}000$.

Notice that each time as we increase the price we sell fewer units, but total revenue increases. Let's do a few more. You do the calculations.

Let     $p = \$1{,}000$,

then     $q = f(1{,}000)$

Revenue =

Let     $p = \$10{,}000$,

then     $q = f(10{,}000)$

Revenue =

Let $p = \$100,000,$

then $q = f(100,000)$

Revenue =

What happened? We were doing fine until the last price increase. At a price of \$10,000 per unit, we would have sold 90,000 units with total revenue of \$900,000,000. But at a price of \$100,000 per unit no one was willing to buy any of the product. We had priced ourselves out of the market.

Where is all of this leading? The real question to be answered is what price should we charge for the product so as to maximize total revenue. This problem is what is typically called a classical optimization problem. The next example will illustrate how to use the concepts of calculus to solve it rather than trial-and-error as we were doing above.

## EXAMPLE 2

Let the demand function for a particular product be

$$q = f(p) = 120,000 - 3p$$

where $q =$ the quantity sold at price $p$

and $p =$ price in dollars.

a. What price will result in maximum total revenue?

b. What is the maximum revenue that can be expected?

c. How many units will be sold when revenue is maximized?

Solution:  1.  What is to be maximized in the problem? Revenue
Write the equation for revenue, in words first if necessary.

$$\text{Revenue} = \text{price} \cdot \text{quantity}$$

$$= p \cdot q$$

2.  Use the constraints of the problem to write the function in terms of a single variable and simplify.

Use $q = 120,000 - 3p$ and substitute.

$$\text{Revenue} = p \cdot q$$
$$R(p) = p(120,000 - 3p)$$
$$= 120,000p - 3p^2$$

3.  Find critical points and classify.

$$R(p) = 120,000p - 3p^2$$
$$R'(p) = 120,000 - 6p$$
$$0 = 120,000 - 6p$$
$$6p = 120,000$$
$$p = 20,000$$

If $p^* = 20{,}000$,

$$R(20{,}000) = 20{,}000[120{,}000 - 3(20{,}000)]$$

$$= 20{,}000[120{,}000 - 60{,}000]$$

$$= 20{,}000[60{,}000]$$

$$= 1{,}200{,}000{,}000$$

Test:   $R(p) = -3p^2 + 120{,}000p$

The function is a quadratic with $a = -3$.

The parabolic curve opens down.

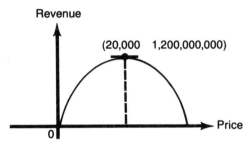

The point (20,000, $1,200,000,000) is an absolute maximum.

4.   Answer questions posed in original problem.

Answers:   a.   A price of $20,000 per unit will result in maximum total revenue.

b.   At a price of $20,000, maximum revenue of $1,200,000,000 can be expected.

c.   At a price of $20,000 per unit, revenue will be maximized, and 60,000 units will be sold.

$$q = f(p) = 120{,}000 - 3p$$

$$q = f(20{,}000) = 120{,}000 - 3(20{,}000)$$

$$= 120{,}000 - 60{,}000$$

$$= 60{,}000$$

---

Several observations need to be made about the above example. Could we have found the answer by trial-and-error? Probably, but using calculus was much faster. How do we know that if we increased the price, to say $20,025, that total revenue would not increase? Because the revenue function is a quadratic with an absolute maximum which is the value we are using.

# EXAMPLE 3

Sketch the revenue function from Example 2.

Determine the restricted domain.

Solution:    The revenue function is $R(p) = p(120{,}000 - 3p)$

$$= 120{,}000p - 3p^2$$

$$= -3p^2 + 120{,}000$$

The function is a quadratic with $a = -3$.

The parabolic curve opens down.

The $y$-intercept is at 0. (Recall this is actually the $q$-axis here.)

The $x$-intercepts are at 0 and 40,000 because

$$0 = p(120{,}000 - 3p)$$

$$p = 0 \qquad 120{,}000 - 3p = 0$$

$$120{,}000 = 3p$$

$$40{,}000 = p$$

(As before, these are really the $p$-intercepts.)

The vertex is located at $(20{,}000, 1{,}200{,}000{,}000)$ from Example 2.

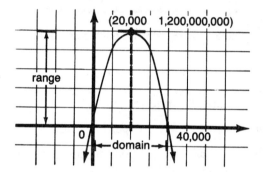

Answer:

Regarding the restricted domain, the question to be answered is, are there any limitations on the values that can be used for $p$?

Because $p$ represents price, a negative number would be meaningless. And because $R(p)$ represents total revenue, a negative number would also be meaningless. Therefore the relevant portion of the graph must be in the first quadrant. From the graph, that means the restricted domain for $p$ is $0 \le p \le 40{,}000$. If $p > 40{,}000$ the total revenue would be negative and that would imply a negative number for quantity, which would be meaningless.

Answer: Restricted domain:    $0 \le p \le 40{,}000$

You try the next one. Use Example 2 as your guide.

## Problem 1

Suppose that the number of cookbooks $q$ sold in a small community is given by $q = 100 - 25p$ where $p$ is the price in dollars of the cookbooks and $q$ is the quantity sold at price $p$. Use calculus to find the price that will lead to maximum revenue.

Solution:   1.   What is to be maximized in the problem?

Write the equation, in words first if necessary.

2.   Use the constraints of the problem to write the function in terms of a single variable and simplify.

3.   Find critical points and classify.

Test:

4.   Answer question posed in original problem.

Answer:   A price of $2 per cookbook would lead to maximum revenue.

---

You might have found the following additional information from the previous problem even though it was not asked for. At a price of $2 per cookbook, the maximum revenue would be $100, and 50 cookbooks would be sold.

The next few examples will look at maximizing profit.

# EXAMPLE 4

Paul is a free-lance TV cameraman. To supplement his income between jobs, he started the Spaniel Manufacturing Company, a small firm making sunglasses. The sunglasses are priced at $8 and the company can sell all it produces. Paul has determined that the firm's total cost is given by

$$C(x) = 20 + 2x + .01x^2$$

where $x = $ the number of sunglasses

and $C(x) = $ the total cost in dollars of producing $x$ sunglasses.

How many sunglasses should the Spaniel Manufacturing Company produce to maximize profit? What is the maximum profit?

---

Solution:   Before actually solving the problem, we are going to do some preliminary computations again so that I am sure you understand the various bits of information given in the problem and what we want to eventually find.

Total Revenue $=$ selling price $\cdot$ quantity sold

Total Cost $\quad = 20 + 2x + .01x^2$

Profit $\qquad\quad = $ Total Revenue $-$ Total Cost

Suppose $x = 10$:

Total Revenue $= 8(10)$ $\qquad\qquad\qquad\qquad\qquad = 80$

Total Cost $\quad = 20 + 2(10) + .01(100) = 20 + 20 + 1 = 41$

Profit $\qquad\quad =$ $\qquad\qquad\qquad\qquad\qquad\qquad\quad = 39$

The interpretation of the above is that if Paul produces 10 sunglasses and sells them for $8 each, he would take in as revenue $80. At the same time it will cost him $41 to produce the glasses, leaving the company with a profit of $39.

Suppose $x = 100$:

Total Revenue $= 8(100)$ $= 800$

Total Cost $= 20 + 2(100) + .01(10,000) = 20 + 200 + 100 = 320$

Profit $=$ $= 480$

If Paul decided instead to produce 100 sunglasses, his revenue would now be $800. The total cost to produce the 100 pairs would be $320, leaving the company with a profit of $480.

Now that you see how the various pieces of information fit together, we can proceed with solving the original problem.

1.  What is to be maximized in the problem? Profit

    Write the equation for profit, in words first if necessary.

    $$\text{Profit} = \text{Total Revenue} - \text{Total Cost}$$

    $$= (\text{price})(\text{quantity}) - \text{Total Cost}$$

2.  Use the constraints of the problem to write the function in terms of a single variable and simplify.

    Substitute $20 + 2x + .01x^2$ for total cost, $8 for price, and $x$ for quantity.

    $$\text{Profit} = (\text{price})(\text{quantity}) - \text{Total Cost}$$

    $$P(x) = 8x - (20 + 2x + .01x^2)$$

    $$= 8x - 20 - 2x - .01x^2$$

    $$= -.01x^2 + 6x - 20$$

3.  Find critical points and classify.

    $$P(x) = -.01x^2 + 6x - 20$$

    $$P'(x) = -.02x + 6$$

    $$0 = -.02x + 6$$

    $$.02x = 6$$

    $$2x = 600$$

    $$x = 300$$

    If $x^* = 300$,

    $$P(300) = -.01(300)^2 + 6(300) - 20$$

    $$= -.01(90,000) + 1,800 - 20$$

    $$= -900 + 1,800 - 20$$

    $$= 880$$

Test:    $P(x) = -.01x^2 + 6x - 20$

The function is a quadratic with $a = -.01$.

The parabolic curve opens down.

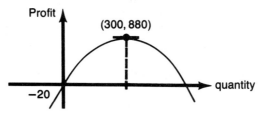

The point (300, 880) is an absolute maximum.

4.   Answer questions posed in original problem.

Answer:    Paul should manufacture 300 pairs of sunglasses to achieve his maximum profit of $880.

---

The last example was okay, but the answer was somewhat predictable. I prefer examples where the answers are not so obvious.

## EXAMPLE 5

During a week, a manufacturer can sell $q$ units of a product at a price $p$ cents per unit where $p = 300 - .05q$. The total cost of making $q$ units is $c$ cents where $c = 200 + q$.

a.   How many units should be manufactured per week so as to maximize profit?

b.   What price per unit should be used so as to maximize profit?

c.   What will the maximum profit be?

---

Solution:    This time we will do an even briefer analysis of the information before starting to work the problem.

$$\text{Suppose } q = 100,$$
$$\text{then} \qquad p = 300 - .05(100) = 300 - 5 = 295$$
$$\text{and} \qquad c = 200 + 100 \qquad\qquad = 300$$

The interpretation thus far is that if 100 units were manufactured, the selling price would be 295 cents or $2.95 per unit and the total cost of making the 100 units would be 300 cents or $3.00.

And continuing with the analysis,

Total Revenue = price · quantity = 2.95(100)  = $295

Total Cost     = (calculated above)            = $3

Profit            = Total Revenue − Total Cost = $292

In this problem we are being asked to determine the quantity to produce so as to maximize profit. The analysis above merely showed the profit if 100 units were produced.

1.   What is to be maximized in the problem? Profit

     Write the equation for profit, in words first if necessary.

$$\text{Profit} = \text{Total Revenue} - \text{Total Cost}$$

$$= (\text{price})(\text{quantity}) - \text{Total Cost}$$

2.   Use the constraints of the problem to write the equation in terms of a single variable and simplify.

     Substitute $200 + q$ for total cost, $300 - .05q$ for price, and $q$ for quantity.

$$\text{Profit} = (\text{price})(\text{quantity}) - \text{Total Cost}$$

$$P(q) = (300 - 0.5q)q - (200 + q)$$

$$= 300q - 0.5q^2 - 200 - q$$

$$= -0.5q^2 + 299q - 200$$

3.   Find the critical points and classify.

$$P(q) = -0.5q^2 + 299q - 200$$

$$P'(q) = -1.0q + 299$$

$$0 = -q + 299$$

$$q = 299$$

If $q^* = 299$,

$$P(299) = -0.5(299)^2 + 299(299) - 200$$

$$= -0.5(89{,}401) + 89{,}401 - 200$$

$$= -44{,}700.5 + 89{,}401 - 200$$

$$= 44{,}500.5 \text{ cents or } \$445.005$$

Test:   $P(q) = -0.5q^2 + 299q - 200$

The function is a quadratic with $a = -0.5$.

The parabolic curve opens down.

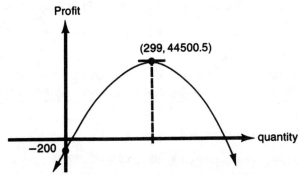

The point (299, 44500.5) is an absolute maximum.

4.   Answer questions posed in original problem.

     Answers:  a.  The company should produce 299 units per week so as to maximize profit.

b.   price $= p = 300 - 0.5q$

$$= 300 - 0.5(299)$$

$$= 300 - 149.5$$

$$= 150.5$$

To maximize profit, the manufacturer should sell the product at a price of 150.5 cents or $1.505 per unit.

c.   Maximum profit is expected to be $445.005.

---

Ready to try a problem similar to the last example?

## Problem 2

The total cost in dollars to produce $q$ units is given by $C = 3q^2 + 15q + 45$. The company sells all it produces at $87 per unit. How many units should be produced so as to maximize profit? What is the maximum profit?

---

Solution:   1.   What is to be maximized in the problem? Profit
Write the equation, in words first if necessary.

$$\text{Profit} = \text{Total Revenue} - \text{Total Cost}$$

$$= (\text{price})(\text{quantity}) - (\text{total cost})$$

2.   Use the constraints of the problem to write the equation in terms of a single variable and simplify.

3.   Find the critical points and classify.

Test:

4.   Answer questions posed in original problem.

Answer:   The company should produce 12 units to achieve maximum profit of $387.00.

---

## EXAMPLE 6

Pen is the leading stylist at the top beauty salon in the city. Pen is paid primarily on commission and enjoys shopping. His most profitable service is a haircut because it does not require a great deal of time and he is allowed to set his own prices within reason. Pen, being a bright person, knows that his profit is a function of the price he charges. In fact, he has determined that his profit per day on haircuts is:

$$y = P(x) = 25x - x^2 - 100$$

where $P(x) =$ profit in dollars

and        $x =$ the price in dollars for a haircut.

The owner, however, has stipulated that Pen may not charge more than $14 nor less than $6 for haircuts. What is the optimal price Pen should charge so as to maximize his profits and still stay within the owner's limitations?

Solution:   What is to be maximized in the problem? Profit

Write the equation for profit.

The equation for profit was given as:

$$y = P(x) = 25x - x^2 - 100 \text{ with } 6 \le x \le 14$$

Do you notice something different? We are being asked to find a maximum on a closed interval. Recall from Unit 18, a three-step procedure is used.

1.   Find the $x$-coordinates of critical points within the interval.

$$y = P(x) = 25x - x^2 - 100 \qquad 6 \le x \le 14$$

$$P'(x) = 25 - 2x$$

$$0 = 25 - 2x$$

$$2x = 25$$

$$x = 12.50$$

2.   Determine the $y$-coordinate for each of the critical points within the interval and the lower and upper limits of interval.

$$y = P(12.50) = 25(12.5) - (12.5)^2 - 100 = 312.5 - 156.25 - 100 = 56.25$$

$$y = P(6) = 25(6) - (6)^2 - 100 = 150 - 36 - 100 \qquad\qquad = 14$$

$$y = P(14) = 25(14) - (14)^2 - 100 = 350 - 196 - 100 \qquad\quad = 54$$

3.   The absolute maximum is 56.25 and occurs when $x = 12.50$.

Answer:   Even though the owner would allow Pen to charge as much as $14 for haircuts, he should charge only $12.50 thereby maximizing his daily profit to $56.25.

---

I especially like that last example. So often the tendency is to charge as much as possible, whereas the problem illustrates that Pen would actually achieve greater profits by charging less.

Next is a review problem for you.

## Problem 3

Suppose the total cost in dollars of manufacturing $x$ items is given by the function $C(x) = 3x^2 + x + 48$. Find $C(2)$, $C(5)$, $C(10)$, $C(0)$ and interpret.

---

Solution:

Answers:   The total cost of manufacturing 2 items is $62.

The total cost of manufacturing 5 items is $128.

The total cost of manufacturing 10 items is $358.

Fixed costs are $48.

# MINIMIZING AVERAGE COSTS

In the last problem total costs for manufacturing various numbers of units were computed. It is not possible to minimize total cost because costs continue to increase when more and more units are manufactured. However, it is possible to minimize the average cost per unit which is what the next example shows. We will use the same total cost function.

## EXAMPLE 7

The total cost in dollars of manufacturing $x$ items is given by the function $C(x) = 3x^2 + x + 48$. For what value of $x$ is the average cost per unit a minimum? What is the minimum average cost?

Solution:   Again let us first analyze what we are attempting to find.

Suppose $x = 2$.

From Problem 3, the total cost of producing 2 units was found to be $62. Therefore the average cost for the 2 units is $62/2 = $31$ per unit.

Suppose $x = 5$.

The total cost of producing 5 units was found to be $128. Now the average cost becomes $128/5 = $25.60$ per unit.

Suppose $x = 10$.

The total cost of producing 10 units was found to be $358 and the average cost per unit becomes $358/10 = $35.80$.

The question being asked in this problem is how many units should be produced so that the average cost per unit is the lowest?

1.   What is to be minimized in the problem? Average Cost

Write the equation for average cost, in words first.

$$\text{Average Cost} = \frac{\text{Total Cost}}{\text{quantity}}$$

2.   Use the constraints of the problem to write the equation in terms of a single variable and simplify.

Substitute $3x^2 + x + 48$ for total cost and $x$ for quantity.

$$A(x) = \frac{3x^2 + x + 48}{x}$$

3.  Find critical points and classify.

$$A(x) = \frac{3x^2 + x + 48}{x}$$

$$A'(x) = \frac{x\dfrac{d(3x^2 + x + 48)}{dx} - (3x^2 + x + 48)\dfrac{d(x)}{dx}}{x^2}$$

$$= \frac{x(6x + 1) - (3x^2 + x + 48)(1)}{x^2}$$

$$= \frac{6x^2 + x - 3x^2 - x - 48}{x^2}$$

$$= \frac{3x^2 - 48}{x^2}$$

$$0 = \frac{3x^2 - 48}{x^2}$$

$$0 = 3x^2 - 48$$

$$3x^2 = 48$$

$$x^2 = 16$$

$$x^* = 4 \qquad \cancel{x^* = -4}$$

(−4 is not within the domain and
should not be considered further.)

If $x^* = 4$,

$$A(4) = \frac{3(4)^2 + 4 + 48}{4}$$

$$= \frac{48 + 4 + 48}{4}$$

$$= 25$$

Test:  First-derivative test with $A'(x) = \dfrac{3x^2 - 48}{x^2}$

For the point (4, 25)                          the curve is:

On the left:      let $x = 3$, $A'(3) = -$        decreasing

At the point:    (4, 25)                          levels off

On the right:    let $x = 5$, $A'(5) = +$        increasing

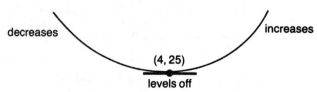

The point (4, 25) is a relative minimum.

4.  Answer questions posed in original problem.

Answers:    The average cost per unit is minimized if 4 units are produced, in which case the average cost is $25 per unit.

---

Thus far we have looked at maximizing total revenue, maximizing profit, and minimizing average cost per unit. There are a whole host of optimization problems covering such topics as ordering costs, inventory expenses, volumes, material requirements, surfaces, and so forth. But rather than introduce some of those that require a fair amount of background material, we will conclude the unit with one last type. I call them fence problems for lack of a better title. I like fence problems because they cover the concept of optimization, they are applicable to most everyone, and best of all they only require two familiar formulas from geometry: finding the area and the perimeter of a rectangle.

# FORMULAS FROM GEOMETRY

Area of a Rectangle = (length)(width) = $lw$

Perimeter of a Rectangle = 2(length) + 2(width) = $2l + 2w$

Remember the area of a rectangle is the number of square units contained within the rectangle, whereas the perimeter is the distance measured around the outside.

Keep the two formulas in mind and the basic technique stated at the beginning of this unit. We are about to tackle some fence problems.

## EXAMPLE 8

Ben, a farmer, with a field adjacent to a straight river wishes to fence a rectangular region for grazing. He has available 1,600 feet of fencing with none needed along the river. Being a bit tight with his money, Ben does not want to buy any more fencing but he does want to design the region so that he achieves the maximum possible area.

What should be the dimensions of the field in order that it have maximum area?

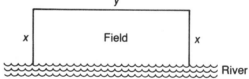

Solution:    Again let's look at some of the possibilities before starting right in with the solution. Ben has 1,600 feet of fencing and is going to fence a rectangular region on three sides. Below are some extreme possibilities.

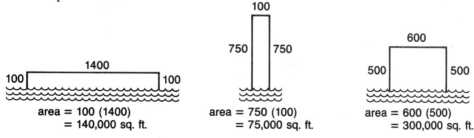

The problem is asking us to determine the dimensions so that the field will have maximum area given the constraint of 1,600 feet of fencing for the three sides.

1.  What is to be minimized in the problem? Area of Rectangle

    Write the equation for area of rectangle, in words first.

    $$\text{Area} = (\text{length})(\text{width})$$

    $$= y \cdot x$$

2.  Use the constraints of the problem to write the equation in terms of a single variable and simplify.

    What are the constraints of this problem?

    There is only 1,600 feet of fencing and it must go around three sides, or stated in symbols:

    $$x + y + x = 1{,}600$$

    $$y + 2x = 1{,}600$$

    You have a choice, either solve the above equation for $x$ or $y$. Which variable is easier to solve for? I would have to say $y$.

    $$y = 1{,}600 - 2x$$

    Now use substitution to rewrite the equation in Step 1 in terms of $x$ and simplify.

    $$\text{Area} = y \cdot x$$

    $$A(x) = (1{,}600 - 2x)x$$

    $$= 1{,}600x - 2x^2$$

    $$= -2x^2 + 1{,}600x$$

3.  Find critical points and classify.

    $$A(x) = -2x^2 + 1{,}600x$$

    $$A'(x) = -4x + 1{,}600$$

    $$0 = -4x + 1{,}600$$

    $$4x = 1{,}600$$

    $$x^* = 400$$

    Test: $A(x) = -2x^2 + 1{,}600x$ is a quadratic with $a = -2$. The parabolic curve opens downward. The critical point in an absolute maximum.

4.  Answer questions posed in original problem.

    The question asked for the dimensions of the field, meaning $x$ and $y$.

    If $x = 400$ and $y = 1{,}600 - 2x$,

    $$\text{then } y = 1{,}600 - 2(400)$$

    $$= 800$$

Answer: To maximize the area, given 1,600 feet of fencing for three sides, Ben should lay out the field to be 400 × 800 feet with the longest side parallel to the river as shown in the sketch. The area of the field will be 320,000 sq ft.

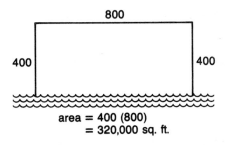

area = 400 (800)
= 320,000 sq. ft.

The last example is probably the most difficult we have attempted, but I know you can do it.

## EXAMPLE 9

A rectangular playground must contain 1,050 square feet and is to be enclosed by a fence, then divided down the middle by another piece of fence. A sketch is provided below. The cost of the interior fencing for the middle section is $.50 per foot. The exterior fencing must be heavier and it costs $1.50 per foot.

How should the playground be designed
so as to minimize the cost of the fencing?

What is the minimum cost?

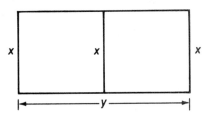

Solution: Let's consider some of the possibilities again so that we are sure we understand the problem. The rectangular area is to be 1,050 square feet with fencing as described above.

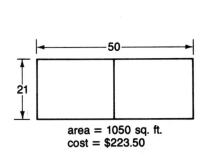

area = 1050 sq. ft.
cost = $223.50

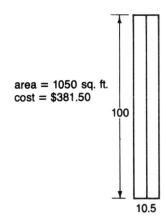

area = 1050 sq. ft.
cost = $381.50

The problem is asking us to determine the dimensions so that the playground's area will be 1,050 square feet, while at the same time the cost of the fencing will be minimized.

1.  What is to be minimized in the problem? Cost of fencing

    Write the equation for the cost of fencing, in words first.

Total Cost = cost of interior fence + cost of exterior fence

$$= .50 \text{ (length of interior} + 1.50 \text{ (length of exterior}$$
$$\text{fencing required)} \qquad \text{fencing required)}$$

$$= .50x + 1.50(2x + 2y)$$

$$= .50x + 3x + 3y$$

$$= 3.5x + 3y$$

2. Use the constraints of the problem to write the equation in terms of a single variable and simplify.

What are the constraints of this problem?
The area of the rectangle must be 1,050 square feet, or stated in symbols:

$$xy = 1,050$$

Again you have a choice, either solve the above equation for $x$ or $y$. Which variable is easier to solve for? It doesn't really matter, but I will solve for $y$.

$$y = \frac{1,050}{x}$$

Now use substitution to rewrite the equation in Step 1 in terms of $x$ and simplify.

$$\text{Total Cost} = 3.5x + 3y$$

$$C(x) = 3.5x + 3\left(\frac{1,050}{x}\right)$$

$$= 3.5x + \frac{3,150}{x}$$

3. Find critical points and classify.

$$C(x) = 3.5x + \frac{3,150}{x}$$

$$= 3.5x + 3,150x^{-1}$$

$$C'(x) = 3.5 + 3,150(-1x^{-1-1})$$

$$= 3.5 - 3,150x^{-2}$$

$$= 3.5 - \frac{3,150}{x^2}$$

$$0 = 3.5 - \frac{3,150}{x^2}$$

$$0 = 3.5x^2 - 3,150$$

$$3.5x^2 = 3,150$$

$$x^2 = 900$$

$$x^* = 30 \qquad \cancel{x^* = -30}$$

($-30$ is not within the domain and should not be considered further.)

Test:   Second-derivative test with $C''(x) = \dfrac{6,300}{x^3}$

At $x = 30$, $C''(30) = +$, concave up, minimum point

4.  Answer questions posed in original problem.

The question asked for the dimensions of the playground, meaning $x$ and $y$, that minimizes cost.

$$\text{If } x = 30 \text{ and } y = 1,050/x,$$

$$\text{then } y = 1,050/30$$

$$= 35$$

Answer:  To minimize the cost for the fencing, the area should be $30 \times 35$ feet with the fence in the middle being parallel to the 30 foot side as shown in the sketch.

Total Cost $= 3.5x + 3y$

$\qquad = 3.5(30) + 3(35)$

$\qquad = 105 + 105$

$\qquad = 210$

The minimum cost for the fencing would be \$210.

---

There, that wasn't so bad was it?

You should now be familiar with the general technique for optimization problems and be able to apply it to a variety of applications.

Briefly stated the general technique is:

1.  Decide what quantity is to be maximized or minimized and write the equation for it.

2.  Use the constraints of the problem to write the equation in terms of a single variable.

3.  Find all critical points and classify.

4.  Answer the original question.

Before beginning the next unit you should solve the following problems. Do not get discouraged if you are unable to solve all of them. Remember the complete solutions are provided at the back of the book if you need help.

## EXERCISES

1.  $P(x) = 50x - 0.01x^2 - 1,000$ with $x \geq 0$ is the profit in dollars when $x$ units are produced. How many units should be produced to maximize profit? What would the maximum profit be?

2.  The total cost in dollars of manufacturing $x$ items is given by the function $C(x) = x^2 + 4x + 16$ with $x \geq 0$.

    a.  Determine $C(10)$ and interpret.

    b.  Find the average cost per unit if 10 units are produced.

c. For what value of $x$ is the average cost per unit a minimum?

d. What is the minimum average cost?

3. Let the demand function for a particular product be

$$q = f(p) = 1,152 - 5p$$

where $q$ = the quantity sold at price $p$

and $p$ = price in dollars.

a. What price will result in maximum total revenue?

b. What is the maximum revenue that can be expected?

c. How many units will be sold when revenue is maximized?

d. Sketch the revenue function.

e. Determine the restricted domain for the revenue function.

4. A manufacturing company can sell all it produces of a certain product at a price of $50. The company's total cost is given by

$$C(x) = 110 - 2.5x^2 + x^3 \text{ with } x \geq 0$$

where $x$ = the number of units produced

and $C(x)$ = total cost in dollars of producing $x$ units.

How many units should the company produce to maximize profit? What is the maximum profit?

5. A machine shop has determined that their cost is a function of the number of machines in operation and is given by

$$C(x) = 20x + \frac{1,280}{x}$$

where $x$ = the number of machines in operation

and $C(x)$ = total cost in dollars.

How many machines should the company operate so as to minimize their cost?

6. A park manager has $100 to buy fencing to enclose a rectangular garden along the side of a building. No fence is needed along the building. The fence along the ends costs $1.25 per foot and the fence along the front costs $2 per foot. Use calculus to determine the dimensions $x$ and $y$ that maximize the enclosed area. What will the maximum area be?

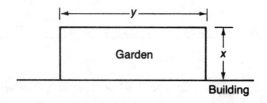

If additional practice is needed:

Barnett and Ziegler, pages 236–241, problems 1–42
Budnick, pages 515–518, problems 1–17; pages 534–537, problems 1–13
Hoffmann, pages 188–193, problems 1–36
Piascik, pages 205–207, problems 1–14

# UNIT 20

## Implicit Differentiation

In this unit implicit differentiation will be explained. When you have completed the unit, you will be able to differentiate an implicit equation without first having to rewrite the equation explicitly.

All of our work with derivatives so far has been with explicit functions or equations. Each equation has been written as $y = f(x)$ where the dependent variable is a function of the independent variable. Stated another way, an explicit equation has one variable expressed in terms of the other variables.

Examples of explicit functions:

$$y = f(x) = x^2 - 2x - 3$$

$$y = f(x) = 2x^5 - 7$$

$$y = f(x) = x^3 + x - 6$$

$$y = f(x) = \frac{x^2 + 2}{x - 1}$$

Recall from Unit 2, an implicit equation is one that does *not* directly express one variable in terms of the others.

Examples of implicit equations:

$$x^2 + x + y = 5$$

$$2x^2 + 3y = 10$$

$$y^2 = 2x^2 + 1$$

$$x^2 + y^2 = 5$$

$$xy = 1$$

$$y^3 + y^2 = 3x - x^7$$

With the exception of the last explicit function, which requires the use of the Quotient Rule, I am confident you could differentiate each of the explicit equations without even needing paper and pencil. But how do we differentiate implicit equations like the second group of examples? There are two choices: rewrite the equation explicitly, or use a technique called implicit differentiation.

# Differentiating Implicit Equations

To find the derivative, $\dfrac{dy}{dx}$, the implicit equation may be solved for the dependent variable and the derivative determined as usual. This method is illustrated by the next example.

## EXAMPLE 1

Find $\dfrac{dy}{dx}$ for $x^2 + x + y = 5$.

Solution:    Given.                                                $x^2 + x + y = 5$

Solve for the dependent variable.                $y = 5 - x^2 - x$

Differentiate as usual.                                   $\dfrac{dy}{dx} = -2x - 1$

In the above example, rewriting the equation explicitly for $y$ in terms of $x$ was easy. However in other situations, it may be difficult or impossible to solve for $y$. I doubt that any of us could solve $y^3 + y^2 = 3x - x^7$ for $y$.

**Implicit differentiation** is a technique for finding the derivative of an implicit equation without first having to rewrite the equation explicitly.

# Procedure for Implicit Differentiation

To find the derivative, $\dfrac{dy}{dx}$, of an implicit equation:

1.   Differentiate the equation term by term, regarding $y$ as a function of $x$.

2.   Solve the resulting equation for $\dfrac{dy}{dx}$. If $\dfrac{dy}{dx}$ appears in more than one term, factoring it out as a common factor will be required.

We will redo Example 1 using implicit differentiation so that you can contrast the methods.

## EXAMPLE 2

Find $\dfrac{dy}{dx}$ for $x^2 + x + y = 5$.

Solution:    1.   Differentiate the equation term by term, regarding $y$ as a function of $x$.

As I have done in the past, there is an extra step to show you what is intended before actually taking the derivative.

$$x^2 + x + y = 5$$

$$\frac{d(x^2)}{dx} + \frac{d(x)}{dx} + \frac{d(y)}{dx} = \frac{d(5)}{dx}$$

$$2x + 1 + \frac{dy}{dx} = 0$$

2.  Solve resulting equation for $\frac{dy}{dx}$. $\qquad \frac{dy}{dx} = -1 - 2x$

---

Don't let that first step confuse you. We are differentiating and regarding $y$ as a function of $x$. What you need to keep in mind is:

a.  $\frac{d(x)}{dx} = \frac{dx}{dx} = 1$, because it is the derivative of $x$ with respect to $x$.

b.  but $\frac{d(y)}{dx} = \frac{dy}{dx}$, because it is the derivative of $y$ with respect to $x$.

There is nothing more that can be done to simplify the expression.

## EXAMPLE 3

Find the derivative of $2x^2 + 3y = 10$ using implicit differentiation.

Solution: $\qquad\qquad\qquad\qquad\qquad\qquad 2x^2 + 3y = 10$

1.  Differentiate term by term, regarding $y$ as a function of $x$.

$$\frac{d(2x^2)}{dx} + \frac{d(3y)}{dx} = \frac{d(10)}{dx}$$

$$4x + 3\frac{dy}{dx} = 0$$

2.  Solve for $\frac{dy}{dx}$.

$$3\frac{dy}{dx} = -4x$$

$$\frac{dy}{dx} = \frac{-4x}{3}$$

# The Chain Rule Revisited

From Unit 12, the Chain Rule is repeated here:

**Chain Rule**

If $f(x) = [u(x)]^n$, then $f'(x) = n[u(x)]^{n-1}u'(x)$.

In words, the derivative of a function raised to a power is equal to the power times the function raised to the power minus one, and all of that is multiplied times the derivative of the function.

With implicit differentiation we regard $y$ as a function of $x$ and use the Chain Rule to differentiate term by term. To make it clearer, think of replacing $u(x)$ in the above Rule with the letter $y$.

For example;        If $f(x) = [u(x)]^3$, then $f'(x) = 3[u(x)]^2 u'(x)$.

Replace with $y$:   If $f(x) = y^3$, then $f'(x) = 3y^2 \dfrac{dy}{dx}$.

We will use the above result in the next example.

## EXAMPLE 4

Find the derivative of $y^3 = 2x^2 + 1$ using implicit differentiation.

Solution:                                                    $$y^3 = 2x^2 + 1$$

1.  Differentiate term by term.          $$\frac{d(y^3)}{dx} = \frac{d(2x^2)}{dx} + \frac{d(1)}{dx}$$

$$3y^2 \frac{dy}{dx} = 4x + 0$$

2.  Solve for $\dfrac{dy}{dx}$.          $$3y^2 \frac{dy}{dx} = 4x$$

$$\frac{dy}{dx} = \frac{4x}{3y^2}$$

Do you notice something different about this example? Previously all derivatives have been in terms of $x$, the independent variable. With this example the derivative is in terms of both variables, $x$ and $y$. However, the interpretation remains the same. The derivative can be thought of as a formula for the slope of the tangent line at any point on the curve. Except with implicit differentiation, to evaluate the formula often requires both the $x$ and $y$ values.

Are you feeling somewhat comfortable with the technique? Try the next two.

## Problem 1

Find the derivative for $2y + 3x + 15 = 0$ using implicit differentiation.

Solution:

Answer:  $\dfrac{dy}{dx} = \dfrac{-3}{2}$

You just used implicit differentiation to find the slope of the line.

## Problem 2

Differentiate $x^2 + y^2 = 5$ using implicit differentiation.

Solution:

Answer: $\dfrac{dy}{dx} = \dfrac{-x}{y}$

## EXAMPLE 5

Find the slope of the tangent line to the curve $y^3 = x^2$ at the point (8, 4).

Solution: The first derivative is a formula for the slope of the tangent line at any given point on the curve. Use implicit differentiation to find $\dfrac{dy}{dx}$.

$$y^3 = x^2$$

$$\frac{d(y^3)}{dx} = \frac{d(x^2)}{dx}$$

$$3y^2 \frac{dy}{dx} = 2x$$

$$\frac{dy}{dx} = \frac{2x}{3y^2}$$

At the point (8, 4)    $\dfrac{dy}{dx} = \dfrac{2(8)}{3(4)^2} = \dfrac{16}{48} = \dfrac{1}{3}$

Answer: At the point (8, 4) the slope of the tangent line to the curve is 1/3.

Observe that to evaluate the derivative in this example, both the $x$- and $y$-coordinates were needed for the calculations.

The next example looks easier than it is, which is why I am doing it.

## EXAMPLE 6

Differentiate $xy = 1$ using implicit differentiation.

Solution: The $xy$ term is a product. It is $x$ times $y$. Both are factors, and as such must be differentiated using the Product Rule.

$$xy = 1$$

Differentiate term by term. $$\frac{d(xy)}{dx} = \frac{d(1)}{dx}$$

Use Product Rule on left side. $$x\frac{d(y)}{dx} + y\frac{d(x)}{dx} = 0$$

$$x\frac{dy}{dx} + y\,(1) = 0$$

Simplify. $$x\frac{dy}{dx} = -y$$

Solve for $\frac{dy}{dx}$. $$\frac{dy}{dx} = \frac{-y}{x}$$

---

Don't worry, we will do a few more with factors and the Product Rule.

====

## EXAMPLE 7

Differentiate the equation: $x^3y^2 = 3$.

---

Solution:   Again the left side of the equation is a product and must be differentiated using the Product Rule.

$$x^3y^2 = 3$$

$$\frac{d(x^3y^2)}{dx} = \frac{d(3)}{dx}$$

$$x^3\frac{d(y^2)}{dx} + y^2\frac{d(x^3)}{dx} = 0$$

$$x^3(2y)\frac{dy}{dx} + y^2(3x^2) = 0$$

$$2x^3y\frac{dy}{dx} + 3x^2y^2 = 0$$

$$2x^3y\frac{dy}{dx} = -3x^2y^2$$

$$\frac{dy}{dx} = \frac{-3x^2y^2}{2x^3y}$$

$$\frac{dy}{dx} = \frac{-3y}{2x}$$

---

Thus far all of the examples could have been solved for $y$ and differentiated as usual. That is not the case with the next example. Without implicit differentiation we would be unable to work the problem.

The major difference from the earlier examples will occur in line three where $\frac{dy}{dx}$ will appear in more than one term. When $\frac{dy}{dx}$ appears in more than one term, it must be factored out as a common factor. Then to solve for $\frac{dy}{dx}$, divide the entire equation by the coefficient of $\frac{dy}{dx}$.

# EXAMPLE 8

Differentiate $y^3 + y^2 = 3x - x^7$.

Solution:

$$y^3 + y^2 = 3x - x^7$$

Differentiate term by term.

$$\frac{d(y^3)}{dx} + \frac{d(y^2)}{dx} = \frac{d(3x)}{dx} - \frac{d(x^7)}{dx}$$

$$3y^2 \frac{dy}{dx} + 2y \frac{dy}{dx} = 3 - 7x^6$$

Factor out $\dfrac{dy}{dx}$.

$$\frac{dy}{dx}(3y^2 + 2y) = 3 - 7x^6$$

Divide by coefficient.

$$\frac{dy}{dx} \frac{\cancel{(3y^2 + 2y)}}{\cancel{(3y^2 + 2y)}} = \frac{3 - 7x^6}{3y^2 + 2y}$$

$$\frac{dy}{dx} = \frac{3 - 7x^6}{3y^2 + 2y}$$

That wasn't too bad, was it?

We will conclude this unit with an example to illustrate another advantage of implicit differentiation. First the problem will be worked as an explicit function using the Chain Rule.

# EXAMPLE 9

If $y = \sqrt{1 - x^2}$, find $\dfrac{dy}{dx}$ using the Chain Rule.

Solution:

$$y = \sqrt{1 - x^2}$$

Rewrite as a power of $x$.

$$y = (1 - x^2)^{\frac{1}{2}}$$

Use Chain Rule.

$$\frac{dy}{dx} = \frac{1}{2}(1 - x^2)^{\frac{1}{2} - 1} \frac{d(1 - x^2)}{dx}$$

Simplify.

$$= \frac{1}{2}(1 - x^2)^{-\frac{1}{2}}(-2x)$$

Rewrite with positive exponents.

$$= \frac{1}{\cancel{2}} \cdot \frac{1}{(1 - x^2)^{\frac{1}{2}}} \cdot (-\cancel{2}x)$$

Rewrite without fractional exponents.

$$= \frac{-x}{\sqrt{1 - x^2}}$$

Now compare how much shorter, and hence easier, the problem becomes by using implicit differentiation. The solution changes at the first step. Instead of rewriting the equation as a power of $x$ as we did in Example 9, square both sides of the equation to remove the radical sign. Then differentiate implicitly.

# EXAMPLE 10

If $y = \sqrt{1 - x^2}$, find $\dfrac{dy}{dx}$ using implicit differentiation.

Solution:

|  |  |
|---|---|
|  | $y = \sqrt{1 - x^2}$ |
| Square both sides of the equation. | $y^2 = (\sqrt{1 - x^2})^2$ |
|  | $y^2 = 1 - x^2$ |
| Use implicit differentiation. | $\dfrac{d(y^2)}{dx} = \dfrac{d(1)}{dx} - \dfrac{d(x^2)}{dx}$ |
|  | $2y \dfrac{dy}{dx} = 0 - 2x$ |
| Solve for $\dfrac{dy}{dx}$. | $\dfrac{dy}{dx} = \dfrac{-2x}{2y}$ |
| Simplify. | $\dfrac{dy}{dx} = \dfrac{-x}{y}$ |

Are the answers equivalent? Yes. If you were to substitute $\sqrt{1 - x^2}$ for $y$ into the second answer, the two derivatives would be identical.

To my way of thinking, the second way was much easier than having to deal with fractional exponents and radicals. But as always, the final choice is yours. Use whichever method gives you the best results.

You should now be able to differentiate an implicit equation without having to first rewrite the equation explicitly.

Remember to differentiate an implicit equation, the equation is differentiated term by term, regarding $y$ as a function of $x$. Then the resulting equation is solved for $\dfrac{dy}{dx}$. If $\dfrac{dy}{dx}$ appears in more than one term, factoring it out as a common factor will be required.

Use implicit differentiation to find the derivatives of the following equations before continuing to the next unit.

# EXERCISES

Differentiate implicitly.

1.  $x^2 + 5y = 10$

2.  $y^2 = x^3$

3.  $x^2 + y^2 = 9$

4.  $x^2 - y^2 + 3y = 7$

5.  Find the slope of the tangent line to the curve $x^2 - y^2 - x = 1$ at the point (2, 1).

6.  $x^3 y = 2$

7.  Evaluate the derivative of $xy = 15$ at (3, 5).

8.  $y = \sqrt{x}$

9.  For a certain product, the demand and unit price are related by

$$q = p^2 - 40p + 2{,}150 \qquad 0 \le p \le 20$$

where $p$ = price in dollars

and     $q$ = quantity demanded at price $p$.

a.   Find $\dfrac{dp}{dq}$.

b.   Evaluate $q$ at $p = 10$.

c.   Evaluate $\dfrac{dp}{dq}$ at $p = 10$.

d.   Interpret the results of b and c.

---

If additional practice is needed:

Hoffmann, page 124, problems 1–23
Piascik, pages 142–143, problems 1–10

# UNIT 21

## Exponential Functions

Exponential functions will be introduced and defined in this unit. When you have finished it you will be able to identify an exponential function, differentiate it, and find and classify all its critical points.

We've gone this far together so it is only fair to warn you. Typically this is the least favorite topic in calculus. If I had to guess why, it is probably because the material is brand new. Many of you will have never seen an exponential function before, much less the ones we will be dealing with. Added to that, we use an irrational number as the base and it all becomes a bit much for some students. My advice is don't give up. It is only one unit. You can do it.

Thus far in the book we have been working with power functions. A **power function** has the independent variable appearing only in the base and the exponent is a real number. In contrast, an **exponential function** has the independent variable appearing in the exponent.

---

**Power Function:** $f(x) = x^n$, $n$ a real number

**Exponential Function:** $f(x) = b^x$, $b$ a real number, $b \neq 0$

---

We have spent many units learning how to graph, to differentiate, and to locate and classify critical points for power functions. In this unit you will learn how to do all of those same things for exponential functions. Hopefully it will not take us as long.

## THE NUMBER $e$

Although any number may be used as the base in an exponential function, one of the most commonly used bases is the number $e$. From geometry you most likely remember the irrational number $\pi$, that is approximately equal to 3.14, and used in formulas for the area and circumference of a circle. Like $\pi$, $e$ is an irrational number, meaning it is a nonrepeating and nonterminating decimal, with a value approximately equal to 2.718.

This book will deal exclusively with exponential functions to the base $e$.

# THE GRAPH OF $f(x) = e^x$

The simplest exponential function to examine is $f(x) = e^x$. Again we will use another picture.

## EXAMPLE 1

Graph:   $f(x) = e^x$.

Solution:   Since this is a new type of function and its graph is probably unfamiliar to you, we will plot a reasonable number of points.

Many calculators today contain a key for $e$. If yours does, use it. Otherwise use Table I, which is provided in the Appendix, to obtain the various values for $e^x$. Practice reading the table by verifying that the values listed below are correct.

Let $x = -3$, then $f(-3) = e^{-3} = 0.050$       Let $x = .5$,   then $f(.5) = e^{.5} = 1.649$

Let $x = -2$, then $f(-2) = e^{-2} = 0.135$       Let $x = 1$,   then $f(1) = e^1 = 2.718$

Let $x = -1$, then $f(-1) = e^{-1} = 0.368$       Let $x = 2$,   then $f(2) = e^2 = 7.389$

Let $x = -.5$, then $f(-.5) = e^{-.5} = 0.607$     Let $x = 3$,   then $f(3) = e^3 = 20.086$

Let $x = 0$,    then $f(0) = e^0 = 1.0$

For the majority of our work after this graph, the answers will be left in terms of $e$.

The above nine points have been plotted on the grid below and connected with a smooth curve. The curve has no breaks because the domain for the function is the set of all real numbers. Why?

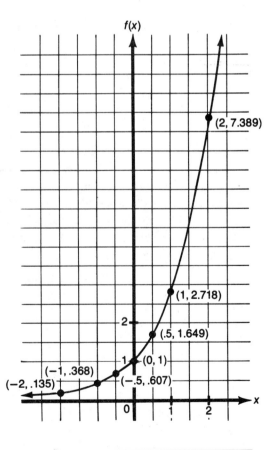

Answer:

# FACTS ABOUT $e^x$

There are several observations that can be made regarding the function $f(x) = e^x$. Refer to Example 1, expecially the graph pictured above, to verify each of the following statements.

1.  $e^x$ is always positive.

2.  $e^x = 0$ has no solution. That is the same as saying the curve has no $x$-intercept, but we will use it in the form of no solution more frequently than regarding the $x$-intercept.

    *To generalize further, e raised to any power can never equal 0.*

3.  $e^0 = 1$.

4.  As $x$ gets very large (say as $x = 4$, 100, or 1,000), $e^x$ gets very, very large. (In Example 1, the point where $x = 3$ was already too large to be included on the graph.)

5.  As $x$ gets very small (say as $x = -4$, $-100$, or $-1,000$), $e^x$ gets closer and closer to zero. Mathematically, that would be stated as $e^x$ is **asymptotic** to the $x$-axis on the left. This means the value of $e^x$ is approaching, but never reaches, the value of 0 as $x$ is getting smaller and smaller. (In Example 1, the point where $x = -3$ was already too close to 0 to be shown on the graph.)

6.  The domain of $f(x) = e^x$ is the set of real numbers. Remember the only time values need to be excluded from the domain are for functions with even-rooted radicals or variables in the denominator, neither of which exist here.

7.  The function is increasing over its entire domain.

While all of the above statements are important, for our purposes the first two are the ones you need to remember.

# APPLICATIONS

Exponential functions to the base $e$ are found throughout applied mathematics. For example, if interest is compounded continuously, the resulting bank balance is determined by an exponential function. In the absence of any environmental constraints, population increases exponentially. Sales of many products decrease exponentially when advertising is discontinued. Learning curves, which are exponential functions, describe the relationship between the efficiency with which an individual performs a task and the amount of training the individual has had. And logistic curves, which are exponential functions, describe the spread of epidemics or rumors through a community.

We will do one example and let it go at that.

## EXAMPLE 2

The total number of hamburgers sold by a national fast-food chain is growing exponentially according to the function

$$Q(t) = 4e^{2t-3}$$

where $t$ = the years since 1980 ($t = 0$ corresponds to 1980)

and $Q(t)$ = the number of hamburgers in hundreds sold in $t$ years.

The company is considering using the number of hamburgers sold in a future advertising campaign, somewhat like the way McDonald's did once. Just to see how the number would look in the ads, how many hamburgers had been sold by 1986?

Solution:   Since $t$ = the years since 1980, 1986 would correspond to $t = 6$.

$$Q(t) = 4e^{2t-3}$$

$$Q(6) = 4e^{2(6)-3}$$

$$= 4e^9$$

$$= 4(8,103.1) \quad \text{From Table I}$$

$$= 32,412.4$$

Answer:   By 1986, the fast-food chain had sold 32,412.4 hundred or 3,241,240 hamburgers.

## Problem 1

Continuing with Example 2, how many hamburgers had been sold in 1980?

Solution:

Answer:   .2 hundred or 20

Answers might vary due to rounding.

# DIFFERENTIATION

By now I know you are able to take the derivative of all polynomial functions, as well as of the product of two polynomials, of the quotient of two polynomials, and of a polynomial raised to a power.

A brief summary of the rules thus far are:

| | |
|---|---|
| Power: | If $f(x) = x^n$, then $f'(x) = nx^{n-1}$ |
| Constant: | If $f(x) = c$, then $f'(x) = 0$ |
| Constant · Function: | If $f(x) = c \cdot u(x)$, then $f'(x) = c \cdot u'(x)$ |
| Special Case: | If $f(x) = c \cdot x^n$, then $f'(x) = cnx^{n-1}$ |
| Sum: | If $f(x) = u(x) + v(x)$, then $f'(x) = u'(x) + v'(x)$ |
| Difference: | If $f(x) = u(x) - v(x)$, then $f'(x) = u'(x) - v'(x)$ |
| Product Rule: | If $f(x) = F(x) \cdot S(x)$, then $f'(x) = F(x)S'(x) + S(x)F'(x)$ |
| Quotient Rule: | If $f(x) = \dfrac{N(x)}{D(x)}$, then $f'(x) = \dfrac{D(x)N'(x) - N(x)D'(x)}{[D(x)]^2}$ |
| Chain Rule: | If $f(x) = [u(x)]^n$, then $f'(x) = n[u(x)]^{n-1}u'(x)$ |

The differentiation of exponential functions requires the addition of only one more rule.

**Exponential Rule**

If $f(x) = e^{u(x)}$ where $u(x)$ is a function of $x$, then $f'(x) = u'(x)e^{u(x)}$.

In words, the derivative of $e$ raised to some exponent containing $x$ equals the derivative of the exponent times the original function.

Notice that we are working with a function of $x$ so all the previous rules still apply.

# EXAMPLE 3

If $f(x) = e^{5x}$, find $f'(x)$.

Solution:   To differentiate $e$ raised to some exponent containing $x$, take the derivative of the exponent times the original function.

$$f(x) = e^{5x}$$
$$f'(x) = \frac{d(5x)}{dx}e^{5x}$$
$$= 5e^{5x}$$

Keep in mind all that is happening here is that we are considering another group of functions, called exponentials. The derivative remains a formula for finding the slope of the tangent line at any point along the curve. Critical points are located and classified the same way as before. And exponential functions are maximized or minimized using the same general technique presented in an earlier unit.

# EXAMPLE 4

If $f(x) = e^{5x}$, find and interpret a. $f'(0)$, b. $f'(1)$, and c. $f'(-1)$.

Solution:   a.

$$f(x) = e^{5x}$$

$$f(0) = e^0 = 1, \text{ which is the point } (0, 1) \text{ on the curve.}$$

$$f'(x) = 5e^{5x}$$

$$f'(0) = 5e^0 = 5(1) = 5$$

Answer:   The slope of the tangent line to the curve at the point $(0, 1)$ is 5.

b.

$$f(x) = e^{5x}$$

$$f(1) = e^{5(1)} = e^5, \text{ which is the point } (1, e^5) \text{ on the curve.}$$

Note:   Leaving the answer as $e^5$ is perfectly acceptable. This is not an engineering course. We do not need a decimal approximation for any further use in the problem. Mentally I think of $e$ as being almost 3 and let it go at that. If more precision is required, from Table I, $e^5 = 148.4$ and the point would be about $(1, 148.4)$.

$$f'(x) = 5e^{5x}$$

$$f'(1) = 5e^{5(1)} = 5e^5$$

Answer:   The slope of the tangent line to the curve at the point $(1, e^5)$ is $5e^5$.

c.

$$f(x) = e^{5x}$$

$$f(-1) = e^{5(-1)} = e^{-5}, \text{ which is the point } (-1, e^{-5}).$$

$$f'(x) = 5e^{5x}$$

$$f'(-1) = 5e^{5(-1)} = 5e^{-5}$$

Answer:   The slope of the tangent line to the curve at the point $(-1, e^{-5})$ is $5e^{-5}$.

If the answer is to be written with positive exponents only, the slope of the tangent line to the curve at the point

$$\left(-1, \frac{1}{e^5}\right) \text{ is } \frac{5}{e^5}.$$

Again, as stated in part b, if you are going to do something more with the answer and need an approximate value, from Table I, $e^{-5} = .00674$ and $5(.00674) = .0337$. The answer would now read, the slope of the tangent line to the curve at the point $(-1, .00674)$ is approximately .0337.

---

In case you are curious, the graph of $f(x) = e^{5x}$ is shaped about the same as the function in Example 1 except the curve is going up faster on the right and dropping down to the $x$-axis faster on the left.

Unfortunately, the graphs of exponential functions as a whole are not predictable. Although some generalizations can be made about specific types of exponential functions, graphing of exponential functions will not be discussed in this book.

## EXAMPLE 5

If $f(x) = e^{x^2+1}$, find $f'(x)$.

Solution:   The derivative of $e$ raised to some exponent containing the independent variable equals the derivative of the exponent times the original function.

$$f(x) = e^{x^2+1}$$

$$f'(x) = \frac{d(x^2+1)}{dx} e^{x^2+1}$$

$$= 2x e^{x^2+1}$$

derivative of exponent        original function

## EXAMPLE 6

If $f(x) = e^x$, find $f'(x)$.

Solution:    $f(x) = e^x$

$$f'(x) = \frac{d(x)}{dx} e^x$$

$$= 1 \cdot e^x$$

$$= e^x$$

Example 6 illustrates that the derivative of $e^x$ is $e^x$. It is a special case of the Exponential Rule.

> Special Case:    If $f(x) = e^x$, then $f'(x) = e^x$.

By now you are probably anxious to try a few yourself before they get harder.

## Problem 2

Find $f'(x)$ if $f(x) = e^{6x+5}$.

Solution:

Answer:    $f'(x) = 6e^{6x+5}$

## Problem 3

Find $f'(x)$ if $f(x) = e^{x^2+3x-1}$.

---

Solution:

Answer: $f'(x) = (2x + 3)\, e^{x^2+3x-1}$

The parentheses are required to show that the entire quantity is a factor.

---

The following examples require the use of more than one rule.

## EXAMPLE 7

If $f(x) = \dfrac{e^x}{x}$, find $f'(x)$.

---

Solution: I hope you recognized that $f(x)$ is a quotient and as such we must use the Quotient Rule.

$$f(x) = \frac{e^x}{x}$$

Use Quotient Rule. $\quad f'(x) = \dfrac{x\dfrac{d(e^x)}{dx} - e^x\dfrac{d(x)}{dx}}{x^2}$

Differentiate. $\quad = \dfrac{x(e^x) - e^x(1)}{x^2}$

Simplify. $\quad = \dfrac{xe^x - e^x}{x^2}$

Or factored, if you prefer. $\quad = \dfrac{e^x(x - 1)}{x^2}$

---

That wasn't too bad, but the next one is not quite so obvious.

## EXAMPLE 8

If $f(x) = 4xe^x$, find $f'(x)$.

Solution:   This time $f(x)$ is a product. Do you see why?

$$f(x) = \underbrace{4x}_{}\overbrace{e^x}^{}$$

first factor    second factor

Use Product Rule.

$$f'(x) = (4x)\frac{d(e^x)}{dx} + (e^x)\frac{d(4x)}{dx}$$

Differentiate.

$$= 4x(e^x) + e^x(4)$$

Simplify.

$$= 4xe^x + 4e^x$$

Or, if you like factoring.

$$= 4e^x(x + 1)$$

Here is another example with the Product Rule.

## EXAMPLE 9

If $f(x) = (1 - 3x)e^{x^2}$, find $f'(x)$.

Solution:

$$f(x) = \underbrace{(1 - 3x)}_{}\overbrace{e^{x^2}}^{}$$

first factor    second factor

Use Product Rule.

$$f'(x) = (1 - 3x)\frac{d(e^{x^2})}{dx} + (e^{x^2})\frac{d(1 - 3x)}{dx}$$

Differentiate.

$$= (1 - 3x)(2xe^{x^2}) + (e^{x^2})(-3)$$

Simplify.

$$= 2xe^{x^2} - 6x^2e^{x^2} - 3e^{x^2}$$

$$= -6x^2e^{x^2} + 2xe^{x^2} - 3e^{x^2}$$

Or factored, if you prefer.

$$= e^{x^2}(-6x^2 + 2x - 3)$$

# LOCATING AND CLASSIFYING CRITICAL POINTS

Recall critical points, if they exist, are located by taking the first derivative, setting it equal to zero, solving for $x$, and then determining the corresponding $y$-coordinate. Once located, a critical point may be classified using either the original-function test, the first-derivative test, or the second-derivative test.

The same procedures are used for exponential functions. Do not let that irrational number $e$ confuse you. It is nothing more than a commonly accepted symbol used to represent a number approximately equal to

2.718. And exponential functions are just another group of functions with different characteristics and applications than polynomial functions, but the concepts and techniques remain the same.

To save ourselves some work, I will use functions for which we already have found the derivative.

# EXAMPLE 10

Find and classify all critical points for $f(x) = 4xe^x$.

Solution:
$$f(x) = 4xe^x$$

$$f'(x) = 4e^x(x + 1) \quad \text{from Example 8}$$

Find all critical points. Critical points occur where $f'(x) = 0$.

$$f'(x) = 4e^x(x + 1)$$

$$0 = 4e^x(x + 1)$$

To solve an equation of this type, as before, we set each factor equal to zero and solve individually.

$$x + 1 = 0 \qquad 4e^x = 0$$

$$x^* = -1 \qquad e^x = 0,$$

but recall from Example 1, this equation has no solution.

If $x^* = -1$,

$$f(-1) = -4e^{-1} = \frac{-4}{e}, \text{ thus } \left(-1, \frac{-4}{e}\right) \text{ is the only critical point.}$$

If the $e$ makes you uncomfortable, use Table I and rewrite it as $(1, -1.472)$, but it is actually more work and unnecessary.

Classify the critical point.

My choice would be to use the first-derivative test using $f'(x) = 4e^x(x + 1)$.

To save ourselves time, remember we need only determine the sign, positive or negative, of the first derivative. Again recall from Example 1, that $e^x$ is always positive, so $4e^x$ is always positive.

For the point $\left(-1, \dfrac{-4}{e}\right)$ with $f'(x) = 4e^x(x + 1)$

On the left:  Let $x = -2$, then $f'(-2) = (+)(-2 - 1) = (+)(-) = -$     decreasing

At the point:  $\left(-1, \dfrac{-4}{e}\right)$ the curve     levels off

On the right:  Let $x = 0$,   then $f'(0) = (+)(0 + 1) = (+)(+) = +$     increasing

At the critical point, the curve is decreasing, leveling off, then increasing.

Answer:   The critical point $\left(-1, \dfrac{-4}{e}\right)$ is a relative minimum.

In fact, the critical point $\left(-1, \dfrac{-4}{e}\right)$ is an absolute minimum. Do you understand why? The domain is the entire set of real numbers meaning the curve has no breaks. Since there is one and only one relative minimum, it must be the absolute minimum.

That one was a bit long; let's try a shorter one.

## EXAMPLE 11

Find and classify all critical points for $f(x) = e^x$.

Solution:
$$f(x) = e^x$$
$$f'(x) = e^x$$

Find all critical points. Critical points occur where $f'(x) = 0$.

$$0 = e^x$$

but this equation has no solution.

Answer:   $f(x) = e^x$ has no critical points.

Before we do a really long one, I want you to try one.

## Problem 4

Given:   $f(x) = e^{x^2 + 1}$.

a.   Find $f'(x)$.

b.   Find all critical points for $f(x)$.

c.   Classify each critical point.

Solution:

Answers:   a.   Refer to Example 5.

b.   $(0, e)$

c.   Relative minimum

# EXAMPLE 12

Find and classify all critical points for $f(x) = \dfrac{e^x}{x}$.

Solution:   $f(x) = \dfrac{e^x}{x}$

$f'(x) = \dfrac{xe^x - e^x}{x^2}$   from Example 7

Find all critical points. Critical points occur where $f'(x) = 0$.

$0 = \dfrac{xe^x - e^x}{x^2}$

$0 = xe^x - e^x$        Multiply entire equation by $x^2$.

$0 = e^x(x - 1)$        Factor out common factor of $e^x$.

Set each factor equal to zero and solve.

$x - 1 = 0$     $e^x = 0$

$\quad x^* = 1$     From Example 1, $e^x$ has no solution.

If $x^* = 1$,

$f(1) = \dfrac{e^1}{1} = e$, thus $(1, e)$ is the only critical point.

Classify the critical point.

My choice would be to use the first-derivative test using

$$f'(x) = \frac{xe^x - e^x}{x^2} = \frac{e^x(x - 1)}{x^2}$$

To save ourselves time, we need only determine the sign. Also notice that the denominator will always be positive, because it is being squared and $e^x$ is always positive.

For the point $(1, e)$

| | | | |
|---|---|---|---|
| On the left: | Let $x = .5$, then $f'(.5) = (+)(.5 - 1)/ + = -$ | decreasing |
| At the point: | $(1, e)$ the curve | levels off |
| On the right: | Let $x = 2$, then $f'(2) = (+)(2 - 1)/ + = +$ | increasing |

At the critical point, the curve is decreasing, leveling off, then increasing.

Answer: The critical point $(1, e)$ is a relative minimum.

# OPTIMIZATION REVISITED

As you well know, the real purpose behind all of this is to be able to determine where an exponential function achieves its maximum and minimum values. The four-step general technique in condensed form is repeated below for reference.

The general technique is:

1.  Decide what quantity is to be maximized or minimized and write the equation for it.

2.  Use the constraints of the problem to write the equation in terms of a single variable.

3.  Find all critical points and classify.

4.  Answer the original question.

One example is about all you are likely to want to do.

# EXAMPLE 13

Edward, a publisher's rep, has observed that his sales of college textbooks is a function of the number of years the book has been on the market. Typically a new book reaches its peak in sales a few years after it is first published and then sales level off and remain fairly constant. This year the publishing company is introducing a new economics text. Edward estimates his sales for the book will be given by

$$f(x) = x^2 e^{-x} \qquad x \geq 0$$

where $f(x)$ is the number of books sold in thousands in year $x$

and $x$ is the amount of years since the book was published.

In what year should Edward expect to maximize his sales of the new economics book? What is the maximum number of copies of the book it is estimated he will sell in any year?

Solution:  1.  What is to be maximized in the problem? Number of books sold

Write the equation for number of books sold.

The equation for number of books sold was given as

$$f(x) = x^2 e^{-x} \text{ with } x \geq 0$$

2.  Write function in terms of a single independent variable.

The function was given in terms of a single variable.

3.  Find critical points and classify.

$$f(x) = x^2 e^{-x}$$

$$f'(x) = x^2 \frac{d(e^{-x})}{dx} + e^{-x} \frac{d(x^2)}{dx}$$

$$= x^2(-e^{-x}) + e^{-x}(2x)$$

$$= -x^2 e^{-x} + 2xe^{-x}$$

$$= -xe^{-x}(x - 2)$$

$$0 = -xe^{-x}(x - 2)$$

$$-x = 0 \quad x - 2 = 0 \quad e^{-x} = 0$$

$$x^* = 0 \qquad x^* = 2 \quad \text{This has no solution.}$$

If $x^* = 0$,

$f(0) = 0$ and is of little interest to the problem.

If $x^* = 2$,

$$f(2) = (2)^2 e^{-2}$$

$$= 4(.135)$$

$$= .54$$

Test:   First-derivative test with $f'(x) = -xe^{-x}(x - 2)$

Remember $e^{-x}$ is always positive.

At the point $(2, .54)$

On the left:    Let $x = 1$, $f'(1) = (-)(+)(-) = +$          increasing

At the point:    $(2, .54)$                                              levels off

On the right:    Let $x = 3$, $f'(3) = (-)(+)(+) = -$          decreasing

The curve is increasing, then leveling off, then decreasing. The point $(2, .54)$ is a relative maximum. Stated another way, the maximum value for $f(x)$ is .54 and occurs when $x = 2$.

4.   Answer the questions posed in original problem.

Answers:   Edward should expect to maximize the sale of the new economics textbook in its second year on the market, at which time, it is estimated he will sell .54 thousand books or 540 books.

---

You should now be able to identify an exponential equation, differentiate it, and be able to find and classify all of its critical points.

Before continuing to the next unit, practice with the following exercises.

## EXERCISES

Find and classify all critical points for the following three problems:

1.   $f(x) = e^{x^2}$

2.   $f(x) = \dfrac{e^x}{x^2}$

3.   $f(x) = x^2 e^x$

4.   Jude manages the concession stand at one of the local theatres. She is planning the introduction and promotion of a new popcorn line. After testing the new line for several months, she found that the demand is given approximately by

$$p = 5e^{-x} \qquad 0 \le x \le 2$$

where $x$ thousand boxes of popcorn were sold per week
at a price of $p$ dollars each.

At what price will the weekly revenue be maximum?

---

If additional practice is needed:    Barnett and Ziegler, page 285, problems 3–4, 11–14, 33–34, 39–40, 57–58, 61–62    Budnick, page 555, problems 4–17; pages 569–570, problems 1–25 Hoffmann, pages 256–258, problems 1–14, 31–47   Piascik, pages 281–282, problems 1–5

# UNIT 22

## Logarithmic Functions

The purpose of this unit is to provide you with a brief overview of logarithms, especially natural logarithms. First the definition and notation used with natural logarithms will be introduced. Then you will learn to differentiate logarithmic functions and find and classify all their critical points.

Keep in mind that all we are doing in this unit is considering still another group of functions called logarithmic functions. The derivative remains a formula for finding the slope of the tangent line at any point along the curve. Critical points are located and classified the same way as before. And logarithmic functions will be maximized or minimized using the same general technique as presented in an earlier unit.

Although some generalizations can be made about the graphs of specific types of logarithmic functions, graphing of them will not be covered in this book.

The equation $\log_b N = x$ is read "the logarithm of $N$ to the base $b$ is $x$."

---

Definition:  $x$ is called the **logarithm of $N$ to the base $b$** if $b^x = N$, where $N$ and $b$ are both positive numbers, $b \neq 1$.

---

In other words:

$$\log_b N = x \text{ if and only if } b^x = N.$$

Examples of logarithms are:

$$\log_3 9 = 2 \text{ because } 3^2 = 9$$

$$\log_2 8 = 3 \text{ because } 2^3 = 8$$

$$\log_7 7 = 1 \text{ because } 7^1 = 7$$

$$\log_2 16 = 4 \text{ because } 2^4 = 16$$

$$\log_{15} 1 = 0 \text{ because } 15^0 = 1$$

$$\log_2\left(\frac{1}{2}\right) = -1 \text{ because } 2^{-1} = \frac{1}{2}$$

Notice that the logarithm of a positive number $N$ is the exponent to which the base must be raised to produce the number $N$.

# THE NATURAL LOGARITHM

Although any number may be used as the base in a logarithmic function, one of the most commonly used bases is the number $e$. You remember $e$ from the previous unit. It is an irrational number with a value approximately equal to 2.718. When $e$ is used as the base of a logarithmic function, the function is referred to as a natural logarithm. Typically ln is used to denote a natural logarithm, rather than log, and the $e$ is omitted.

The equation $\ln x = y$ is read "the natural logarithm of $x$ is $y$." Often "the natural logarithm of $x$" is shortened and read as "natural log of $x$."

---

Definition:    $y$ is called the **natural logarithm of $x$** if $e^y = x$, where $x$ is a positive number.

---

In words:

$$\ln x = y \text{ if and only if } e^y = x.$$

Examples of natural logarithms are:

$$\ln e = 1 \text{ because } e^1 = e$$

$$\ln 1 = 0 \text{ because } e^0 = 1$$

$$\text{From Table I, } \ln 7.3891 = 2 \text{ because } e^2 = 7.3891$$

$$\text{From Table I, } \ln 20.086 = 3 \text{ because } e^3 = 20.086$$

You need to remember the first two examples. For the last two examples and any other values, Table II is provided in the Appendix. Practice using the table by first verifying that the values listed in the Examples are correct and then doing the Problems on your own.

---

EXAMPLE 1    $\ln 4 = 1.38629$ and $e^{1.38629} = 4$

EXAMPLE 2    $\ln 1.2 = .18232$ and $e^{.18232} = 1.2$

EXAMPLE 3    $\ln .1 = -2.30259$ and $e^{-2.30259} = .1$

---

Cover the answers and find the natural logs of the following numbers using the table.

---

Problem 1    $\ln 1.49 =$

Problem 2    $\ln .50 =$

Problem 3    $\ln 0.3 =$

Problem 4    $\ln 0.61 =$

Answers:    .39878, $-0.69315$, $-1.20397$, $-0.49430$

As with the earlier definition, the natural logarithm of a positive number $x$ is the exponent to which the base $e$ must be raised to produce the number $x$.

$\ln x = y$ is called the logarithmic form,

$b^y = x$ is called the exponential form,

and the two statements are equivalent.

This book will deal exclusively with natural logarithms.

The tendency at this point is to be a bit confused about what logarithmic functions are and what this unit is really about. Recall I said that logarithmic functions are merely another group of functions with properties and applications that are unique to their group. We are not going to graph logarithmic functions, but you will learn how to differentiate them and how to locate and classify their critical points.

# FACTS ABOUT $y = \ln u(x)$

There are several observations to be remembered concerning the function $y = f(x) = \ln u(x)$. Refer to the above material to verify each of the following statements.

1.  $\ln 1 = 0$ because $e^0 = 1$.

2.  $\ln e = 1$ because $e^1 = e$.

3.  The function $u(x)$ must be positive because logarithms are defined for positive numbers only. This fact is often overlooked when we start working with logarithms of algebraic functions rather than of numbers, so you are being warned early on.

# DIFFERENTIATION

Because we are still working with a function of $x$, all of our previous rules apply.

A brief summary of the rules thus far are:

Power: If $f(x) = x^n$, then $f'(x) = nx^{n-1}$

Constant: If $f(x) = c$, then $f'(x) = 0$

Constant · Function: If $f(x) = c \cdot u(x)$, then $f'(x) = c \cdot u'(x)$

Special Case: If $f(x) = c \cdot x^n$, then $f'(x) = cnx^{n-1}$

Sum: If $f(x) = u(x) + v(x)$, then $f'(x) = u'(x) + v'(x)$

Difference: If $f(x) = u(x) - v(x)$, then $f'(x) = u'(x) - v'(x)$

Product: If $f(x) = F(x) \cdot S(x)$, then $f'(x) = F(x)S'(x) + S(x)F'(x)$

Quotient: If $f(x) = \dfrac{N(x)}{D(x)}$, then $f'(x) = \dfrac{D(x)N'(x) - N(x)D'(x)}{[D(x)]^2}$

Chain: If $f(x) = [u(x)]^n$, then $f'(x) = n[u(x)]^{n-1}u'(x)$

Exponential: If $f(x) = e^{u(x)}$, then $f'(x) = u'(x)e^{u(x)}$

Special Case: If $f(x) = e^x$, then $f'(x) = e^x$

To differentiate a natural logarithmic function requires the addition of only one more rule.

---

**Logarithmic Rule**

If $f(x) = \ln u(x)$, then $f'(x) = \dfrac{u'(x)}{u(x)}$

Verbally, the derivative of the ln $u(x)$ is the derivative of $u(x)$ divided by $u(x)$ itself.

---

# EXAMPLE 4

If $f(x) = \ln x$, find $f'(x)$.

---

Solution:   The derivative of ln $u(x)$ is simply the derivative of $u(x)$ divided by $u(x)$ itself.

$$f(x) = \ln x$$

$$f'(x) = \frac{\dfrac{d(x)}{dx}}{x}$$

$$= \frac{1}{x}$$

---

Example 4 illustrates that the derivative of ln $x$ is $\dfrac{1}{x}$. It is a frequently encountered function and can be considered a special case of the logarithmic rule.

---

Special Case:   If $f(x) = \ln x$, then $f'(x) = \dfrac{1}{x}$.

---

# EXAMPLE 5

If $f(x) = \ln (x^2 + 5x)$, find $f'(x)$.

---

Solution:   The derivative of ln $u(x)$ is the derivative of $u(x)$ divided by $u(x)$.

$$f(x) = \ln (x^2 + 5x)$$

$$f'(x) = \frac{\dfrac{d(x^2 + 5x)}{dx}}{(x^2 + 5x)}$$

$$= \frac{2x + 5}{x^2 + 5x}$$

## EXAMPLE 6

If $f(x) = \ln x^3$, find $f'(x)$.

---

Solution:   The derivative of $\ln u(x)$ is the derivative of $u(x)$ divided by $u(x)$.

$$f(x) = \ln x^3$$

$$f'(x) = \frac{\dfrac{d(x^3)}{dx}}{x^3}$$

$$= \frac{3x^2}{x^3}$$

$$= \frac{3}{x}$$

---

Aren't these easy compared to some of the functions we have been doing? Try the next few problems on your own. Be sure to simplify your answers.

## Problem 5

Find $f'(x)$ if $f(x) = \ln x^2$.

---

Solution:

Answer:   $f'(x) = 2/x$

---

## Problem 6

Find $f'(x)$ if $f(x) = \ln (x^2 + 7x + 1)$.

---

Solution:

Answer:   $f'(x) = \dfrac{2x + 7}{x^2 + 7x + 1}$

---

## Problem 7

Find $f'(x)$ if $f(x) = \ln x^2 + 7x + 1$. Be careful.

Solution:

Answer: $f'(x) = \dfrac{2}{x} + 7$

Did I catch you on that last one? I really was not trying to trick you, but merely trying to show you the importance of parentheses. Problem 6 was the ln of the entire quantity in parentheses whereas Problem 7 was the ln of only $x^2$ with the $7x$ and the 1 being separate terms. In Problem 7 you had to take the derivative term by term. The following example also illustrates having to take the derivative term by term.

## EXAMPLE 7

Find $f'(x)$ if $f(x) = \ln x + \ln 3x$.

Solution:   $f(x)$ is a function of two separate terms, much like Problem 7.

$$f(x) = \ln x + \ln 3x$$

$$f'(x) = \frac{1}{x} + \frac{\dfrac{d(3x)}{dx}}{3x}$$

$$= \frac{1}{x} + \frac{3}{3x}$$

$$= \frac{1}{x} + \frac{1}{x}$$

$$= \frac{2}{x}$$

## EXAMPLE 8

Find $f'(x)$ if $f(x) = \ln 3$.

Solution:   What kind of function is $f(x)$? It is a constant. And the derivative of a constant is zero.

$$f(x) = \ln 3$$

$$f'(x) = 0$$

272 LOGARITHMIC FUNCTIONS

Having trouble with that? Let's rewrite $f(x)$ and try it again.

From Table II, if $f(x) = \ln 3$

$$\text{then } f(x) = 1.09861$$

$$\text{and } f'(x) = 0$$

---

There is no way to avoid it; the next examples require the Product Rule and the Quotient Rule.

========================================

## EXAMPLE 9

Find $f'(x)$ if $f(x) = \dfrac{\ln x}{x}$.

---

Solution:  $f(x)$ is a quotient and as such we must use the Quotient Rule.

$$f(x) = \frac{\ln x}{x}$$

Use Quotient Rule.
$$f'(x) = \frac{x \dfrac{d(\ln x)}{dx} - (\ln x) \dfrac{d(x)}{dx}}{x^2}$$

Special Case with $\ln x$.
$$= \frac{x\left(\dfrac{1}{x}\right) - (\ln x)(1)}{x^2}$$

Simplify.
$$= \frac{1 - \ln x}{x^2}$$

========================================

## EXAMPLE 10

Find $f'(x)$ if $f(x) = x^2 \ln x$.

---

Solution:   This time $f(x)$ is a product.

$$f(x) = x^2 \underbrace{\ln x}$$
$$\text{first factor} \qquad \text{second factor}$$

Use Product Rule.
$$f'(x) = (x^2)\frac{d(\ln x)}{dx} + (\ln x)\frac{d(x^2)}{dx}$$

Special Case with $\ln x$.
$$= (x^2)\left(\frac{1}{x}\right) + (\ln x)(2x)$$

Simplify.
$$= x + 2x \ln x$$

Or, if you like it factored.
$$= x(1 + 2 \ln x)$$

# LOCATING AND CLASSIFYING CRITICAL POINTS

Critical points, as you well know by now, occur where the first derivative equals zero. They are located by setting the first derivative equal to zero and solving the resulting equation and then determining the corresponding $y$-coordinate. Once located, a critical point may be classified as either a relative maximum, a relative minimum, or a stationary inflection point. One of three tests is used to make this determination: the original-function test, the first-derivative test, or the second-derivative test.

The same procedures are used for logarithmic functions.

## EXAMPLE 11

Find and classify all critical points for $f(x) = x - \ln x$ for $x > 0$.

Solution:
$$f(x) = x - \ln x$$

$$f'(x) = 1 - \frac{1}{x} \quad \text{from Special Case.}$$

Find all critical points. Critical points occur where $f'(x) = 0$.

$$0 = 1 - \frac{1}{x}$$

$$0 = x - 1 \quad \text{(Cleared of fractions)}$$

$$x^* = 1$$

If $x^* = 1$,

$$f(1) = 1 - \ln 1 = 1 - 0 = 1,$$

thus $(1, 1)$ is the only critical point.

Classify the critical point.

For variety, we will use the second-derivative test.

$$f'(x) = 1 - \frac{1}{x}$$

$$= 1 - x^{-1}$$

$$f''(x) = 0 - (-1x^{-2})$$

$$= 0 + x^{-2}$$

$$= \frac{1}{x^2}$$

At the point $(1, 1)$

$f''(x) = +$, curve is concave up, $\smile$ , point is minimum.

Answer: The critical point $(1, 1)$ is a relative minimum. Or stated another way, the minimum value of $f(x)$ is 1 and occurs when $x = 1$.

Here is a short problem for you to do, then I will do one more. Don't look at the answer until you have finished the problem.

## Problem 8

Find and classify all critical points for $f(x) = \ln x$.

Solution:

Answer:   $f(x)$ has no critical points.

## EXAMPLE 12

Find and classify all critical points for $f(x) = \ln x - \frac{1}{2} x^2$ for $x > 0$.

Solution:
$$f(x) = \ln x - \frac{1}{2} x^2$$

$$f'(x) = \left(\frac{1}{x}\right) - \frac{1}{2}(2x) \quad \text{from Special Case.}$$

$$= \frac{1}{x} - x$$

Find all critical points. Critical points occur where $f'(x) = 0$.

$$0 = \frac{1}{x} - x$$

$$0 = 1 - x^2 \quad \text{(Cleared of fractions)}$$

$$x^2 = 1$$

$$x^* = 1 \qquad x^* = -1$$

($-1$ is not within the domain and should not be considered further.)

If $x^* = 1$,

$$f(1) = \ln 1 - \frac{1}{2}(1)^2 = 0 - \frac{1}{2} = -\frac{1}{2},$$

thus $\left(1, -\frac{1}{2}\right)$ is the only critical point.

Classify the critical point.

This time we will use the first-derivative test for practice.

For the point $\left(1, -\dfrac{1}{2}\right)$ with $f'(x) = \dfrac{1}{x} - x$      the curve is:

On the left:     Let $x = \dfrac{1}{2}$, then $f'\left(\dfrac{1}{2}\right) = 2 - \dfrac{1}{2} = +$     increasing

At the point:     $\left(1, -\dfrac{1}{2}\right)$     levels off

On the right:     Let $x = 2$, then $f'(2) = \dfrac{1}{2} - 2 = -$     decreasing

The curve is increasing, leveling off, then decreasing.

Answer:     The critical point $\left(1, -\dfrac{1}{2}\right)$ is a relative maximum. Or stated another way, the maximum value of $f(x)$ is $-\dfrac{1}{2}$ and occurs when $x = 1$.

---

You should now be able to identify a natural logarithmic function, differentiate it, and be able to find and classify all of its critical points.

This concludes the section on differential calculus of a single variable. By now you should be able to take the derivative of all polynomial functions, exponential functions to the base $e$, and natural logarithmic functions. A complete listing of the rules for differentiation presented in this book are provided below:

## A Brief Summary of All the Rules

Power:      If $f(x) = x^n$, then $f'(x) = nx^{n-1}$

Constant:      If $f(x) = c$, then $f'(x) = 0$

Constant · Function:      If $f(x) = c \cdot u(x)$, then $f'(x) = c \cdot u'(x)$

    Special Case:      If $f(x) = c \cdot x^n$, then $f'(x) = cnx^{n-1}$

Sum:      If $f(x) = u(x) + v(x)$, then $f'(x) = u'(x) + v'(x)$

Difference:      If $f(x) = u(x) - v(x)$, then $f'(x) = u'(x) - v'(x)$

Product:      If $f(x) = F(x) \cdot S(x)$, then $f'(x) = F(x)S'(x) + S(x)F'(x)$

Quotient:      If $f(x) = \dfrac{N(x)}{D(x)}$, then $f'(x) = \dfrac{D(x)N'(x) - N(x)D'(x)}{[D(x)]^2}$

Chain:      If $f(x) = [u(x)]^n$, then $f'(x) = n[u(x)]^{n-1}u'(x)$

Exponential:      If $f(x) = e^{u(x)}$, then $f'(x) = u'(x)e^{u(x)}$

    Special Case:      If $f(x) = e^x$, then $f'(x) = e^x$

Logarithmic:      If $f(x) = \ln u(x)$, then $f'(x) = \dfrac{u'(x)}{u(x)}$

    Special Case:      If $f(x) = \ln x$, then $f'(x) = \dfrac{1}{x}$

Before beginning the next unit, you should solve the following problems involving logarithmic functions. Reduce answers to lowest terms, but you need not change them to decimal approximations.

## EXERCISES

1.  Differentiate $f(x) = \ln 3x$.

2.  If $f(x) = \ln (x^2 - 5)$, find $f'(x)$.

3.  Find $\dfrac{dy}{dx}$ if $y = f(x) = \ln x^2 - 5$.

4.  If $f(x) = \dfrac{2x}{\ln x}$, find $f'(x)$.

5.  If $f(x) = e^x \ln x$, find $f'(x)$.

6.  Find all critical points for $f(x) = \ln (x^2 - 6x + 25)$.

7.  Find and classify all critical points for $f(x) = 4 \ln x - 2x$.

---

If additional practice is needed:

Barnett and Ziegler, pages 285–288, problems 1–64
Budnick, pages 578–579, problems 11–30, 47–64
Hoffmann, pages 256–258, problems 15–26, 48–51
Piascik, page 270, problems 5, 8; page 281, problems 1–2

# UNIT 23

## Integration—Indefinite Integrals

This unit starts with the beginning of what is traditionally called integral calculus. The purpose of the unit is to provide you with an understanding of an indefinite integral. When you have finished the unit, you will be able to integrate a variety of functions.

In this unit we perform an operation that is the reverse of differentiation. It is called antidifferentiation or integration. Typically we will be given a function, $f(x)$, and be asked to find a function, $F(x)$, such that $F'(x), = f(x)$. Such a function is called an antiderivative of $f(x)$.

## NOTATION AND TERMINOLOGY

Like differentiation, integration has its own notation and terminology.

$$\int f(x)\, dx = F(x) + C \qquad \text{where } F'(x) = f(x)$$

$\int f(x)\, dx$  is called the **indefinite integral** of $f(x)$.

$\int$  is called an integral sign.

$f(x)$  is the quantity to be integrated.

$dx$  indicates that $x$ is the variable with respect to which the integration is to take place.

$F(x)$  is the antiderivative.

$C$  is called the constant of integration

$\int f(x)\, dx$ is read "the indefinite integral" of $f$ of $x\, dx$ and the process for finding $\int f(x)\, dx$ is called **indefinite integration**. Basically we will be starting with the derivative of a function and working to find the function itself.

We will start with a review differentiation problem from the previous unit, which I can then use as an example to illustrate the above notation.

## Problem 1

If $f(x) = x \ln x - x$, find $f'(x)$. Hint: The first term is a product.

Solution:

Answer:   $f'(x) = \ln x$

From Problem 1 you found that if $f(x) = x \ln x - x$, then $f'(x) = \ln x$. In terms of our new notation, we could write $\int \ln x \, dx = x \ln x - x + C$. That is, if we were asked to find a function whose derivative is $\ln x$, it is $x \ln x - x + C$. I will explain more about $C$, the constant of integration, after a few more examples. For the time being, accept it as a separate term added on at the end.

In effect, we have just derived our first integration formula.

> Indefinite Integral of $\ln x$
>
> $\int \ln x \, dx = x \ln x - x + C$ where $C$ is an arbitrary constant.
>
> In words, the indefinite integral of $\ln x$ is $x \ln x - x$, plus the constant of integration.

# INTEGRATION OF POWER FUNCTIONS

Next we will work with one of the major rules, that for integrating power functions, and do a few examples to get you started with this topic.

> Indefinite Integral of $x^n$
>
> $\int x^n \, dx = \dfrac{x^{n+1}}{n+1} + C$ for $n \neq -1$ and $C$ an arbitrary constant.
>
> In words, the indefinite integral of $x$ to the $n$th power equals $x$ to the $n+1$ power divided by $n+1$, plus the constant of integration.

Stated still another way, to integrate $x$ to the $n$th power, we increase the exponent by 1 and then divide by the new exponent, and finally $C$, the arbitrary constant term, is added as a separate term.

## EXAMPLE 1

Find: $\displaystyle\int x^5 \, dx.$

Solution:   We are being asked to find a function whose derivative is $x^5$.

According to the above rule, we increase the exponent by 1 and then divide by the new exponent. Then $C$ is added as a separate term.

$$\int x^5 \, dx = \frac{x^6}{6} + C$$

The nice part about integration is that we can always check our answer by differentiating.

Check:   Let   $F(x) = \dfrac{x^6}{6} + C$

$$= \frac{1}{6}x^6 + C$$

then $F'(x) = \dfrac{1}{6}(6x^5) + 0$

$$= x^5$$

Therefore the answer to the integration problem above is correct.

## EXAMPLE 2

Find: $\displaystyle\int x^3 \, dx.$

Solution:   We are being asked to find a function, whose derivative is $x^3$.

According to the rule for integrating powers of $x$, we increase the exponent by 1 and then divide the term by the new exponent.

$$\int x^3 \, dx = \frac{x^4}{4} + C$$

Check:   Let   $F(x) = \dfrac{1}{4}x^4 + C$

then $F'(x) = \dfrac{1}{4}(4x^3) + 0$

$$= x^3, \text{ and the answer checks.}$$

## EXAMPLE 3

Find: $\displaystyle\int x\,dx.$

Solution:    We are being asked to find a function whose derivative is $x$.

According to the rule, we increase the exponent by 1 and then divide by the new exponent.

$$\int x\,dx = \frac{x^2}{2} + C$$

Check:   Let   $F(x) = \dfrac{1}{2}x^2 + C$

then $F'(x) = \dfrac{1}{2}(2x)$

$= x$, and the answer checks.

Let's talk a bit about the constant of integration, $C$. In Example 3 we were being asked to find a function whose derivative was $x$. There are actually infinitely many such functions. To mention just a few:

$$F_1(x) = \frac{1}{2}x^2 + 1$$

$$F_0(x) = \frac{1}{2}x^2$$

$$F_3(x) = \frac{1}{2}x^2 - 3$$

$$F_5(x) = \frac{1}{2}x^2 + 5$$

Verify for yourself that each of the above functions has $x$ as its derivative. In summary, all antiderivatives of $f(x) = x$ are functions of the form

$F(x) = \dfrac{1}{2}x^2 + C$ where $C$ is an arbitrary constant.

If we were to graph the antiderivatives of $f(x)$, we would have a family of curves. Each is a parabola, differing only in its $y$-intercept. A few of them are illustrated.

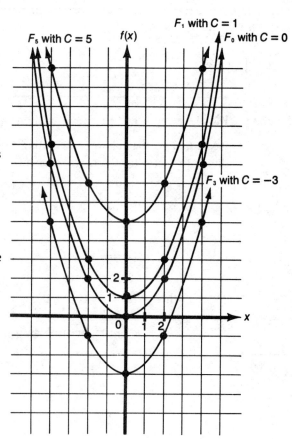

# EVALUATION OF THE CONSTANT OF INTEGRATION

If the problem states an additional property, often called an initial condition, it is then possible to find $C$. That is, we are able to determine the one member of the family of curves that is an antiderivative as well as satisfying the given initial condition.

## EXAMPLE 4

Given $F'(x) = x$, find the function, $F(x)$, that passes through the point $(2, 1)$.

Solution:   If      $F'(x) = x$

then   $F(x) = \displaystyle\int x\, dx$

$= \dfrac{x^2}{2} + C$

The problem is asking for the one curve that passes through the point $(2, 1)$, or if $x = 2$, $y = 1$. Substitute and solve for $C$.

$$F(x) = \frac{x^2}{2} + C$$

$$1 = \frac{(2)^2}{2} + C$$

$$1 = 2 + C$$

$$C = -1$$

Answer:   $F(x) = \dfrac{1}{2}x^2 - 1$

Notice that with an initial condition the family of antiderivatives is narrowed to only the one function that also satisfies the initial condition.

# ADDITIONAL RULES FOR INTEGRATION

Remember with derivatives when there was more than one term, we took the derivative term by term. With integration the rule states that if there is more than one term, we also integrate term by term.

Indefinite Integral of a Sum (or Difference)

$$\int [f(x) + g(x)]\, dx = \int f(x)\, dx + \int g(x)\, dx$$

In words, the indefinite integral of a sum (or difference) of two functions is the sum (or difference) of their individual integrals.

## EXAMPLE 5

Find:   $\displaystyle\int (x^2 - x^3 + x)\, dx$.

Solution:   To integrate more than one term, we integrate term by term.

$$\int (x^2 - x^3 + x)\, dx = \int x^2\, dx - \int x^3\, dx + \int x\, dx$$

$$= \left(\frac{x^3}{3} + c_1\right) - \left(\frac{x^4}{4} + c_2\right) + \left(\frac{x^2}{2} + c_3\right)$$

$$= \frac{x^3}{3} - \frac{x^4}{4} + \frac{x^2}{2} + C$$

where $C = c_1 - c_2 + c_3$

Note:   In the future it is not necessary to write the arbitrary constants associated with each integral as I have done above. Instead simply add a single constant of integration as the last term of the final answer.

---

Indefinite Integral of a Constant Times a Function

$$\int k\, f(x)\, dx = k \int f(x)\, dx$$

In words, the indefinite integral of a constant $k$ times a function $f(x)$ is equal to the constant $k$ times the indefinite integral of the function $f(x)$.

---

This rule states that if the function is being multipled by some constant the constant can be pulled out in front of the integral sign.

## EXAMPLE 6

Find:   $\displaystyle\int 6x^2\, dx$.

Solution:   $\displaystyle\int 6x^2\, dx = 6 \int x^2\, dx$

$$= 6\left(\frac{x^3}{3} + c_1\right)$$

$$= 2x^3 + 6c_1$$

$$= 2x^3 + C$$

where $6c_1$ is rewritten as the arbitrary constant $C$.

Note:   As in the previous example, in the future simply add the constant of integrations as a term to the final answer.

Keep in mind that integration is the inverse of differentiation. We are given the derivative and asked to find the original function. Therefore a rule for integration can be obtained by stating in reverse the rule for differentiation, much like we did with Problem 1.

For the next two rules the differentiation rules will be stated followed by the corresponding rule for integration. The $\frac{dy}{dx}$ notation will be used so as to make the similarity as obvious as possible.

# LOGARITHMIC RULE: SPECIAL CASE

$$\frac{d(e^x)}{dx} = e^x$$

Indefinite Integral of $e^x$

$$\int e^x \, dx = e^x + C$$

In words, the indefinite integral of $e^x$ is $e^x$.

And the last rule of this unit relates back to the integration of powers of $x$. That rule was valid for all $n$, provided $n \neq -1$. This last rule demonstrates how to integrate when $n$ does equal $-1$.

# LOGARITHMIC RULE: SPECIAL CASE

$$\frac{d(\ln x)}{dx} = \frac{1}{x}$$

Indefinite Integral of $\frac{1}{x}$

$$\int \frac{1}{x} \, dx = \ln |x| + C$$

or

$$\int x^{-1} \, dx = \ln |x| + C$$

In words, the indefinite integral of $\frac{1}{x}$ is $\ln |x|$.

The absolute value signs are required because recall that logarithms are defined for non-negative numbers only. The absolute value signs, therefore, guarantee that the number will be non-negative.

Observe in the last two rules that each one is exactly the reverse of the differentiation rule. If we read the differentiation rule from left to right, it follows to read the integration rule from right to left.

I was wrong, we have one more rule in this unit.

Power Rule: Special Case
$$\frac{d(x)}{dx} = 1$$

---

Indefinite Integral of 1

$$\int dx = x + C$$

In words, the indefinite integral of 1 is $x$.

---

This one often causes trouble. Since the derivative of $x$ is 1, it follows that the integral of 1 is $x$.

You might want to make a list of the various integration rules for handy reference as we do a variety of examples.

# EXAMPLE 7

Find: $\int (3x^2 - 5x + 2)\, dx.$

Solution: $\int (3x^2 - 5x + 2)\, dx = \int 3x^2\, dx - \int 5x\, dx + \int 2\, dx$

$$= 3\int x^2\, dx - 5\int x\, dx + 2\int dx$$

$$= 3\left(\frac{x^3}{3}\right) - 5\left(\frac{x^2}{2}\right) + 2x + C$$

$$= x^3 - \frac{5}{2}x^2 + 2x + C$$

# EXAMPLE 8

Find: $\int \frac{1}{x^2}\, dx.$

Solution: $\int \frac{1}{x^2}\, dx = \int x^{-2}\, dx$

$$= \frac{x^{-2+1}}{-1} + C$$

$$= -x^{-1} + C$$

$$= \frac{-1}{x} + C$$

Did you notice in Example 8 the similarity to differentiation? The function was written first as a power of $x$, then integrated, and finally rewritten with a positive exponent.

From now on I am going to shorten the process by omitting the first step.

## EXAMPLE 9

Find: $\int \left( 2e^x + \dfrac{6}{x} - 7x^2 \right) dx.$

Solution: $\int \left( 2e^x + \dfrac{6}{x} - 7x^2 \right) dx = 2 \int e^x \, dx + 6 \int \dfrac{1}{x} dx - 7 \int x^2 \, dx$

$$= 2e^x + 6 \ln |x| - \dfrac{7 \, x^3}{3} + C$$

Ready to try a few? Remember you can always check your answer by differentiating.

## Problem 2

Find: $\int (2x + 3) \, dx.$

Solution:

Answer: $x^2 + 3x + C$

## Problem 3

Find: $\int (x^3 - 3x^2 + 5x - 3x^{-1}) \, dx.$

Solution:

Answer: $\dfrac{x^4}{4} - x^3 + \dfrac{5x^2}{2} - 3 \ln |x| + C$

The next two examples are to illustrate the importance of the $dx$ in the indefinite integral.

## EXAMPLE 10

Find:  $\displaystyle\int 3px^2\,dx$.

Solution:  $\displaystyle\int 3px^2\,dx = 3p\int x^2\,dx$

$$= \cancel{3}p\,\frac{x^3}{\cancel{3}} + C$$

$$= px^3 + C$$

## EXAMPLE 11

Find:  $\displaystyle\int 3px^2\,dp$.

Solution:  $\displaystyle\int 3px^2\,dp = 3x^2\int p\,dp$

$$= 3x^2\,\frac{p^2}{2} + C$$

$$= \frac{3p^2x^2}{2} + C$$

In Example 10 the $dx$ indicates that $x$ is the variable and the integration is to take place with respect to $x$, and $p$ is considered a constant; whereas in Example 11 the $dp$ indicates that $p$ is the variable and the integration is to take place with respect to $p$, and $x$ is considered a constant.

Thus far all of the examples of integration of powers have been whole numbers, but as you well know, often we have to work with radicals or fractional exponents.

## EXAMPLE 12

Find:  $\displaystyle\int \sqrt{x}\,dx$.

Solution:  $\displaystyle\int \sqrt{x}\,dx = \int x^{\frac{1}{2}}\,dx$   $\displaystyle = \frac{x^{\frac{1}{2}+1}}{\dfrac{3}{2}} + C$

$$= \frac{2x^{\frac{3}{2}}}{3} + C$$

$$= \frac{2\sqrt{x^3}}{3} + C$$

We will conclude the unit with an application problem.

## EXAMPLE 13

A firm produces posters. At a production level of $x$ posters, the marginal cost is $MC = C'(x) = 5 - 0.001x$. Find the cost function, $C(x)$, if fixed cost is known to be \$200.

Solution:   $C(x) = \int C'(x)\, dx$

$\qquad\qquad = \int (5 - 0.001x)\, dx$

$\qquad\qquad = 5 \int dx - 0.001 \int x\, dx$

$\qquad\qquad = 5x - 0.001 \dfrac{x^2}{2} + C$

$\qquad\qquad = 5x - 0.0005x^2 + C$

Since the fixed cost is \$200, then $C(0) = 200$,

and $\qquad\qquad C(x) = 5x - 0.0005x^2 + C,$

therefore $\qquad 200 = 5(0) - 0.0005(0)^2 + C$

$\qquad\qquad\quad 200 = C$

Answer:   $C(x) = 5x - 0.0005x^2 + 200$

You should now be able to identify an indefinite integral. In addition, you should be able to integrate a select group of functions such as $e^x$, $\ln x$, and powers of $x$, using the seven rules of integration. And if an initial condition is specified, you should be able to evaluate the constant of integration.

Before continuing to the next unit, integrate the following functions. Be sure to simplify your answers.

## EXERCISES

Find the indefinite integral (if possible).

1.  $\displaystyle\int (2x^4 + x^2)\, dx$

2.  $\displaystyle\int \left( x + \dfrac{1}{x} \right) dx$

3.  $\displaystyle\int x^{\frac{1}{3}}\, dx$

4.  $\displaystyle\int 2x^{-5}\, dx$

5. $\displaystyle\int (15x^2 - \sqrt{2})\, dx$

6. $\displaystyle\int 3e^x\, dx$

7. $\displaystyle\int \ln 3\, dx$

8. $\displaystyle\int 2 \ln x\, dx$

9. $\displaystyle\int 5ax^4 q\, dq$

10. $\displaystyle\int \left( x^3 - \frac{2}{x^2} + 2 \right) dx$

11. $\displaystyle\int (1 - \ln x)\, dx$

12. $\displaystyle\int \frac{e^x - 2x}{2}\, dx$

13. $\displaystyle\int (3x^{-2} + x^{-1})\, dx$

14. $\displaystyle\int \frac{\sqrt{x}}{2}\, dx$

15. $\displaystyle\int x^3 \left( \frac{2}{x} + 4x \right) dx$

16. Find the equation of the curve, given that $f'(x) = 2x^3 - x + 5$ and that the point $(3, 38)$ satisfies $f(x)$.

17. Determine $f(x)$ given that $f'(x) = 2x - 5$ with an initial condition of $f(5) = 4$.

18. The marginal cost function is           $MC = 6x + 1$

    where $x$ equals the number of units produced and MC is the marginal cost measured in dollars. Fixed cost is \$50. What is the total cost of producing 10 units?

19. The function describing the marginal profit from producing and selling a product is

    $$MP = 100 - 2x$$

    where $x$ equals the number of units and $MP$ is the marginal profit measured in dollars. When 10 units are produced and sold, total profit equals \$150. Determine the total profit function.

---

If additional practice is needed:

Barnett and Ziegler, pages 312–316, problems 1–82
Budnick, pages 660–661, problems 1–20
Hoffmann, pages 284–286, problems 1–38

# UNIT 24

## Integration by Substitution

This unit will continue our discussion of integration. Specifically, you will learn a technique called integration by substitition.

Recall the seven rules of integration from the previous unit:

1. $\displaystyle\int x^n\, dx = \frac{x^{n+1}}{n+1} + C, \quad n \neq -1$

2. $\displaystyle\int dx = x + C$

3. $\displaystyle\int [f(x) + g(x)]\, dx = \int f(x)\, dx + \int g(x)\, dx$

4. $\displaystyle\int c\, f(x)\, dx = c \int f(x)\, dx$

5. $\displaystyle\int e^x\, dx = e^x + C$

6. $\displaystyle\int \frac{1}{x}\, dx = \ln|x| + C$

7. $\displaystyle\int \ln x\, dx = x \ln x - x + C$

As you probably guessed, the seven rules apply directly to only a limited number of integration problems. Fortunately, however, the use of the rules can be extended by a rather straightforward technique called integration by substitution. The technique is a procedure whereby certain functions can be integrated by substituting a new variable for a function and integrating with respect to the substituted variable.

The technique can be summarized by the following six steps.

1.  Let $u$ equal the quantity *inside* the parentheses and/or the most complicated algebraic expression.

2.  Take the derivative $\dfrac{du}{dx}$.

3.  "Pretend" to solve for $dx$ by

    a.  first cross multiplying and

    b.  then dividing by the coefficient of $dx$.

4.  Substitute the $u$ and $du$ terms into the original problem and simplify.

5.  If possible, integrate with respect to $u$.

6.  Change back to the original variable.

## EXAMPLE 1

Find: $\displaystyle\int (3x - 5)^5 \, dx.$

Solution:   If we had unlimited time and nothing better to do, we could multiply the quantity out and integrate term by term using the rules from Unit 23. But lacking the time and inclination to do that, integration by substitution provides an easier and certainly faster method.

Using the six steps:

1.  Let $u$ equal the quantity *inside* the parentheses.          Let $u = 3x - 5$

2.  Take the derivative $\dfrac{du}{dx}$.                                              $\dfrac{du}{dx} = 3$

3.  "Pretend" to solve for $dx$ by

    a.  first cross multiply,                                                               $du = 3 \, dx$

    b.  then divide by coefficient of $dx$.                                     $\dfrac{du}{3} = dx$

4.  Substitute $u$ and $du$ terms into the original problem and simplify.

    Substitute:                                                            $\displaystyle\int (3x - 5)^5 \, dx = \int u^5 \, \frac{du}{3}$

    Simplify:                                                                          $= \dfrac{1}{3} \displaystyle\int u^5 \, du$

5.  Integrate with respect to $u$.                                          $= \dfrac{1}{3} \cdot \dfrac{u^6}{6} + C$

                                                                                                      $= \dfrac{u^6}{18} + C$

6.  Change back to the original variable.                               $= \dfrac{(3x - 5)^6}{18} + C$

## EXAMPLE 2

Find: $\displaystyle\int (2 - 7x)^{\frac{2}{3}}\, dx.$

---

Solution:

1. Let $u$ equal the quantity *inside* the parentheses. $\qquad$ Let $u = 2 - 7x$

2. Take the derivative $\dfrac{du}{dx}$. $\qquad\qquad\qquad\qquad\qquad \dfrac{du}{dx} = -7$

3. "Pretend" to solve for $dx$ by

   a. first cross multiply, $\qquad\qquad\qquad\qquad\qquad\qquad du = -7\, dx$

   b. then divide by coefficient of $dx$. $\qquad\qquad\qquad \dfrac{du}{-7} = dx$

4. Substitute $u$ and $du$ terms into the original problem and simplify.

   Substitute: $\qquad\qquad\qquad\qquad \displaystyle\int (2 - 7x)^{\frac{2}{3}}\, dx = \int u^{\frac{2}{3}} \dfrac{du}{-7}$

   Simplify: $\qquad\qquad\qquad\qquad\qquad\qquad = \dfrac{-1}{7} \displaystyle\int u^{\frac{2}{3}}\, du$

5. Integrate with respect to $u$. $\qquad\qquad\qquad\quad = -\dfrac{1}{7} \cdot \dfrac{u^{\frac{5}{3}}}{\dfrac{5}{3}} + C$

   $\qquad\qquad\qquad\qquad\qquad\qquad\qquad\qquad\qquad = \dfrac{-3}{35} u^{\frac{5}{3}} + C$

6. Change back to the original variable. $\qquad\qquad = \dfrac{-3}{35} (2 - 7x)^{\frac{5}{3}} + C$

---

By now I am sure you are ready to try a problem on your own.

## Problem 1

Find: $\displaystyle\int (2x + 3)^{4}\, dx.$

---

Solution:

1. Let $u$ equal the quantity *inside* the parentheses.

2. Take the derivative $\dfrac{du}{dx}$.

3. "Pretend" to solve for $dx$ by

   a. first cross multiply,

   b. then divide by coefficient of $dx$.

4.   Substitute $u$ and $du$ terms into the original problem and simplify.

Substitute:

Simplify:

5.   Integrate with respect to $u$.

6.   Change back to the original variable.

Answer:   $\dfrac{(2x+3)^5}{10} + C$

Although the next example appears more difficult, you will soon see it really isn't.

# EXAMPLE 3

Find:   $\displaystyle\int (x^2 - x + 7)^3 (2x - 1)\, dx.$

Solution:   Recall that the first step of this technique states to let $u$ equal the quantity inside the parentheses and/or the most complicated algebraic expression. Since we have two quantities in parentheses, let $u$ equal the more complicated of the two, $x^2 - x + 7$.

1.   Let $u$ equal the quantity.                                     Let $u = x^2 - x + 7$

2.   Take the derivative $\dfrac{du}{dx}$.                          $\dfrac{du}{dx} = 2x - 1$

3.   "Pretend" to solve for $dx$ by

a.   first cross multiply,                                      $du = (2x - 1)\, dx$

b.   then divide by coefficient of $dx$.                 $\dfrac{du}{(2x - 1)} = dx$

4.   Substitute $u$ and $du$ terms into the original problem and simplify.

Substitute:   $\displaystyle\int (x^2 - x + 7)^3 (2x - 1)\, dx = \int u^3 (2x - 1)\, \dfrac{du}{(2x - 1)}$

Simplify:   $= \displaystyle\int u^3 \cancel{(2x - 1)}\, \dfrac{du}{\cancel{(2x - 1)}}$

Note:   the common factors of $(2x - 1)$ cancel out.        $= \displaystyle\int u^3\, du$

5.   Integrate with respect to $u$.                                 $= \dfrac{u^4}{4} + C$

6.  Change back to the original variable.                              $= \dfrac{(x^2 - x + 7)^4}{4} + C$

Integration by substitution works for a variety of problems. We will go through a few more examples and then stop while you try some.

From now on I will put the first three steps off to the left to provide a more condensed format for working the problems.

## EXAMPLE 4

Find:  $\displaystyle\int \dfrac{8x}{(4x^2 + 1)^2}\, dx.$

Solution:                          $\displaystyle\int \dfrac{8x}{(4x^2 + 1)^2}\, dx = \int \dfrac{8x}{u^2}\dfrac{du}{8x}$

Let $u = 4x^2 + 1$                        $= \displaystyle\int \dfrac{8x}{u^2}\dfrac{du}{8x}$

$\dfrac{du}{dx} = 8x$                              $= \displaystyle\int \dfrac{du}{u^2}$

$du = 8x\, dx$                              $= \displaystyle\int u^{-2}\, du$

$\dfrac{du}{8x} = dx$                              $= \dfrac{u^{-2+1}}{-2+1} + C$

$= \dfrac{u^{-1}}{-1} + C$

$= \dfrac{-1}{u} + C$

$= \dfrac{-1}{4x^2 + 1} + C$

The next example contains a quantity under a radical sign. As when we were taking derivatives, we will need to first rewrite the expression using a fractional exponent, then proceed as usual with integration by substitution.

## EXAMPLE 5

Find:  $\displaystyle\int x\sqrt{x^2 + 5}\, dx.$

Solution:                          $\displaystyle\int x\sqrt{x^2 + 5}\, dx = \int x(x^2 + 5)^{\frac{1}{2}}\, dx$

$$= \int x\, u^{\frac{1}{2}}\, \frac{du}{2x}$$

$$\text{Let } u = x^2 + 5 \qquad\qquad = \int \cancel{x}\, u^{\frac{1}{2}}\, \frac{du}{2\cancel{x}}$$

$$\frac{du}{dx} = 2x \qquad\qquad = \frac{1}{2} \int u^{\frac{1}{2}}\, du$$

$$du = 2x\, dx \qquad\qquad = \frac{1}{2} \cdot \frac{u^{\frac{3}{2}}}{\frac{3}{2}} + C$$

$$\frac{du}{2x} = dx \qquad\qquad = \frac{1}{3}\, u^{\frac{3}{2}} + C$$

$$= \frac{(x^2 + 5)^{\frac{3}{2}}}{3} + C$$

Now it is time for you to try a few. We will start with some easy ones and work up to more difficult problems.

## Problem 2

Find: $\displaystyle\int (10x - 3)(5x^2 - 3x - 17)^{\frac{1}{7}}\, dx.$

Solution:

Answer: $\displaystyle\frac{7}{8}(5x^2 - 3x - 17)^{\frac{8}{7}} + C$

## Problem 3

Find: $\displaystyle\int x(3x^2 + 2)^3\, dx.$

Solution:

Answer: $\displaystyle\frac{(3x^2 + 2)^4}{24} + C$

## Problem 4

Find: $\displaystyle\int \frac{dx}{(4x+2)^2}$.

Solution:

Answer: $\displaystyle\frac{-1}{4(4x+2)} + C$

Let's try two more; then I will explain some additional examples of the technique. The next problem looks harder than it really is. I am confident you will be able to solve it without any difficulty.

## Problem 5

Find: $\displaystyle\int [(x-1)^5 + 3(x-1)^2 + 6(x-1) + 5]\,dx$.

Solution:

Hint: Let $u = x - 1$

Answer: $\displaystyle\frac{1}{6}(x-1)^6 + (x-1)^3 + 3(x-1)^2 + 5(x-1) + C$

## Problem 6

Find: $\displaystyle\int x\sqrt[3]{3x^2 + 5}\,dx$.

Solution:

Answer: $\displaystyle\frac{1}{8}(3x^2+5)^{\frac{4}{3}} + C$

The next example involves the integration of a logarithm and, while the technique remains the same, simplifying the final answer has an additional step.

## EXAMPLE 6

Find: $\displaystyle\int (6x - 1) \ln (3x^2 - x + 7)\, dx.$

Solution: $\displaystyle\int (6x - 1) \ln (3x^2 - x + 7)\, dx = \int \cancel{(6x - 1)} \ln u\, \frac{du}{\cancel{(6x - 1)}}$

Let $u = 3x^2 - x + 7$ $\qquad = \displaystyle\int \ln u\, du$

$\dfrac{du}{dx} = 6x - 1$ $\qquad\qquad = u \ln u - u + C$

$du = (6x - 1)\, dx$ $\qquad = (3x^2 - x + 7) \ln (3x^2 - x + 7) - (3x^2 - x + 7) + C$

$\dfrac{du}{(6x + 1)} = dx$ $\qquad = (3x^2 - x + 7) \ln (3x^2 - x + 7) - 3x^2 + x - 7 + C$

$\qquad\qquad\qquad\qquad = (3x^2 - x + 7) \ln (3x^2 - x + 7) - 3x^2 + x + C$

Be sure to notice what happened in the final steps of this problem. The parentheses were removed and the $-7 + C$ were combined into a new constant of integration, but still denoted simply by $C$.

Integration by substitution can be used with exponential functions as well, the only difference being that it is usually desirable to let $u$ equal the exponent of $e$ rather than the quantity in parentheses. As you will note, the exponent is the more complicated expression.

## EXAMPLE 7

Find: $\displaystyle\int (2x + 3)\, e^{x^2 + 3x + 5}\, dx.$

Solution: $\displaystyle\int (2x + 3)\, e^{x^2 + 3x + 5}\, dx = \int \cancel{(2x + 3)} e^u\, \frac{du}{\cancel{(2x + 3)}}$

Let $u = x^2 + 3x + 5$ $\qquad = \displaystyle\int e^u\, du$

$\dfrac{du}{dx} = 2x + 3$ $\qquad\qquad = e^u + C$

$du = (2x + 3)\, dx$ $\qquad = e^{x^2 + 3x + 5} + C$

$\dfrac{du}{(2x + 3)} = dx$

## Problem 7

Find: $\displaystyle\int 2x\, e^{x^2 - 1}\, dx.$

Solution:

Answer:  $e^{x^2-1} + C$

---

## Problem 8

Find:  $\int x\, e^{x^2}\, dx.$

---

Solution:

Answer:  $\dfrac{1}{2} e^{x^2} + C$

---

Now we will consider one last situation, that of the integration of a rational function, which is the quotient of two polynomials. With a quotient there are no parentheses appearing, but if we consider the numerator and denominator as separate functions, the procedure remains the same. In other words, let $u$ be the more complicated expression, either the numerator or the denominator, and proceed as usual.

## EXAMPLE 8

Find:  $\int \dfrac{2}{x+1}\, dx.$

---

Solution:
$$\int \frac{2}{x+1}\, dx = \int \frac{2}{u}\, du$$

Let $u = x + 1$
$$= 2 \int \frac{du}{u}$$

$$\frac{du}{dx} = 1$$
$$= 2 \ln |u| + C$$

$$du = dx$$
$$= 2 \ln |x + 1| + C$$

---

## Problem 9

Find:  $\int \dfrac{\ln 3x}{x}\, dx.$

Solution:

Answer: $\dfrac{1}{2}(\ln 3x)^2 + C$

By now you should be wondering what happens if the factors involving the original variable do *not* cancel out so conveniently after the substitution step. The next and final example should answer the question.

# EXAMPLE 9

Find: $\displaystyle\int (x^3 - 1)^{\frac{1}{2}}\, dx.$

Solution: $\displaystyle\int (x^3 - 1)^{\frac{1}{2}}\, dx = \int u^{\frac{1}{2}} \dfrac{du}{3x^2}$

Let $u = x^3 - 1$

$\dfrac{du}{dx} = 3x^2$

$du = 3x^2\, dx$

$\dfrac{du}{3x^2} = dx$

Observe that after substituting we are left with a factor of $x^2$ in the denominator and no matching factor in the numerator with which to cancel it. And $\dfrac{1}{3x^2}$ cannot be taken outside the integral because $x^2$ is *not* a constant. Thus, we are left with an integration problem containing both variables. To finish the problem requires methods beyond the scope of this unit. A method of approach will be discussed in Unit 25.

You should be able by now to integrate a variety of problems—many which contain radicals, exponentials, logarithms, or quotients—using a technique called integration by substitution.

Remember the technique can be summarized by the following six steps.

1.  Let $u$ equal the quantity *inside* the parentheses and/or the most complicated algebraic expression.
2.  Take the derivative.
3.  "Pretend" to solve for $dx$.
4.  Substitute and simplify.
5.  Integrate with respect to $u$.
6.  Change back to the original variable.

Before beginning the next unit you should integrate the following problems. When you have completed them, check your answers against those at the back of the book.

## EXERCISES

Find:

1. $\displaystyle\int e^{3-5x}\, dx$

2. $\displaystyle\int \frac{x}{(5+x^2)^4}\, dx$

3. $\displaystyle\int \frac{2x^3}{x^4+1}\, dx$

4. $\displaystyle\int (2x-2)\, e^{x^2-2x+1}\, dx$

5. $\displaystyle\int 3x^2 \ln(x^3-2)\, dx$

6. $\displaystyle\int 36x^2\sqrt{6x^3+1}\, dx$

7. $\displaystyle\int 2x(5x^3-2x)^7\, dx$

8. $\displaystyle\int \frac{4x^2}{\sqrt{x^3-7}}\, dx$

If additional practice is needed:

Barnett and Ziegler, pages 322–325, problems 1–61
Budnick, pages 665–666, problems 1–24
Hoffmann, pages 293–294, problems 1–39
Piascik, pages 324–325, problems 1–4

# UNIT 25

## Integration Using Formulas

This unit continues our discussion of integration. When you have finished the unit you will be able to integrate a variety of functions using integration formulas.

At the conclusion of the last unit, the final example illustrated a problem that we were unable to integrate with the methods and rules available to us at that point. If your future plans include the need to do extensive integration, I would suggest you obtain a good set of integration formulas. At last count, there were over seven hundred rules for integration. These can be found in most any mathematical handbook. Do understand, even with a complete table of integration formulas, not all functions can be integrated.

A condensed version of the more frequently used formulas is provided below for your convenience. The first seven rules from the previous units appear at the beginning. In the list, $x$ is the variable and all other letters are constants. The constant of integration has been omitted. The numbers at the left are for referencing purposes in the examples to follow.

## SELECTED INTEGRATION FORMULAS

1. $$\int x^n \, dx = \frac{x^{n+1}}{n+1} \quad \text{provided } n \neq -1$$

2. $$\int dx = x$$

3. $$\int [f(x) + g(x)] \, dx = \int f(x) \, dx + \int g(x) \, dx$$

4. $$\int c \, f(x) \, dx = c \int f(x) \, dx$$

5. $\int e^x \, dx = e^x$

6. $\int \dfrac{dx}{x} = \ln |x|$

7. $\int \ln x \, dx = x \ln x - x$

8. $\int \dfrac{dx}{x \ln x} = \ln |\ln x|$

9. $\int x \, e^x \, dx = (x - 1) \, e^x$

10. $\int \dfrac{dx}{a^2 - x^2} = \dfrac{1}{2a} \ln \left| \dfrac{a + x}{a - x} \right|$

11. $\int \dfrac{dx}{x^2 - a^2} = \dfrac{1}{2a} \ln \left| \dfrac{x - a}{x + a} \right|$

12. $\int \dfrac{dx}{\sqrt{x^2 - a^2}} = \ln |x + \sqrt{x^2 - a^2}|$

13. $\int \dfrac{dx}{\sqrt{x^2 + a^2}} = \ln (x + \sqrt{x^2 + a^2})$

14. $\int \dfrac{dx}{x^2 \sqrt{x^2 \pm a^2}} = \mp \dfrac{\sqrt{x^2 \pm a^2}}{a^2 x}$

Note: If the plus sign appears under the radical to be integrated, which is the top sign, the answer uses the top sign, which is the minus in front of the radical and the plus sign under the radical.

15. $\int \dfrac{dx}{(x^2 \pm a^2)^{\frac{3}{2}}} = \dfrac{\pm x}{a^2 \sqrt{x^2 \pm a^2}}$

16. $\int x^n \ln x \, dx = x^{n+1} \left( \dfrac{\ln x}{n + 1} - \dfrac{1}{(n + 1)^2} \right)$ for $n \neq 1$

17. $\int \dfrac{x^n \, dx}{\sqrt{a + bx}} = \dfrac{2x^n \sqrt{a + bx}}{b(2n + 1)} - \dfrac{2an}{b(2n + 1)} \int \dfrac{x^{n-1} \, dx}{\sqrt{a + bx}}$

18. $\int \dfrac{dx}{x^n \sqrt{a + bx}} = -\dfrac{\sqrt{a + bx}}{a(n - 1)x^{n-1}} - \dfrac{b(2n - 3)}{2a(n - 1)} \int \dfrac{dx}{x^{n-1} \sqrt{a + bx}}$

19. $\int \sqrt{(x^2 \pm a^2)^n} \, dx = \dfrac{x\sqrt{(x^2 \pm a^2)^n}}{n + 1} \pm \dfrac{na^2}{n + 1} \int \sqrt{(x^2 \pm a^2)^{n-2}} \, dx$

20.  $\displaystyle\int x^n\, e^x\, dx = x^n e^x - n \int x^{n-1} e^x\, dx$

21.  $\displaystyle\int \frac{e^x\, dx}{x^n} = -\frac{e^x}{(n-1)x^{n-1}} + \frac{1}{n-1} \int \frac{e^x\, dx}{x^{n-1}},\ n \neq 1$

22.  $\displaystyle\int x^n (\ln x)^m\, dx = \frac{x^{n+1}}{n+1}\,(\ln x)^m - \frac{m}{n+1} \int x^n\,(\ln x)^{m-1}\, dx$

23.  $\displaystyle\int e^{ax}\, dx = \frac{e^{ax}}{a}$

24.  $\displaystyle\int x^n e^{ax}\, dx = \frac{x^n e^{ax}}{a} - \frac{n}{a} \int x^{n-1} e^{ax}\, dx$

25.  $\displaystyle\int (ax + b)^n\, dx = \frac{(ax + b)^{n+1}}{(n+1)a}$

26.  $\displaystyle\int x(ax + b)^{\frac{m}{2}}\, dx = \frac{2(ax + b)^{\frac{m+4}{2}}}{a^2(m+4)} - \frac{2b(ax + b)^{\frac{m+2}{2}}}{a^2(m+2)}$

27.  $\displaystyle\int \frac{dx}{x(ax + b)} = \frac{1}{b}\,\ln\left|\frac{x}{ax + b}\right|$

As I stated earlier, if you are required to do a great deal of integration, extensive lists of formulas are available in mathematical handbooks. The above list of twenty-seven are the ones I believe that you are most apt to encounter. In most sets of tables, the formulas are grouped in similar types. That was not quite possible with this list due to its limited size.

## USING INTEGRATION FORMULAS

To integrate using formulas involves finding the rule which matches the given function to be integrated, determining the values of the various constants, and finally substituting those values into the right-hand side of the formula. Often it is a one-step procedure as some of the following examples will illustrate.

## EXAMPLE 1

Find:  $\displaystyle\int \frac{dx}{x(2x - 3)}$.

Solution: The integral fits the pattern of Rule 27.

$$\int \frac{dx}{x(ax+b)} = \frac{1}{b}\ln\left|\frac{x}{ax+b}\right|$$

$$\int \frac{dx}{x(2x+3)}$$

With $a = 2$

and $b = 3$,

substitute into the right-hand side of the formula.

Answer: $\frac{1}{3}\ln\left|\frac{x}{2x+3}\right| + C$

## EXAMPLE 2

Find: $\int \frac{dx}{\sqrt{x^2+25}}$.

Solution: The integral fits the pattern of Rule 13.

$$\int \frac{dx}{\sqrt{x^2+a^2}} = \ln\left(x + \sqrt{x^2+a^2}\right)$$

$$\int \frac{dx}{\sqrt{x^2+25}}$$

With $a^2 = 25$,

substitute into the right-hand side of the formula.

Answer: $= \ln\left(x + \sqrt{x^2+25}\right)$

## EXAMPLE 3

Find: $\int \frac{dx}{x^2\sqrt{x^2-9}}$.

Solution: The integral fits the pattern of Rule 14 using the bottom signs.

$$\int \frac{dx}{x^2\sqrt{x^2\pm a^2}} = \mp\frac{\sqrt{x^2\pm a^2}}{a^2x}$$

$$\int \frac{dx}{x^2\sqrt{x^2-9}}$$

With $a^2 = 9$,

substitute into the right-hand side of the formula.

Answer: $= \frac{\sqrt{x^2-9}}{9x} + C$

You should find these problems relatively short. Let me help you with one and then do the second one completely on your own.

## Problem 1

Find: $\displaystyle\int \frac{dx}{4-x^2}.$

Solution:   The integral fits the pattern of Rule 10.

$$\int \frac{dx}{a^2-x^2} = \frac{1}{2a}\ln\left|\frac{a+x}{a-x}\right|$$

$$\int \frac{dx}{4-x^2}$$

With $a = 2$, substitute.

Answer:   $\displaystyle\frac{1}{4}\ln\left|\frac{2+x}{2-x}\right| + C$

This time you're on your own.

## Problem 2

Find: $\displaystyle\int \frac{dx}{x(5x+7)}.$

Solution:

Answer:   $\displaystyle\frac{1}{7}\ln\left|\frac{x}{5x+7}\right| + C$

As you probably noticed, some of the formulas contain integral signs on the right-hand side of the formula as well. When using one of those formulas, you must continue to integrate the right-hand side until all the integral signs are removed.

# EXAMPLE 4

Find: $\int xe^x\,dx$.

---

**Solution:** The integral fits the pattern of Rule 20.

$$\int x^n e^x\,dx \qquad = x^n e^x - n \int x^{n-1} e^x\,dx$$

$$\int xe^x\,dx$$

With $n = 1$,
substitute. $\qquad = xe^x - \int x^0 e^x\,dx$

$$= xe^x - \int e^x\,dx$$

Use Rule 5. $\quad = xe^x - e^x + C$

---

We will do another one requiring additional integration steps.

# EXAMPLE 5

Find: $\int (\ln x)^2\,dx$.

---

**Solution:** Although it might not be obvious at first glance, the integral fits the pattern of Rule 22 because the function can be rewritten using $x^0$, which equals 1.

$$\int x^n(\ln x)^m\,dx \qquad = \frac{x^{n+1}(\ln x)^m}{n+1} - \frac{m}{n+1}\int x^n(\ln x)^{m-1}\,dx$$

$$\int x^0(\ln x)^2\,dx$$

With $n = 0$
and $m = 2$,
substitute. $\qquad = \dfrac{x(\ln x)^2}{1} - \dfrac{2}{1}\int x^0(\ln x)^{2-1}\,dx$

$$= x\,(\ln x)^2 - 2\int (\ln x)\,dx$$

Use Rule 7. $\qquad = x(\ln x)^2 - 2(x\ln x - x) + C$

$$= x(\ln x)^2 - 2x\ln x + 2x + C$$

---

A word of warning, now that you have choices regarding which formula to use; the required number of steps and the final form of your answer can vary greatly. Consider the next problem.

## Problem 3

Find: $\displaystyle\int \frac{x\,dx}{\sqrt{1+x}}$ using the specified rules.

Solution:   Step 1:   Use Rule 17 with $n = 1$, $a = 1$, and $b = 1$.

Step 2:   Use Rule 25 with $n = -\dfrac{1}{2}$, $a = 1$, and $b = 1$.

Answer:   $\dfrac{2x\sqrt{1+x} - 4\sqrt{1+x}}{3} + C$

For a change of pace I had you do the hard problem; I will take the easy one.

## EXAMPLE 6

Find:   $\displaystyle\int \frac{x\,dx}{\sqrt{1+x}}$

Solution:   By way of showing you your choices, I am going to solve the same problem, but using Rule 26 after rewriting the function.

$$\int x(ax + b)^{\frac{m}{2}}\,dx = \frac{2(ax + b)^{\frac{m+4}{2}}}{a^2(m + 4)} - \frac{2b(ax + b)^{\frac{m+2}{2}}}{a^2(m + 2)}$$

$$\int x(x + 1)^{\frac{-1}{2}}\,dx$$

With $a = 1$,

and $b = 1$,

and $m = -1$,

substitute.   $= \dfrac{2(x + 1)^{\frac{-1+4}{2}}}{(-1 + 4)} - \dfrac{2(x + 1)^{\frac{-1+2}{2}}}{(-1 + 2)} + C$

$= \dfrac{2(x + 1)^{\frac{3}{2}}}{3} - 2(x + 1)^{\frac{1}{2}} + C$

Observe two things. One, notice how much shorter the second method was than the way you worked the problem. Second, notice the apparent different answers. Actually the two answers are equivalent, but their forms are different. The same situation might very well happen with the exercises at the end of the unit. Remember you can always check by differentiating.

You should now be able to integrate a variety of functions using a table of integrals. A condensed version of some of the more commonly used formulas was provided in the unit. Remember that even with extensive formulas available, not all functions can be integrated.

Before beginning the next unit, use the formulas provided in this unit to integrate the following functions. Check you answers with those in the back of the book. If you used a different formula than I did, verify that your answer is correct by differentiating.

## EXERCISES

Find each of the following:

1. $\displaystyle\int xe^x\, dx$

2. $\displaystyle\int \frac{dx}{x^2 - 1}$

3. $\displaystyle\int (2x - 3)^3\, dx$

4. $\displaystyle\int \frac{dx}{x \ln x}$

5. $\displaystyle\int \frac{dx}{\sqrt{x^2 - 100}}$

6. $\displaystyle\int x^5 \ln x\, dx$

7. $\displaystyle\int x^2 e^x\, dx$

8. $\displaystyle\int x(x - 6)^{\frac{1}{2}}\, dx$

9. $\displaystyle\int x^3 (\ln x)^2\, dx$

10. $\displaystyle\int x^2 e^{3x}\, dx$

If additional practice is needed:

Budnick, page 683, problems 23–36
Hoffmann, page 304, problems 1–20
Piascik, page 334, problems 1–2

# UNIT 26

## Integration—Definite Integrals

In this unit you will be introduced to the concept of a definite integral. Upon completion of this unit you will be able to evaluate a definite integral and determine the area under a curve or between two curves.

---

Definition:   The **definite integral** of a continuous function $f(x)$ from $x = a$ to $x = b$ is defined as:

$$\int_a^b f(x)\, dx = F(x) \Big|_a^b = F(b) - F(a)$$

where $F'(x) = f(x)$.

---

Stated another way, the definite integral is the difference of the antiderivative evaluated at the lower limit subtracted from the antiderivative evaluated at the upper limit.

$\int_a^b f(x)\, dx$ is read "the definite integral of $f(x)$ from $x = a$ to $x = b$." The definite integral results in a numerical value with $a$ and $b$ called the limits of integration.

## PROCEDURE FOR DETERMINING THE DEFINITE INTEGRAL

To calculate a definite integral, which as stated above results in a numerical value, requires a straight-forward four-step procedure.

1. Integrate the function.

2. Evaluate the antiderivative at the upper limit.

3. Evaluate the antiderivative at the lower limit.

4. Subtract the value found in Step 3 from the value found in Step 2.

The next few examples will illustrate the procedure.

## EXAMPLE 1

Find: $\displaystyle\int_1^2 2x \, dx$.

Solution:
$$\int_1^2 f(x) \, dx = F(x) \Big|_1^2 = F(2) - F(1)$$

$$\int_1^2 2x \, dx = \frac{\cancel{2}x^2}{\cancel{2}} + C \Big|_1^2 = [(2)^2 + C] - [(1)^2 + C]$$

$$= [4 + C] - [1 + C]$$

$$= 4 + C - 1 - C$$

$$= 3$$

Observe that the constant of integration cancels out, which will always happen. Therefore it is acceptable to omit the C entirely from the calculations when determining a definite integral.

## EXAMPLE 2

Find: $\displaystyle\int_{-1}^2 x^4 \, dx$.

Solution:
$$\int_{-1}^2 f(x) \, dx = F(x) \Big|_{-1}^2 = F(2) - F(-1)$$

$$\int_{-1}^2 x^4 \, dx = \frac{x^5}{5} \Big|_{-1}^2 = \frac{[(2)^5]}{5} - \frac{[(-1)^5]}{5}$$

$$= \frac{32}{5} - \frac{(-1)}{5}$$

$$= \frac{32 + 1}{5}$$

$$= \frac{33}{5}$$

The operation we are performing is referred to as integrating between limits.

## EXAMPLE 3

Find: $\displaystyle\int_0^1 (1 - x^2) \, dx$.

Solution:     $$\int_0^1 f(x)\,dx = F(x)\Big|_0^1 = F(1) - F(0)$$

$$\int_0^1 (1 - x^2)\,dx = x - \frac{x^3}{3}\Big|_0^1 = \left(1 - \frac{1}{3}\right) - [0 - 0]$$

$$= \frac{2}{3}$$

# EXAMPLE 4

Find:    $\displaystyle\int_1^5 (x + 2)\,dx.$

Solution:     $$\int_1^5 f(x)\,dx = F(x)\Big|_1^5 = F(5) - F(1)$$

$$\int_1^5 (x + 2)\,dx = \frac{x^2}{2} + 2x \Big|_1^5 = \left(\frac{25}{2} + 10\right) - \left(\frac{1}{2} + 2\right)$$

$$= \frac{25}{2} + 10 - \frac{1}{2} - 2$$

$$= 12 + 8$$

$$= 20$$

When there are fractions present as in Example 4, I tend to remove the parentheses before collecting terms. Typically it makes the problem easier by allowing you to combine the fractions with the same denominator first. If you are working with decimals, it doesn't make any difference.

Are you ready to try a few? Work *all* of the following problems. Don't skip any because I intend to use the problems as illustrations of basic properties of a definite integral after you have finished.

# Problem 1

Find:    $\displaystyle\int_1^2 x\,dx.$

Solution:

Answer:    1.5

## Problem 2

Find: $\displaystyle\int_{-1}^{1} x^3\, dx$.

Solution:

Answer: 0

## Problem 3

Find: $\displaystyle\int_{1}^{3} (x^3 + x)\, dx$.

Solution:

Answer: 24

## Problem 4

Find: $\displaystyle\int_{3}^{1} (x^3 + x)\, dx$.

Solution:

Answer: −24

## Problem 5

Find: $\displaystyle\int_{1}^{1} (-2x^3 + 5x^2 + 3x + 1)\, dx$.

Solution:

I will do one more example before going on to something new. Why am I doing it rather than you? Because the limits are in terms of a constant, $a$, rather than being numerical values. The technique, however, is the same.

## EXAMPLE 5

Find: $\displaystyle\int_a^{2a} (a + x)\, dx.$

Solution:
$$\int_a^{2a} f(x)\, dx = F(x)\Big|_a^{2a} = F(2a) - F(a)$$

$$\int_a^{2a} (a + x)\, dx = ax + \frac{x^2}{2}\Big|_a^{2a} = \left[a(2a) + \frac{(2a)^2}{2}\right] - \left[a^2 + \frac{a^2}{2}\right]$$

$$= 2a^2 + \frac{4a^2}{2} - a^2 - \frac{a^2}{2}$$

$$= a^2 + \frac{3a^2}{2}$$

$$= \frac{5a^2}{2}$$

# PROPERTIES OF A DEFINITE INTEGRAL

A definite integral has the following basic properties.

1.  $\displaystyle\int_a^b f(x)\, dx = -\int_b^a f(x)\, dx$

2.  $\displaystyle\int_b^a f(x)\, dx = 0$

3.  $\displaystyle\int_a^b f(x)\, dx = \int_a^c f(x)\, dx + \int_c^b f(x)\, dx$   where $a \le c \le b$

The basic properties can, if recognized and used correctly, save time during the calculations for a definite integral.

The first property states that if the limits are interchanged the absolute value of the definite integral remains the same, but the sign changes. If you will refer back to Problem 3, you found that the definite integral of $(x^3 + x)$ integrated from $x = 1$ to $x = 3$ was 24. Whereas in Problem 4 the same function integrated from $x = 3$ to $x = 1$ was $-24$, or the same absolute value, but with opposite signs.

The second property should be obvious. If the limits of integration are equal, the difference between the antiderivative evaluated at the upper limit and lower limit is zero. Problem 5 was an illustration of this property.

The third property indicates that the value of a definite integral integrated from its lower limit to its upper limit is the same as if an intermediate value is used between the two limits, with each definite integral calculated separately and then their values combined.

The next problem will be used to verify the third property.

## Problem 6

Verify: $\displaystyle\int_{-1}^{5} 2x\,dx = \int_{-1}^{1} 2x\,dx + \int_{1}^{5} 2x\,dx.$

Solution:

Answer:  $24 = 0 + 24$

# APPLICATIONS

If this were a book for mathematics majors or pre-engineering students, which it is not, we would just be starting into the real strength of integral calculus—that of application problems. However, for the majority of readers of this book, application problems requiring integration are somewhat limited. Two examples should be sufficient.

## EXAMPLE 6

Professor Haymaker is the director of the art gallery at a private college. Tomorrow morning is the scheduled opening of the student show followed by a luncheon for the artists. He is trying to decide whether or not an attendant need be present while he is attending the luncheon. Professor Haymaker estimates that $t$ hours after the doors open at 9 a.m. visitors will be entering at the rate of $-4t^3 + 54t^2$ people per hour. How many visitors are estimated will visit the gallery during the luncheon between noon and 2 p.m.?

Solution:  The problem states that visitors will be entering at the **rate** of $-4t^3 + 54t^2$, which indicates that the function is a derivative, say $F'(t)$.

We need to determine the antiderivative, $F(t)$, which would represent the total number of visitors entering since 9 a.m. and to estimate the number that will visit between noon and 2 p.m. would be given by evaluating $F(5) - F(3)$. ($x = 3$ corresponds to noon, that is three hours from 9 a.m., and $x = 5$ corresponds to 2 p.m.) But $F(5) - F(3)$ is nothing more than the definite integral evaluated from $x = 3$ to $x = 5$.

What we need to find, then, is

$$\int_3^5 (-4t^3 + 54t^2)\, dt = \frac{-\cancel{4}t^4}{\cancel{4}} + \frac{\overset{18}{\cancel{54}}t^3}{\cancel{3}} \Bigg|_3^5$$

$$= -t^4 + 18t^3 \Bigg|_3^5$$

$$= [-(5)^4 + 18(5)^3] - [-(3)^4 + 18(3)^3]$$

$$= [-625 + 2{,}250] - [-81 + 486]$$

$$= -625 + 2{,}250 + 81 - 486$$

$$= 1{,}220$$

Answer:   It is estimated that 1,220 visitors will enter the gallery between noon and 2 p.m. on opening day.

# EXAMPLE 7

A manufacturing firm knows that its marginal cost is given by the function

$$MC = C'(x) = 12x + 8$$

where $x$ is the number of units manufactured and

$MC$ is the marginal cost per unit in dollars.

Determine the total change in cost if production rises from 20 to 25 units.

Solution:   The total change in cost is determined by evaluating $C(25) - C(20)$, but that is nothing more than the definite integral evaluated from $x = 20$ to $x = 25$.

$$\text{Total change in cost} = \int_{20}^{25} C'(x)\, dx$$

$$= \int_{20}^{25} (12x + 8)\, dx$$

$$= \frac{\overset{6}{\cancel{12}}\, x^2}{\cancel{2}} + 8x \Bigg|_{20}^{25}$$

$$= 6x^2 + 8x \Bigg|_{20}^{25}$$

$$= [6(25)^2 + 8(25)] - [6(20)^2 + 8(20)]$$

$$= (3{,}750 + 200) - (2{,}400 + 160)$$

$$= 1{,}390$$

Answer:   If production rises from 20 to 25 units, the total change in cost will be $1,390.

# AREA UNDER A CURVE

We can now state a theorem that relates the area under a curve to the definite integral.

The definite integral can be interpreted as the area under a continuous and non-negative curve $y = f(x)$, bounded by the $x$-axis on the bottom, and the vertical lines $x = a$ on the left and $x = b$ on the right.

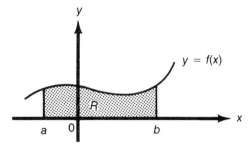

## Fundamental Theorem of Calculus

$$\text{The area of } R = \int_a^b f(x)\, dx$$

$$= F(x) \Big|_a^b$$

$$= F(b) - F(a)$$

provided $f(x)$ is continuous and non-negative over the interval $a \leq x \leq b$.

Observe that the fundamental theorem of calculus requires that $f(x)$ be non-negative over the interval. Therefore graphs will be required to verify that the function is, indeed, non-negative.

## EXAMPLE 8

Find the area between the $x$-axis and the curve $f(x) = 5x$ from $x = 0$ to $x = 4$.

Solution:    A sketch of $f(x)$ and the desired area appear below. The shaded area is computed by the definite integral.

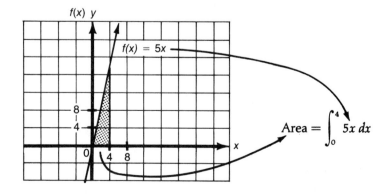

$$= \frac{5x^2}{2} \Big|_0^4$$

$$= \frac{[5(16)]}{2} - 0$$

$$= 40 \text{ square units}$$

Note that the answers to these problems will be in square units because we are determining areas.

---

What would have been the answer to Example 8 if you had used the formula from geometry for finding the area of a triangle?

## EXAMPLE 9

Find the area bounded by the curve $f(x) = 3x^2$, the $x$-axis, and the vertical line $x = 2$.

---

Solution:    A sketch of $f(x)$ and the indicated area is provided. Observe that $f(x)$ is non-negative over the desired interval $0 \leq x \leq 2$ as required. The shaded area can be determined by the definite integral.

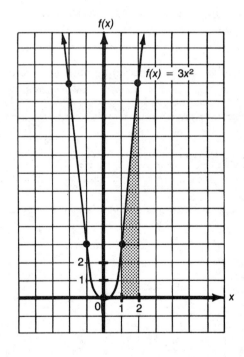

$$\text{Area} = \int_0^2 3x^2 \, dx$$

$$= \frac{\cancel{3}x^3}{\cancel{3}} \Big|_0^2$$

$$= x^3 \Big|_0^2$$

$$= (2)^3 - 0$$

$$= 8 \text{ square units}$$

---

Does the answer to Example 9 appear reasonable? As best you can, count the square units in the shaded area and compare your answers.

Use the graph provided in Example 9 and compute the following areas.

## Problem 7

Find the area between the $x$-axis and the curve $f(x) = 3x^2$ from $x = 1$ to $x = 3$.

Solution:

Answer:   26 square units

## Problem 8

Find the area bounded by the curve $f(x) = 3x^2$, the $x$-axis, $x = -1$, and $x = 1$.

Solution:

Answer:   2 square units

It is time for a review problem on graphing a cubic that we can then use for the next few examples.

## EXAMPLE 10

Sketch:   $y = f(x) = x^3 + x^2 - 6x$.

Solution:    1.    This is a cubic with $a = 1$ and $d = 0$.

2.    The curve opens up:

3.    The $y$-intercept is at 0.

4.    The $x$-intercepts are 0, $-3$, and 2 because

$$0 = x^3 + x^2 - 6x$$
$$= x(x^2 + x - 6)$$
$$= x(x + 3)(x - 2)$$
$$x = 0 \qquad x = -3 \qquad x = 2$$

Answer:

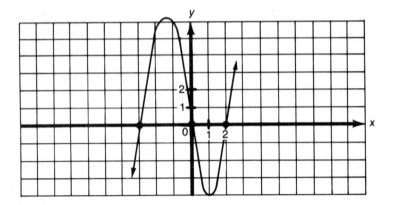

---

As stated earlier, the fundamental theorem requires that the function be non-negative over the interval from $x = a$ to $x = b$. If the function is negative over the interval, the definite integral results in a negative value that is the opposite of the area. In such a case, the area is found by taking the absolute value of the result.

The next example will illustrate what I mean.

## EXAMPLE 11

Find the area between the $x$-axis and the curve $y = f(x) = x^3 + x^2 - 6x$ from $x = 1$ to $x = 2$.

---

Solution:   A sketch of $f(x)$ and the indicated area is repeated below. Note that the curve is negative over the interval $1 \leq x \leq 2$, and thus the shaded area appears below the $x$-axis.

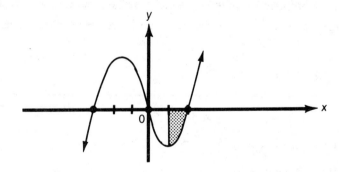

To determine the area, we begin by computing the definite integral:

$$\int_{1}^{2} (x^3 + x^2 - 6x)\, dx = \frac{x^4}{4} + \frac{x^3}{3} - \frac{6x^2}{2}\bigg|_{1}^{2}$$

$$= \left[ \frac{16}{4} + \frac{8}{3} - \frac{24}{2} \right] - \left[ \frac{1}{4} + \frac{1}{3} - \frac{6}{2} \right]$$

$$= 4 + \frac{8}{3} - 12 - \frac{1}{4} - \frac{1}{3} + 3$$

$$= -5 + \frac{7}{3} - \frac{1}{4}$$

$$= \frac{-60 + 28 - 3}{12}$$

$$= -2\frac{11}{12}$$

The definite integral results in a negative number because the area is located below the $x$-axis. The area of the shaded area is found by taking the absolute value of the definite integral.

$$\text{Answer:} \quad \text{Area} = \left| -2\frac{11}{12} \right| = 2\frac{11}{12} \text{ square units.}$$

You might conclude from the above example that anytime the definite integral results in a negative number, you merely take the absolute value of it for the area and then never have to bother with the graph. Unfortunately that won't work all the time as the next example will show.

The next example illustrates the situation where some of the desired area is above the $x$-axis and some is below the $x$-axis.

## EXAMPLE 12

Find the area between the $x$-axis and the curve $f(x) = x^3 + x^2 - 6x$ from $x = -3$ to $x = 3$.

Solution:   A sketch of $f(x)$ and the indicated area shows that some of the area appears above the $x$-axis and some of the area appears below the $x$-axis. In such a situation, each part must be determined separately and then added for the total area.

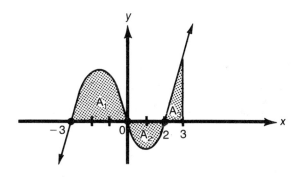

$$\text{Area of } A_1 = \int_{-3}^{0} (x^3 + x^2 - 6x)\, dx$$

$$= \frac{x^4}{4} + \frac{x^3}{3} - 3x^2 \Big|_{-3}^{0}$$

$$= (0 + 0 + 0) - \left[ \frac{81}{4} - \frac{27}{3} - 27 \right]$$

$$= 0 - 20.25 + 9 + 27$$

$$= 15.75 \text{ square units}$$

$$\text{Area of } A_2 = \left| \int_{0}^{2} (x^3 + x^2 - 6x)\, dx \right|$$

$$= \left| \frac{x^4}{4} + \frac{x^3}{3} - 3x^2 \Big|_{0}^{2} \right|$$

$$= \left| \left[ \frac{16}{4} + \frac{8}{3} - 12 \right] - (0 + 0 - 0) \right|$$

$$= |4 + 2.67 - 12|$$

$$= |-5.33|$$

$$= 5.33 \text{ square units}$$

$$\text{Area of } A_3 = \int_{2}^{3} (x^3 + x^2 - 6x)\, dx$$

$$= \frac{x^4}{4} + \frac{x^3}{3} - 3x^2 \Big|_{2}^{3}$$

$$= \left[ \frac{81}{4} + \frac{27}{3} - 27 \right] - \left[ \frac{16}{4} + \frac{8}{3} - 12 \right]$$

$$= 20.25 + 9 - 27 - 4 - 2.67 + 12$$

$$= 7.58 \text{ square units}$$

Thus, the total area is $A = A_1 + A_2 + A_3$

$$= 15.75 + 5.33 + 7.58$$

$$= 28.66 \text{ square units}.$$

---

In effect, when we have areas that appear both above and below the $x$-axis, the total area can be found by adding the areas above the $x$-axis and subtracting the areas that appear below the $x$-axis rather than using absolute values.

Your turn to do an easy one.

## Problem 9

Find: $\displaystyle\int_{-3}^{3} (x^3 + x^2 - 6x)\, dx.$

Solution:

Answer:   **18**

Why does your answer to Problem 9 differ from the answer to Example 12? What does your answer represent? Do you now see the necessity for doing the graph whenever an area is being computed?

# AREA BETWEEN TWO CURVES

Suppose we need to find area between two curves, as shown in the picture. What could we do?

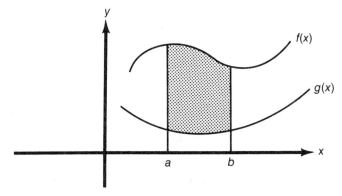

Using the fundamental theorem of calculus, we could find the area under $f(x)$ and above the $x$-axis on the bottom. Then find the area under $g(x)$ and above the $x$-axis. Subtract them, which would result in the shaded area.

That is basically what the next theorem does, except it has you subtract the functions first before calculating the area.

---

If $f(x)$ and $g(x)$ are continuous with $f(x) \geq g(x)$ on the interval $a \leq x \leq b$, then the area between the two curves is equal to

$$\int_{a}^{b} [f(x) - g(x)]\, dx.$$

---

Be sure to notice that the theorem requires that $f(x)$ be above $g(x)$ on the interval. There is no requirement that either function be non-negative.

# EXAMPLE 13

Find the area between $y = f(x) = x^2 + 5$ and $y = g(x) = -x^2$ from $x = 1$ to $x = 2$.

Solution:   A sketch of both functions and the desired area appears below.

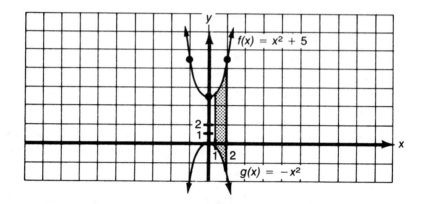

The shaded area is computed by the definite integral of the difference of the functions, $f(x) - g(x)$, with $f(x)$ being the function on the top.

$$\text{Area} = \int_1^2 [f(x) - g(x)]\, dx$$

$$= \int_1^2 [(x^2 + 5) - (-x^2)]\, dx$$

$$= \int_1^2 [x^2 + 5 + x^2]\, dx$$

$$= \int_1^2 [2x^2 + 5]\, dx$$

$$= \frac{2x^3}{3} + 5x \Big|_1^2$$

$$= \left[ \frac{2(8)}{3} + 10 \right] - \left[ \frac{2}{3} + 5 \right]$$

$$= \frac{16}{3} + 10 - \frac{2}{3} - 5$$

$$= \frac{14}{3} + 5$$

$$= 4.67 + 5$$

$$= 9.67 \text{ square units}$$

The final example is the hardest, of course.

# EXAMPLE 14

Find the area bounded by the curves $y = f(x) = x + 1$ and $y = g(x) = x^2 - 2x + 1$.

Solution: $f(x)$ is a linear function with slope of 1 and $y$-intercept of 1.

$g(x)$ is a quadratic with the parabola opening up, $y$-intercept of 1, and $x$-intercept of 1.

A sketch of both functions appears below.

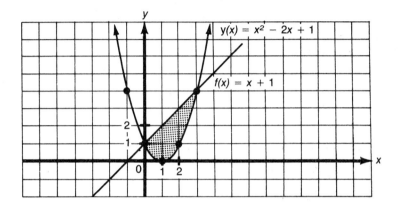

We must determine the $x$-coordinates of the points of intersection.

To find the points of intersections, set

$$f(x) = g(x)$$

$$x + 1 = x^2 - 2x + 1$$

$$0 = x^2 - 3x$$

$$0 = x(x - 3)$$

$$x = 0 \qquad x = 3$$

Using the $x$-coordinates just calculated, the area between the two curves is from $x = 0$ to $x = 3$. This area can be determined by the definite integral of the difference of the two functions integrated from 0 to 3.

$$\text{Area} = \int_0^3 [f(x) - g(x)]\, dx$$

$$= \int_0^3 [(x + 1) - (x^2 - 2x + 1)]\, dx$$

$$= \int_0^3 [x + 1 - x^2 + 2x - 1]\, dx$$

$$= \int_0^3 [-x^2 + 3x]\, dx$$

$$= \frac{-x^3}{3} + \frac{3x^2}{2} \Bigg|_0^3$$

$$= \left( \frac{-(27)}{3} + \frac{27}{2} \right) - 0$$

$$= -9 + 13.5$$

$$= 4.5 \text{ square units}$$

You should now be able to evaluate a definite integral. In addition, you should be able to compute the area bounded by a curve and the $x$-axis, or the area between two curves. Again remember, it is necessary to do a sketch prior to setting up the definite integral when determining areas.

Here are some practice problems to try before going on to the next unit.

## EXERCISES

Evaluate the following four definite integrals:

1.  $\int_{1}^{2} x^2 \, dx$

2.  $\int_{2}^{10} (-x + 10) \, dx$

3.  $\int_{0}^{1} (x - x^2) \, dx$

4.  $\int_{1}^{e} \ln x \, dx$   (This one might surprise you.)

5.  Find the area bounded by the curve $f(x) = x^3 + 8$, the $x$ axis, and the vertical lines at $x = -1$ and $x = 4$.

6.  Find the area between the $x$-axis and the curve $f(x) = x^2 - 4x$.

7.  Find the area between the curve $f(x) = x^3 - 4x$ and the $x$-axis from $x = -1$ to $x = 3$.

8.  Find the area bounded by the curve $f(x) = 4$ and $g(x) = x^2$.

9.  Find the area between $f(x) = x + 1$ and $g(x) = x^2 - 2x + 1$ from $x = -2$ to $x = 0$. Hint: Use the graph from Example 14.

10. Maintenance costs for a home tend to increase as the house gets older. From past records, Penny, a real estate agent, determined that the rate of increase in maintenance costs (in dollars per year) for a small house is given by

$$M'(x) = 9x^2 + 50$$

where $x$ is the age of the house in years and $M(x)$ is the total accumulated cost of maintenance for $x$ years. Write a definite integral that will give the total maintenance costs from 5 to 10 years after the house was built, and evaluate it.

If additional practice is needed:

Barnett and Ziegler, pages 342–346, problems 1–50; pages 355–357, problems 1–43
Budnick, pages 697–698, problems 1–28; pages 706–707, problems 1–16
Hoffmann, pages 313–315, problems 1–36; pages 325–327, problems 1–28
Piascik, pages 306–308, problems 1–20; pages 312–313, problems 1–8

# UNIT 27

## Functions of Several Variables

The purpose of this last unit is to provide you with a working knowledge of differentiating functions of several variables. When you have finished this unit, you will be able to differentiate such functions and, for some, be able to locate and classify their critical points.

## DERIVATIVES REVISITED

Because it has been a number of units since we used derivatives, the formulas have been repeated here. However, the $\frac{dy}{dx}$ notation has been used rather than $f'(x)$ with $u$ and $v$ denoting functions of $x$, $c$ and $n$ are real numbers.

The Rules for Differentiation

$$\frac{d(c)}{dx} = 0$$

$$\frac{d(x^n)}{dx} = nx^{n-1}$$

$$\frac{d(cf(x))}{dx} = c\frac{d(f(x))}{dx}$$

$$\frac{d(u^n)}{dx} = nu^{n-1}\frac{du}{dx}$$

$$\frac{d(uv)}{dx} = u\frac{dv}{dx} + v\frac{du}{dx}$$

$$\frac{d\left(\dfrac{u}{v}\right)}{dx} = \frac{v\dfrac{du}{dx} - u\dfrac{dv}{dx}}{v^2}$$

$$\frac{d(e^u)}{dx} = e^u \frac{du}{dx}$$

$$\frac{d(\ln u)}{dx} = \frac{1}{u} \cdot \frac{du}{dx}$$

# PARTIAL DERIVATIVES

When $f$ is a function of $x$ and $y$, defined by $z = f(x, y)$, it is useful to speak of the instantaneous rate of change in $f$ with respect to $x$ when $y$ is held constant and vice versa.

---

Definition:   The instantaneous rate of change of $f$ with respect to $x$ ($y$ is held constant) is called the **partial derivative with respect to $x$ of $f$**, or simply the first partial.

---

Notation:   $\frac{\partial f}{\partial x} = \frac{\partial z}{\partial x} = f_x(x, y) = f_x$, all denote the partial derivative of $f$ with respect to $x$.

---

The partial of $f$ with respect to $x$ is found by treating $x$ as the only variable, the remaining independent variables as constants, and applying the differentiation rules. The process is known as partial differentiation.

The $f$ and $z$ are interchangeable. So to speak of the partial of $f$ with respect to $x$ is the same as the partial of $z$ with respect to $x$.

---

Definition:   The instantaneous rate of change of $f$ with respect to $y$ ($x$ is held constant) is called the **partial derivative of $f$ with respect to $y$**, or simply the first partial.

Notation:   $\frac{\partial f}{\partial y} = \frac{\partial z}{\partial y} = f_y(x, y) = f_y$, all denote the partial derivative of $f$ with respect to $y$.

---

The partial of $f$ with respect to $y$ is found by treating $y$ as the only variable, the remaining independent variables as constants, and applying the differentiation rules.

The notation expands as you would expect if there are additional independent variables. All of the rules for differentiation we have used remain valid, and you will be happy to learn, there are no new rules. The only difficulty is keeping in mind which variables are being held constant during the differentiation.

## EXAMPLE 1

If $z = f(x, y) = x^2 - y^4$, find $\frac{\partial f}{\partial x}$ and $\frac{\partial f}{\partial y}$.

Solution: To find $\dfrac{\partial f}{\partial x}$, we treat $y$ as a constant and differentiate with respect to $x$.

$$z = f(x, y) = x^2 - y^4$$

$$\frac{\partial f}{\partial x} = f_x(x, y) = 2x$$

To find $\dfrac{\partial f}{\partial y}$, we treat $x$ as a constant and differentiate with respect to $y$.

$$z = f(x, y) = x^2 - y^4$$

$$\frac{\partial f}{\partial y} = f_y(x, y) = -4y^3$$

## EXAMPLE 2

If $z = f(x, y) = 5x + 2x^2y + y^2$, find $f_x$ and $f_y$.

Solution: To find $f_x$, we treat $y$ as a constant and differentiate with respect to $x$. I will put an extra step in before actually differentiating so as to show you what is intended and why.

$$z = f(x, y) = 5x + 2x^2y + y^2$$

$$f_x(x, y) = 5\frac{\partial(x)}{\partial x} + 2y\frac{\partial(x^2)}{\partial x} + \frac{\partial(y^2)}{\partial x}$$

$$= 5(1) + 2y(2x) + 0$$

$$= 5 + 4xy$$

Observe that in the middle term, both the 2 and the $y$ were brought out in front because they are both being considered as constants and only the $x^2$ was differentiated.

To find $f_y$, we treat $x$ as a constant and differentiate with respect to $y$.

$$z = f(x, y) = 5x + 2x^2y + y^2$$

$$f_y(x, y) = \frac{\partial(5x)}{\partial y} + 2x^2\frac{\partial(y)}{\partial y} + \frac{\partial(y^2)}{\partial y}$$

$$= 0 + 2x^2(1) + 2y$$

$$= 2x^2 + 2y$$

Notice the similarity of the notation. With functions of a single independent variable, $\dfrac{dy}{dx}$ is used. With functions of several variables, $\dfrac{\partial f}{\partial x}$ is used.

One more example, and then you can try some. Or if you like, cover up the answer and try the next one yourself.

## EXAMPLE 3

If $z = f(w, x, y) = ye^x + xe^{2y} - 7w^5$, find $f_w$, $f_x$, and $f_y$.

Solution:    To find $f_w$, we treat $x$ and $y$ as constants and differentiate with respect to $w$.

$$z = f(w, x, y) = ye^x + xe^{2y} - 7w^5$$

$$\frac{\partial z}{\partial w} = f_w(w, x, y) = \frac{\partial(ye^x)}{\partial w} + \frac{\partial(xe^{2y})}{\partial w} - 7\frac{\partial(w^5)}{\partial w}$$

$$= 0 + 0 - 7(5w^4)$$

$$= -35w^4$$

To find $f_x$, we treat $w$ and $y$ as constants and differentiate with respect to $x$.

$$z = f(w, x, y) = ye^x + xe^{2y} - 7x^5$$

$$\frac{\partial z}{\partial x} = f_x(w, x, y) = y\frac{\partial(e^x)}{\partial x} + e^{2y}\frac{\partial(x)}{\partial x} - \frac{\partial(7w^5)}{\partial x}$$

$$= y(e^x) + e^{2y}(1) + 0$$

$$= ye^x + e^{2y}$$

To find $f_y$, we treat $w$ and $x$ as constants and differentiate with respect to $y$.

$$z = f(w, x, y) = ye^x + xe^{2y} - 7w^5$$

$$\frac{\partial z}{\partial y} = f_y(w, x, y) = e^x\frac{\partial(y)}{\partial y} + x\frac{\partial(e^{2y})}{\partial y} - \frac{\partial(7w^5)}{\partial y}$$

$$= e^x(1) + x(2e^{2y}) + 0$$

$$= e^x + 2xe^{2y}$$

Ready to try one? I will make it easier.

## Problem 1

If $z = f(w, x, y) = x^2 - 4x + 3y + 20w^3 - 2w^2x^3y^5$, find $f_w$, $f_x$, and $f_y$.

Solution:

Answers:  $f_w = 60w^2 - 4wx^3y^5$

$f_x = 2x - 4 - 6w^2x^2y^5$

$f_y = 3 - 10w^2x^3y^4$

Try one more.

## Problem 2

If $z = f(x, y) = e^{x^2 - 3y}$, find $\dfrac{\partial z}{\partial x}$ and $\dfrac{\partial z}{\partial y}$.

Solution:

Answers: $\dfrac{\partial z}{\partial x} = 2xe^{x^2 - 3y}$

$\dfrac{\partial z}{\partial y} = -3e^{x^2 - 3y}$

# APPLICATIONS

A few application problems might help your understanding of the meaning and the importance of partial derivatives.

Recall that in economics the term marginal analysis refers to the practice of using a derivative to estimate the change in the value of a function resulting from a unit increase in the independent variable. Partial derivatives are used the same way, that is to estimate the change in the value of a function resulting from a one unit increase in one of its variables.

# EXAMPLE 4

W. Spaniel does the production planning for two fibers manufacturing facilities. At the South Carolina plant he estimates that the number of pounds of acetate fibers that can be produced each week is given by the function

$$z = f(x, y) = 12{,}000x + 5{,}000y + 10x^2y - 10x^3$$

where $z =$ the number of pounds of acetate fiber per week,

$x =$ the number of skilled workers at the plant, and

$y =$ the number of unskilled workers at the plant.

Currently the work force at the South Carolina plant is 40 skilled and 60 unskilled workers.

a.  Find and interpret $f(40, 60)$.

b.  Express the rate of change of output with respect to the number of skilled workers.

c.  Find $f_x(40, 60)$.

d.  Interpret the answer to part c.

e.  Compute the actual change in the output that will result if 1 additional skilled worker is hired.

Solution:    a.      $z = f(x, y) = 12{,}000x + 5{,}000y + 10x^2y - 10x^3$

$$z = f(40, 60) = 12{,}000(40) + 5{,}000(60) + (10)(40)^2(60) - 10(40)^3$$

$$= 480{,}000 + 300{,}000 + 960{,}000 - 640{,}000$$

$$= 1{,}100{,}000$$

Answer:    With a work force of 40 skilled and 60 unskilled workers, it is estimated that 1,100,000 pounds of acetate fiber can be produced per week.

b.    The rate of change of output, $z$, with respect to the number of skilled workers, $x$, is the partial derivative $f_x$.

$$z = f(x, y) = 12{,}000x + 5{,}000y + 10x^2y - 10x^3$$

$$\frac{\partial z}{\partial x} = f_x(x, y) = 12{,}000\,\frac{\partial(x)}{\partial x} + \frac{\partial(5{,}000y)}{\partial x} + 10y\,\frac{\partial(x^2)}{\partial x} - 10\,\frac{\partial(x^3)}{\partial x}$$

$$= 12{,}000(1) + 0 + 10y(2x) - 10(3x^2)$$

$$= 12{,}000 + 20xy - 30x^2$$

For any values of $x$ and $y$, this partial derivative is an approximation of the *additional* output that will result from adding one *additional* skilled worker.

c.    $f_x(40, 60)$ denotes that $f_x$ is to be evaluated at $x = 40$ and $y = 60$.

$$f_x(x, y) = 12{,}000 + 20xy - 30x^2$$

$$f_x(40, 60) = 12{,}000 + 20(40)(60) - 30(40)^2$$

$$= 12{,}000 + 48{,}000 - 48{,}000$$

$$= 12{,}000$$

d.    Currently the work force is 40 skilled and 60 unskilled workers. If one additional skilled worker is hired, it is estimated that output would increase by approximately 12,000 pounds of acetate per week.

e.    The actual change is the difference between the output with 41 skilled and 60 unskilled workers and the output with 40 skilled and 60 unskilled workers.

$$z = f(x, y) = 12{,}000x + 5{,}000y + 10x^2y - 10x^3$$

$$z = f(41, 60) = 12{,}000(41) + 5{,}000(60) + 10(41)^2(60) - 10(41)^3$$

$$= 492{,}000 + 300{,}000 + 1{,}008{,}600 - 689{,}210$$

$$= 1{,}111{,}390$$

$$z = f(40, 60) = 1{,}100{,}000$$

$$\text{Actual change} = f(41, 60) - f(40, 60)$$

$$= 1{,}111{,}390 - 1{,}100{,}000$$

$$= 11{,}390$$

In geometric terms, the difference between these two quantities is the difference between using the tangent line to estimate the next point one unit away versus using the actual function to determine the point.

The wording of Example 4 should have sounded familiar. The only difference from the earlier units is that we have expanded our functions to include those with more than one independent variable. The concepts remain the same.

## Problem 3

Continue with W. Spaniel's employment analysis.

a. Find the rate of change of output with respect to the number of unskilled workers.

b. Find $f_y(40, 60)$.

c. Interpret the answer to part b.

d. What would you advise W. Spaniel to do with regard to hiring an additional person?

Solution:

> Answers: a. $f_y(x, y) = 5,000 + 10x^2$; b. $f_y(40, 60) = 21,000$; c. If the number of unskilled workers is increased from 60 to 61 while the number of skilled workers remains at 40, it is estimated that output will increase by 21,000 pounds per week. d. If he decides to hire one more person, the worker should be unskilled rather than skilled.

Here's another short one for you to do.

## Problem 4

At the Virginia fibers plant, W. Spaniel has determined that output is a function of skilled, unskilled, and part-time workers. For this plant, output is estimated by the function

$$z = f(w, x, y) = 1,200x + 500y + x^2y - x^3 - y^2 + 2w^3$$

where $z = $ the number of pounds of acetate fiber per week,

$w = $ the number of part-time workers,

$x = $ the number of skilled workers, and

$y = $ the number of unskilled workers at the plant.

Currently the work force at the Virginia plant is 10 part-time, 31 skilled, and 53 unskilled workers.

Use marginal analysis to estimate the change in output if one additional part-time person was hired.

Solution:

> Answer: $f_w(10, 31, 53) = 6(10)^2 = 600$

Aren't you glad I asked about part time rather than either of the other two?

# SECOND PARTIAL DERIVATIVES

Partial derivatives themselves can be differentiated. The resulting functions are called **second-order partial derivatives**, or simply a **second partial**. If $z = f(x, y)$, then there are four second partials.

1.  The partial of $f_x$ with respect to $x$,

$$\text{denoted by } \frac{\partial^2 z}{\partial x^2} = \frac{\partial^2 f}{\partial x^2} = f_{xx}(x, y) = f_{xx}$$

2.  The partial of $f_y$ with respect to $y$,

$$\text{denoted by } \frac{\partial^2 z}{\partial y^2} = \frac{\partial^2 f}{\partial y^2} = f_{yy}(x, y) = f_{yy}$$

3.  The partial of $f_y$ with respect to $x$,

$$\text{denoted by } \frac{\partial^2 z}{\partial x\, \partial y} = \frac{\partial^2 f}{\partial x\, \partial y} = f_{yx}(x, y) = f_{yx}$$

4.  The partial of $f_x$ with respect to $y$,

$$\text{denoted by } \frac{\partial^2 z}{\partial y\, \partial x} = \frac{\partial^2 f}{\partial y\, \partial x} = f_{xy}(x, y) = f_{xy}$$

# EXAMPLE 5

If $z = f(x, y) = 3xy - x^2 + 5y^3$, find all first and second partials.

Solution:   The two first-order partial derivatives:

$$z = f(x, y) = 3xy - x^2 + 5y^3$$

$$\frac{\partial z}{\partial x} = f_x(x, y) = 3y - 2x$$

$$z = f(x, y) = 3xy - x^2 + 5y^3$$

$$\frac{\partial z}{\partial y} = f_y(x, y) = 3x + 15y^2$$

The four second-order partial derivatives:

1.  To find $f_{xx}(x, y)$, we treat $y$ as a constant and differentiate $f_x$ with respect to $x$.

$$\frac{\partial z}{\partial x} = f_x(x, y) = 3y - 2x$$

$$\frac{\partial^2 z}{\partial x^2} = f_{xx}(x, y) = \frac{\partial(3y)}{\partial x} - 2\frac{\partial(x)}{\partial x}$$

$$= 0 - 2(1)$$

$$= -2$$

2. To find $f_{yy}(x, y)$, we treat $x$ as a constant and differentiate $f_y$ with respect to $y$.

$$\frac{\partial z}{\partial y} = f_y(x, y) \ = 3x + 15y^2$$

$$\frac{\partial^2 z}{\partial y^2} = f_{yy}(x, y) = \frac{\partial(3x)}{\partial y} + 15\frac{\partial(y^2)}{\partial y}$$

$$= 0 + 15(2y)$$

$$= 30y$$

3. To find $f_{xy}(x, y)$, we treat $x$ as a constant and differentiate $f_x$ with respect to $y$.

$$\frac{\partial z}{\partial x} = f_x(x, y) \ = 3y - 2x$$

$$\frac{\partial^2 z}{\partial y \, \partial x} = f_{xy}(x, y) = 3\frac{\partial(y)}{\partial y} - \frac{\partial(2x)}{\partial y}$$

$$= 3(1) - 0$$

$$= 3$$

4. To find $f_{yx}(x, y)$, we treat $y$ as a constant and differentiate $f_y$ with respect to $x$.

$$\frac{\partial z}{\partial y} = f_y(x, y) \ = 3x + 15y^2$$

$$\frac{\partial^2 z}{\partial x \, \partial y} = f_{yx}(x, y) = 3\frac{\partial(x)}{\partial x} + \frac{\partial(15y^2)}{\partial x}$$

$$= 3(1)$$

$$= 3$$

---

The last two partials, $f_{xy}$ and $f_{yx}$, are sometimes referred to as mixed partials or cross partials. The reason should be obvious—the first partial derivative is taken with respect to one variable and the second partial is taken with respect to the other variable. However, if you are like me, I always tend to get the notation for the mixed partials confused in my own mind. The comforting part, though, is that for all of the functions you will ever see, the mixed partials will be equal, as they were in the above example.

Here is a practice problem for you. You might want to cover the answer.

## Problem 5

If $z = f(x, y) = 2x^3y^4$, find all first- and second-order partials.

---

Solution:

Answers: $f_x = 6x^2y^4$, $f_y = 8x^3y^3$, $f_{xx} = 12xy^4$,

$f_{yy} = 24x^3y^2$, $f_{xy} = f_{yx} = 24x^2y^3$

# RELATIVE MAXIMA AND MINIMA FOR BIVARIATE FUNCTIONS

We will conclude this unit with a sketchy discussion of the criteria for maximum and minimum values for bivariate functions. As always, we will allow geometric examples to guide us.

Recall from Unit One, a bivariate function has two independent variables and a multivariate function has more than two independent variables. If $z = f(x, y)$ is a function of two variables $x$ and $y$, it is natural to regard the function as a surface in three-dimensional space. We can use this surface as an aid in studying the properties of the functions, meaning we can continue to draw pictures to help us.

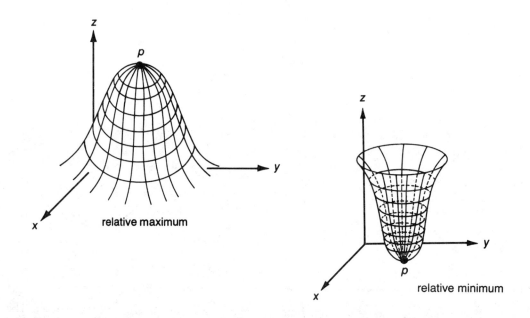

Relative maxima and relative minima of bivariate functions are defined in a manner similar to that used for a single variable function. The graphs illustrate each. Notice that the point marked relative maximum is higher than any of its neighboring points and that the points labeled relative minima are lower than any of their neighboring points. The only difference in terminology is the use of the term **saddle point**, rather than a stationary inflection point, to denote a leveling off point that is neither higher nor lower than all of its neighboring points.

You will be happy to know we are only doing a brief introduction to this material and you will *not* be asked to do any three-dimensional drawings.

For a function $f(x, y)$ to have a relative maximum or minimum, it is necessary that

$$f_x = 0 \quad \text{and}$$

$$f_y = 0 \quad \text{be satisfied simultaneously.}$$

These two conditions are used to determine the critical points of the function.

---

Definition:    A point $(a, b)$ for which both $f_x(a, b) = 0$ and $f_y(a, b) = 0$ is said to be a **critical point** of $f$.

---

One procedure used to classify critical points of a bivariate function is the second-derivative test.

## Second-derivative Test for Classifying Critical Points of a Bivariate Function

Let $D(x, y) = f_{xx}(x, y) f_{yy}(x, y) - [f_{xy}(x, y)]^2$

Evaluate $D$ at the critical point.

1. If $D$ is negative, the critical point is a saddle point.

2. If $D$ is zero, the test in inconclusive.

3. If $D$ is positive, evaluate $f_{xx}$ at the critical point.

   If $f_{xx}$ is negative, the critical point is a maximum, otherwise it is a minimum.

In symbols,   If $D < 0$, it is a saddle point.

   If $D = 0$, test is inconclusive.

   If $D > 0$, it is either a maximum or minimum

   if $f_{xx} < 0$, it is a maximum

   if $f_{xx} > 0$, it is a minimum

Two examples should be sufficient to illustrate the procedure.

## EXAMPLE 6

Find and classify all critical points of $f(x, y) = x^2 + xy + y^2 - 3x + 2$.

Solution:   The first step is to find all the critical points.

Critical points, by definition, occur where $f_x = 0$ and $f_y = 0$ are satisfied simultaneously. Stated another way, it means we take the first partials, set them equal to zero, and solve the equations simultaneously. Notice the similarity to our earlier work.

$$f(x, y) = x^2 + xy + y^2 - 3x + 2$$
$$f_x(x, y) = 2x + y - 3$$
$$f_y(x, y) = x + 2y$$

$$0 = 2x + y - 3$$
$$0 = x + 2y$$

$$3 = 2x + y$$
$$0 = x + 2y$$

$$-6 = -4x - 2y$$
$$\underline{0 = \quad x + 2y}$$
$$-6 = -3x$$

$$x = 2, \text{ and since } 0 = x + 2y$$

by substitution, $0 = 2 + 2y$

$$-2y = 2$$

$$y = -1$$

If $x = 2$ and $y = -1$

with $f(x, y) = x^2 + xy + y^2 - 3x + 2$

then $f(2, -1) = (2)^2 + (2)(-1) + (-1)^2 - 3(2) + 2 = -1$

and the function has a critical point at $(2, -1, -1)$.

To classify the critical point, use the second-derivative test for bivariate functions with

$$f_{xx}(x, y) = 2$$

$$f_{yy}(x, y) = 2$$

$$f_{xy}(x, y) = 1$$

Let $D(x, y) = f_{xx}(x, y)f_{yy}(x, y) - [f_{xy}(x, y)]^2$

$$= (2)(2) - [1]^2$$

$$= 4 - 1$$

$$= 3$$

At the point $(2, -1, -1)$ with

$$D(x, y) = 3$$

$$D(2, -1) = 3$$

Since the value of $D$ is positive, $f$ has either a relative maximum or a relative minimum at the critical point. To determine which one it is, determine the value of $f_{xx}$ at the critical point.

$$f_{xx}(2, -1) = 2$$

Since the value of $f_{xx}$ is positive, the critical point is a relative minimum.

Answer:   The critical point $(2, -1, -1)$ is a relative minimum. Stated another way, $-1$ is the minimum value for $z$ and it occurs when $x = 2$ and $y = -1$.

---

Example 6 was a rather unique situation because all of the second-order partials were constants. That is rarely the case and we will consider one last example where the numerical value of $D$ must be calculated.

I realize that these are long, long, problems, but this is the last one!

# EXAMPLE 7

Find all relative maxima or minima of $f(x, y) = x^2 - 2x + y^3 - 3y^2 - 9y + 6$.

Solution:   The first step is to find all the critical points.

Critical points, by definition, occur where $f_x = 0$ and $f_y = 0$ are satisfied simultaneously. Stated another way, it means we take the first partials, set them equal to zero, and solve the equations simultaneously.

$$f(x, y) = x^2 - 2x + y^3 - 3y^2 - 9y + 6$$
$$f_x(x, y) = 2x - 2$$
$$f_y(x, y) = 3y^2 - 6y - 9$$
$$0 = 2x - 2$$
$$0 = 3y^2 - 6y - 9$$

This time we have two separate equations to be solved. The first one is in terms of $x$ and the second one is in terms of $y$.

$$0 = 2x - 2 \quad 0 = 3y^2 - 6y - 9$$
$$2 = 2x \quad 0 = y^2 - 2y - 3$$
$$x = 1 \quad 0 = (y - 3)(y + 1)$$
$$y - 3 = 0 \qquad y + 1 = 0$$
$$y = 3 \qquad y = -1$$

If $x = 1$ and $y = 3$

and $f(x, y) = x^2 - 2x + y^3 - 3y^2 - 9y + 6$,

then $f(1, 3) = 1 - 2 + 27 - 27 - 27 + 6$

and $(1, 3, -22)$ is a critical point.

If $x = 1$ and $y = -1$

and $f(x, y) = x^2 - 2x + y^3 - 3y^2 - 9y + 6$,

then $f(1, -1) = 1 - 2 - 1 - 3 + 9 + 6 = 10$

and $(1, -1, 10)$ is another critical point.

To classify the critical points, use the second-derivative test for bivariate functions with

$$f_{xx}(x, y) = 2$$
$$f_{yy}(x, y) = 6y - 6$$
$$f_{xy}(x, y) = 0$$

Let $D(x, y) = f_{xx}(x, y)f_{yy}(x, y) - [f_{xy}(x, y)]^2$

$\qquad = [2(6y - 6)] - [0]^2$

$\qquad = 12y - 12$

At the point $(1, 3, -22)$ with

$D(x, y) = 12y - 12$

$D(1, 3) = 12(3) - 12 = 36 - 12 = 24$

Since the value of $D$ is positive, $f$ has either a relative maximum or a relative minimum at the critical point. To determine which one it is, determine the value of $f_{xx}$ at the critical point.

$f_{xx}(1, 3) = 2$

Since the value of $f_{xx}$ is positive, the critical point is a relative minimum.

The critical point $(1, 3, -22)$ is a relative minimum. Stated another way, $-22$ is the minimum value for $z$ and it occurs when $x = 1$ and $y = 3$.

At the point $(1, -1, 10)$ with

$$D(x, y) = 12y - 12$$

$$D(1, -1) = 12(-1) - 12 = -12 - 12 = -24$$

Since the value of $D$ is negative, the critical point is a saddle point.

The critical point $(1, -1, 10)$ is neither a high point nor a low point, but a saddle point.

---

Congratulations, you have completed the book.

You should now be able to find all partial derivatives for multivariate functions and use them to estimate the change in the value of the function for a unit change in one of the independent variables. In addition, for some bivariate functions, you should be able to find and classify all critical points.

I can no longer say, "before continuing to the next unit," but do solve the following problems anyway. There are only five.

# EXERCISES

1.   If $f(x, y) = (3x - 7y)^5$, find $f_x$ and $f_y$.

2.   If $f(w, x, y) = 2xy + 3x^2yw + w^2 + 10$, find $f_w$, $f_x$ and $f_y$.

3.   If $f(x, y) = 3x^4y^3 + 2xy$, find all first- and second-order partials.

4.   Find and classify all critical points for $f(x, y) = 3 - x^2 - y^2$.

5.   The annual profit of a small inn in New England is given by

$$f(x, y) = 10x^2 + 20y^2 - 10xy$$

where $x$ is the weekly advertising budget and $y$ is the number of rooms available for rent. Currently the inn has 10 rooms available and an advertising budget of \$100 per week.

a.   If an additional room is constructed in an unfinished area, how will this affect annual profit?

b.   If an additional dollar is added to the weekly advertising budget, how will this affect annual profit?

---

If additional practice is needed:

Barnett and Ziegler, pages 428–434, problems 1–64; pages 441–442, problems 1–18
Budnick, pages 606–607, problems 1–35; page 620, problems 1–20
Hoffmann, pages 440–442, problems 1–27; pages 467–469, problems 1–12
Piascik, pages 418–420, problems 1–18; pages 427–429, problems 1–11

# Answers to Exercises

## UNIT 1

1.  $f(x) = x^2 - 2x + 3$

    $f(0) = (0)^2 - 2(0) + 3$

    $\quad = 0 - 0 + 3$

    $\quad = 3$

2.  $f(10) = (10)^2 - 2(10) + 3$

    $\quad = 100 - 20 + 3$

    $\quad = 83$

3.  $f(-2) = (-2)^2 - 2(-2) + 3$

    $\quad = 4 + 4 + 3$

    $\quad = 11$

4.  $f(a) = a^2 - 2a + 3$

5.  $f^{-1}(3) = 0$

    The question is asking if the dependent value is 3, what is $x$? From question one above, the answer is 0.

6.  $f(-x) = (-x)^2 - 2(-x) + 3$

    $\quad = x^2 + 2x + 3$

7.  The domain is the set of all reals because nothing to the contrary is specified and $f$ contains neither a quotient nor a radical causing numbers to be excluded.

    $g(x) = \dfrac{1}{x - 3}$

8.  $g(5) = \dfrac{1}{5 - 3}$

    $\quad = \dfrac{1}{2}$

9.  $g(c) = \dfrac{1}{c - 3}$

10. $g(a + 3) = \dfrac{1}{(a + 3) - 3}$

    $\quad = \dfrac{1}{a}$

11. The domain is the set of all real numbers except 3 because 3 would result in the denominator equaling a value of 0.

    $q = h(p) = \sqrt{13 - p}$

12. The independent variable is $p$.

13. To say that the quantity under the radical sign for a square root must be non-negative is the same as saying

    $13 - p \geq 0$

    $13 \geq p$

    or $\quad p \leq 13$;

    therefore the domain is the set of all $p$ such that $p \leq 13$ or, using set notation: $D = \{p : p \leq 13\}$.

14. $h(12) = \sqrt{13 - 12}$

    $\quad = \sqrt{1}$

    $\quad = 1$

15. $h(13) = \sqrt{13 - 13}$

    $\quad = \sqrt{0}$

    $\quad = 0 \qquad$ Recall the square root of 0 is 0.

16. $h(-3) = \sqrt{13 - (-3)}$

    $\quad = \sqrt{16}$

    $\quad = 4$

17. $h(2) = \sqrt{13 - 2}$

    $\quad = \sqrt{11}$

    There is no real advantage to changing this to a decimal approximation at this time.

18. $\quad \alpha(x, y, z) = x - 3z + xy$

    $\alpha(5, 3, -2) = 5 - 3(-2) + (5)(3)$

    $\quad = 5 + 6 + 15$

    $\quad = 26$

19.   $H(7) = 7 - 1$

      $= 6$

20.   $H(0) = 0 + 1 = 1$

21.   $H(-4) = (-4) + 1 = -3$

22.   $H(3) = 3 + 1 = 4$

      $G = \{(5, 6), (5, 7), (8, 5)\}$

23.   $D = \{5, 8\}$

24.   $R = \{5, 6, 7\}$

25.   No, $G$ is not a function.
      When $x$ is 5, there are two different values
      for $y$, 6 and 7.

# UNIT 2

1.   $f(g(5)) = f(3)$      because      $g(5) = 5 - 2 = 3$

     $= 3 + 1$

     $= 4$

2.   $f(H(5)) = f(30)$      because      $H(5) = (5)^2 + 5$

     $= 30 + 1$                          $= 25 + 5$

     $= 31$                              $= 30$

3.   $G(F(2)) = G(-2)$      because      $F(2) = (2)^2 - 3(2)$

     $= 3(-2) - 5$                       $= 4 - 6$

     $= -6 - 5$                          $= -2$

     $= -11$

4.   $g(F(G(2))) = g(F(1))$        $G(2) = 3(2) - 5$

     $= g(-2)$                     $= 6 - 5$

                                   $= 1$

     $= (-2) - 2$

     $= -4$                        $F(1) = (1)^2 - 3(1)$

                                   $= 1 - 3$

                                   $= -2$

5.   Given      $F(x) = x^2 - 3x$

     then    $F(x + 1) = (x + 1)^2 - 3(x + 1)$

             $= x^2 + 2x + 1 - 3x - 3$        Reminder:   $(x + 1)^2 = (x + 1)(x + 1)$

             $= x^2 - x - 2$                              $= x^2 + x + x + 1$

6.   $f(h(x)) = f(2x + 3)$

     $= (2x + 3) + 1$

     $= 2x + 3 + 1$

     $= 2x + 4$

7.  $2f(x) + g(x) - h(x) = 2(x + 1) + (x - 2) - (2x + 3)$

$$= 2x + 2 + x - 2 - 2x - 3$$

$$= x - 3$$

8.  $h(f(x)) = h(x + 1)$

$$= 2(x + 1) + 3$$

$$= 2x + 2 + 3$$

$$= 2x + 5$$

9.  $H(2x) - h(x) = (2x)^2 + (2x) - (2x + 3)$

$$= 4x^2 + 2x - 2x - 3$$

$$= 4x^2 - 3$$

10.  $G(F(x)) = G(x^2 - 3x)$

$$= 3(x^2 - 3x) - 5$$

$$= 3x^2 - 9x - 5$$

11.  $5x + 6y = 72$            implicit

$$6y = -5x + 72$$

$$y = -\frac{5}{6}x + 12 \qquad \text{explicit for } y$$

$$5x + 6y = 72 \qquad \text{implicit}$$

$$5x = -6y + 72$$

$$x = -\frac{6}{5}y + 14.4 \qquad \text{explicit for } x$$

12.  $N(t) = -t^2 + 400t + 50,000$

   a.  $N(0) = -0^2 + 400(0) + 50,000$

$$= 50,000$$

The current population of the community is 50,000 people.

   b.  $N(10) = -(10)^2 + 400(10) + 50,000$

$$= -100 + 4,000 + 50,000$$

$$= 53,900$$

Ten years from now the population of the community is estimated to be 53,900 people.

13.  $C(x) = 6,452 + 72x$    with    $x \geq 0$

   a.  $C(10) = 6,452 + 72(10)$

$$= 6,452 + 720$$

$$= \$7,172$$

b.  $C(100) = 6{,}452 + 72(100)$

$= 6{,}452 + 7{,}200$

$= 13{,}652$

The total cost of producing 100 rugs will be $13,652.

c.  $C(0) = 6{,}452 + 72(0)$

$= 6{,}452$

14.  Total Cost = fixed cost + (variable cost/unit)(number of units)

a.  $C(x) = 13{,}200 + 43x$

b.  $C(0) = 13{,}200 + 43(0)$

$= 13{,}200$

The total cost of producing no braided rugs is $13,200. Or, the company has fixed costs of $13,200.

$C(10) = 13{,}200 + 43(10)$

$= 13{,}200 + 430$

$= 13{,}630$

The total cost of producing 10 braided rugs is $13,630.

$C(100) = 13{,}200 + 43(100)$

$= 13{,}200 + 4{,}300$

$= 17{,}500$

The total cost of producing 100 braided rugs is $17,500.

15.          $R(x) = 153x$

a.  $R(50) = 153(50)$

$= 7{,}650$

b.  $R(100) = 153(100)$

$= 15{,}300$

The total revenue gained from selling 100 crocheted rugs is $15,300.

$R(0) = 153(0) = 0$

The total revenue gained from selling 0 crocheted rugs is $0.

16.  Total Revenue = (selling price per unit)(number of units sold)

a.  $R(x) = 78x$

b.  $R(1{,}000) = 78(1{,}000)$

$= 78{,}000$

The total revenue gained from selling 1,000 braided rugs is $78,000.

c.  **Restricted domain:**  $x \geq 0$

17.  a.  $P(x) = 81x - 6{,}452 \quad$ with $x \geq 0$

$P(200) = 81(200) - 6{,}452$

$\qquad = 16{,}200 - 6{,}452$

$\qquad = 9{,}748$

b.  $P(50) = 81(50 - 6{,}452$

$\qquad = 4{,}050 - 6{,}452$

$\qquad = -2{,}402$

The company would lose \$2,402 on the production and sale of only 50 crocheted rugs.

18.  $\qquad C(x_1, x_2, x_3) = 1{,}750 + 7.5x_1 + 6x_2 + 10.6x_3$

a.  $C(10, 25, 100) = 1{,}750 + 7.5(10) + 6(25) + 10.6(100)$

$\qquad = 1{,}750 + 75 + 150 + 1{,}060$

$\qquad = 3{,}035$

The total cost of producing 10 toy soldiers, 25 teddy bears, and 100 dolls is \$3,035.

b.  Fixed costs are \$1,750.

The variable cost per unit for toy soldiers is \$7.50.

The variable cost per unit for teddy bears is \$6.00.

The variable cost per unit for dolls is \$10.60.

# UNIT 3

1.  $y = f(x) = 2x + 5$

The slope is 2 or $m = \dfrac{\Delta y}{\Delta x} = \dfrac{2}{1}$

The $y$-intercept is 5.

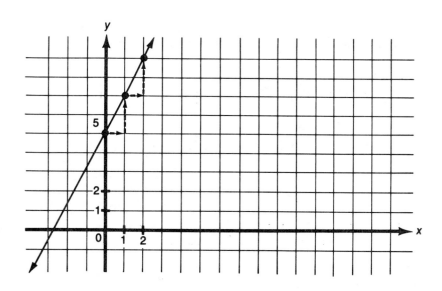

2.   $y = f(x) = \dfrac{7}{3}x - 1$

The slope is $\dfrac{7}{3}$ or $m = \dfrac{\Delta y}{\Delta x} = \dfrac{7}{3}$

The $y$-intercept is $-1$.

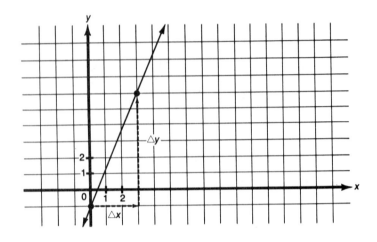

3.   $y = \dfrac{2}{5}x + 3$

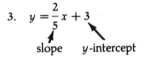
slope    $y$-intercept

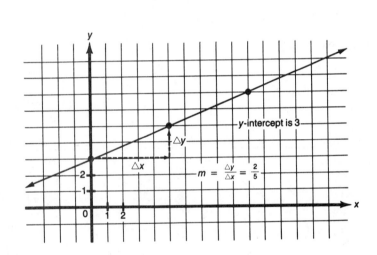

4.   $f(x) = x - 4$

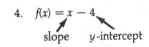
slope    $y$-intercept

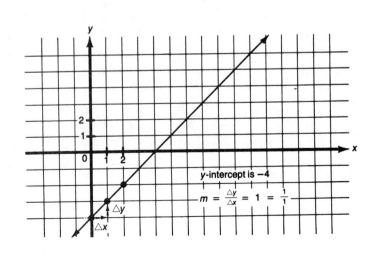

5. $g(x) = -\dfrac{3}{2}x + 7$

   slope    $y$-intercept

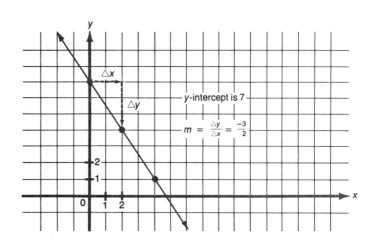

6. $h(x) = -x + 2$

   slope    $y$-intercept

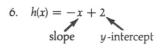

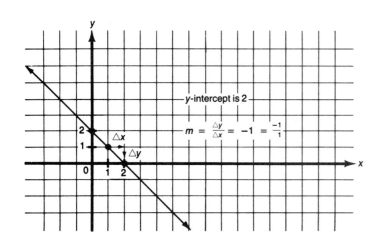

7. $y = f(x) = \dfrac{x}{3} + 1$

   $y = f(x) = \dfrac{1}{3}x + 1$

   slope    $y$-intercept

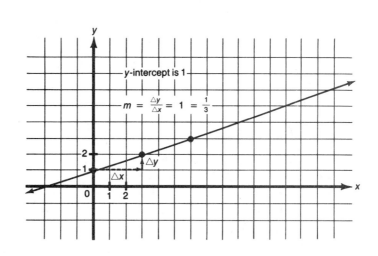

8.  $y = f(x) = .25x + 1.50$

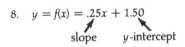

slope    $y$-intercept

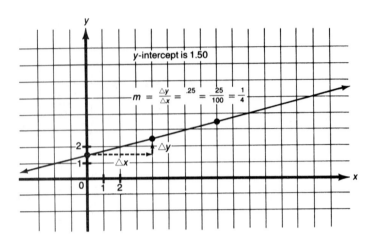

9.  $q = D(p) = -3p + 11$

slope    $y$-intercept

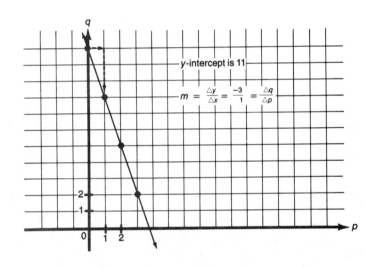

10.  a.  $x - 3y = -12$

   $-3y = -x - 12$

   $y = \dfrac{1}{3}x + 4$    slope-intercept form

   slope    $y$-intercept

   b.  The slope is $\dfrac{1}{3}$ meaning if $x$ increases by 3, the $y$ value will increase by 1.

   c.  The $y$-intercept is 4.

   The $x$-intercept is the value of $x$ when $y = 0$.

   $x - 3y = -12$

   $x - 3(0) = -12$

   $x = -12$

   The $x$-intercept is $-12$ or the point $(-12, 0)$.

d.

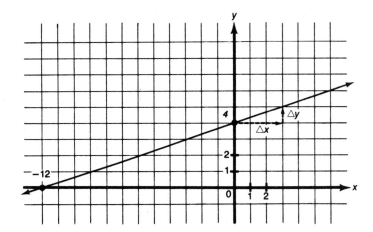

# UNIT 4

1. $y - y_1 = m(x - x_1)$

$y - (-3) = 5(x - 4)$

$y + 3 = 5x - 20$

$y = 5x - 23$

2. $y - y_1 = m(x - x_1)$

$y - 7 = -\dfrac{1}{6}(x - 6)$

$y - 7 = -\dfrac{1}{6}x + 1$

$y = -\dfrac{1}{6}x + 8$

3. $(7, 5)$      $m = \dfrac{\Delta y}{\Delta x} = \dfrac{0}{10} = 0$

$\underline{(-3, 5)}$

$10, 0$

$\Delta x \uparrow \quad \uparrow \Delta y$

To write the function:   $y - y_1 = m(x - x_1)$

$y - 5 = 0(x - 7)$

$y - 5 = 0$

$y = 5$

4. $y - y_1 = m(x - x_1)$

$y - 100 = .15(x - 2.35)$

$y - 100 = .15x - .3525$

$y = .15x + 99.6475$

5. $(-5, -2)$      $m = \dfrac{\Delta y}{\Delta x} = \dfrac{6}{-2} = -3$

$\underline{(-3, -8)}$

$-2, \quad 6$

$y - y_1 = m(x - x_1)$

$y - (-2) = -3(x - (-5))$

$y + 2 = -3(x + 5)$

$y + 2 = -3x - 15$

$y = -3x - 17$

6. $(3600, 1000)$      $m = \dfrac{\Delta y}{\Delta x} = \dfrac{800}{3200} = \dfrac{1}{4}$

$\underline{(400, \quad 200)}$

$3200, \quad 800$

$y - y_1 = m(x - x_1)$

$y - 200 = \dfrac{1}{4}(x - 400)$

$y - 200 = \dfrac{1}{4}x - 100$

$y = \dfrac{1}{4}x + 100$

7. A linear function does not exist, but the equation of the line is $x = -9$.

8. $(3.95, 435)$      $m = \dfrac{\Delta y}{\Delta x} = \dfrac{135}{1.20} = 112.5$

$\underline{(2.75, 300)}$

$1.20, \ 135$

$y - y_1 = m(x - x_1)$

$y - 300 = 112.5(x - 2.75)$

$y - 300 = 112.5x - 309.375$

$y = 112.5x - 9.375$

9.  If the line is to be parallel to $y = 5x + 2$, it must have a slope of 5.

$$y - y_1 = m(x - x_1)$$

$$y - 4 = 5(x - 6)$$

$$y - 4 = 5x - 30$$

$$y = 5x - 26$$

10. If the line is to be perpendicular to $y = 3x - 1$, its slope must be the negative reciprocal of 3, or $m = -\dfrac{1}{3}$.

$$y - y_1 = m(x - x_1)$$

$$y - 2 = -\frac{1}{3}(x - (-7))$$

$$y - 2 = -\frac{1}{3}(x + 7)$$

$$y - 2 = -\frac{1}{3}x - \frac{7}{3}$$

$$y = -\frac{1}{3}x - \frac{7}{3} + 2$$

$$y = -\frac{1}{3}x - \frac{1}{3}$$

# UNIT 5

1.  a.    $L(t) = 1.5t - 60$

$$L(86) = 1.5(86) - 60$$

$$= 129 - 60$$

$$= 69$$

If the temperature is 86°, the street vendor should expect to sell 69 lemonades at lunch.

b.    $L(t) = 1.5t - 60$

$$L(40) = 1.5(40) - 60$$

$$= 60 - 60$$

$$= 0$$

If the temperature is 40° or lower, the street vendor does not expect to sell any lemonade.

c.    $m = 1.5$

For each increase of 1° in the temperature, the number of lemonades sold is expected to increase by 1.5.

However the following is a better answer, since $m = 1.5 = \dfrac{3}{2}$,

for each increase of 2° in the temperature, the number of lemonades sold is expected to increase by 3.

2.  a.  Break even is where   $R(x) = C(x)$

$$10x = 4.5x + 38.5$$

$$5.5x = 38.5$$

$$x = 7$$

Marguerite needs to work 7 hours per week to break even.

b.  $P(x) = R(x) - C(x)$

$$= 10x - (4.5x + 38.5)$$

$$= 10x - 4.5x - 38.5$$

$$= 5.5x - 38.5$$

$$22 = 5.5x - 38.5$$

$$5.5x = 60.5$$

$$x = 11$$

Marguerite needs to work 11 hours next week in order to earn enough profit to pay for a concert ticket of $22.

3.  a.  $p + 2q - 5{,}000 = 0$

$$2q = -p + 5{,}000$$

$$q = -\frac{1}{2}p + 2{,}500$$

$$q = D(p) = -\frac{1}{2}p + 2{,}500$$

b.  $$D(p) = -\frac{1}{2}p + 2{,}500$$

$$D(2{,}000) = -\frac{1}{2}(2{,}000) + 2{,}500$$

$$= -1{,}000 + 2{,}500$$

$$= 1{,}500$$

If the price is $2,000, the demand will be for 1,500 units.

c.  The $y$-intercept is 2,500.

If free, could give away 2,500.

d.  $$D(5{,}000) = -\frac{1}{2}(5{,}000) + 2{,}500$$

$$= 0$$

If the price is $5,000, the demand is 0. And further, at a price of $5,000 or more, there will be no demand.

e.  slope $= \dfrac{-1}{2} = -.5$

For every dollar increase in price, demand will drop by $\dfrac{1}{2}$ unit. Or, for every 2 dollar increase in price, demand will drop by 1 unit.

4.  a.  If $p = 5$, $q = 46$     (5,     46)
        If $p = 9$, $q = 86$    $\underline{(9,\quad 86)}$     $m = \dfrac{\Delta y}{\Delta x} = \dfrac{-40}{-4} = 10$
                            $-4 \quad -40$

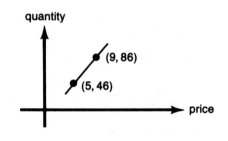

$$y - y_1 = m(x - x_1)$$

$$y - 46 = 10(x - 5)$$

$$y - 46 = 10x - 50$$

$$y = 10x - 4$$

$$q = S(p) = 10p - 4$$

    b.  $q = S(15) = 10(15) - 4$

                $= 150 - 4$

                $= 146$

        If city maps are priced at \$15, suppliers are willing to produce 146 maps.

    c.  The $x$-intercept is where $y = 0$. In this instance we are looking for $p$ when $q = 0$.

        $$0 = 10p - 4$$

        $$10p = 4$$

        $$p = .40$$

        If the price of maps is \$.40 or less, suppliers are unwilling to produce any maps.

    d.  Restricted domain $p \geq .40$

5.  a.  If $p = 5$, then $q = 65$     (5,  65)
        If $p = 10$, then $q = 40$   $\underline{(10,\ 40)}$
                              $-5 \quad 25$     $m = \dfrac{25}{-5} = -5$

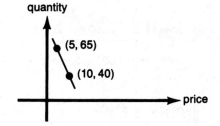

$$y - y_1 = m(x - x_1)$$

$$y - 40 = -5(x - 10)$$

$$y - 40 = -5x + 50$$

$$y = -5x + 90$$

$$q = D(p) = -5p + 90$$

    b.  $y$-intercept is 90

        If the maps are free, could give away 90.

        $x$-intercept:   $0 = -5p + 90$

                    $$5p = 90$$

                    $$p = 18$$

        At a price of \$18 per map, the demand is 0.

    c.  Restricted domain $0 \leq p \leq 18$

    d.  Equilibrium is where          $S(p) = D(p)$

                        $$10p - 4 = -5p + 90$$

                        $$15p = 94$$

                        $$p = 6.27$$

and $q = 10p - 4 = 10(6.27) - 4 = 58.7$

At a price of \$6.27 each, the supply and demand will be equal for 59 city maps.

6.  a.

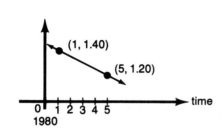

In 1981, cost 1.40  (1, 1.40)

In 1985, cost 1.20  $\underline{(5, 1.20)}$

$\phantom{In 1985, cost 1.20}\;-4,\;\;.20$

$$m = \frac{\Delta y}{\Delta x} = \frac{.20}{-4} = -.05$$

$$y - y_1 = m(x - x_1)$$

$$y - 1.20 = -.05(x - 5)$$

$$y - 1.20 = -.05x + .25$$

$$y = -.05x + 1.45$$

$$C(t) = -.05t + 1.45$$

b.  $m = -.05$

c.    $C(t) = -.05t + 1.45$

$\phantom{c.}\;\;.865 = -.05t + 1.45$

$\phantom{c.}\;\;.05t = .585$

$\phantom{c.}\;\;\;\;t = 11.7$ or in 1991

In late 1991, or 11.7 years from 1980.

7.

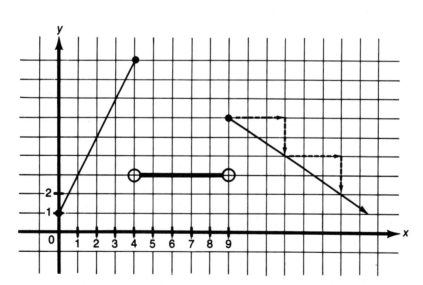

$f(x) = 2x + 1$ if $0 \leq x \leq 4$

| lower limit | if $x = 0$, $f(0) = 2(0) + 1 = 1$ | $(0, 1)$ |
| upper limit | if $x = 4$, $f(4) = 2(4) + 1 = 9$ | $(4, 9)$ |

$f(x) = 3$          if $4 < x < 9$

| lower limit | if $x = 4$, $f(4) = 3$ | $(4, 3)$ open |
| upper limit | if $x = 9$, $f(9) = 3$ | $(9, 3)$ open |

$$f(x) = \frac{-2}{3}x + 12 \qquad \text{if } 9 \leq x$$

limit $\qquad$ if $x = 9$, $f(9) = \frac{-2}{3}(9) + 12 = 6$ $\qquad$ $(9, 6)$

slope is $\frac{-2}{3}$

half line

# UNIT 6

1.  $y = f(x) = x^2 + 2x + 1$

    1.  This is a quadratic with $a = 1$, $b = 2$, $c = 1$.

    2.  The parabola opens up.

    3.  The $y$-intercept is at $1$.

    4.  The $x$-intercept is at $-1$ because

        $0 = x^2 + 2x + 1$

        $0 = (x + 1)^2$

        $x + 1 = 0$

        $\quad x = -1$

    5.  The vertex is located at $(-1, 0)$ because

        axis of symmetry is $x = -1$

        and $f(-1) = (-1)^2 + 2(-1) + 1$

        $\qquad\qquad = 1 - 2 + 1$

        $\qquad\qquad = 0$

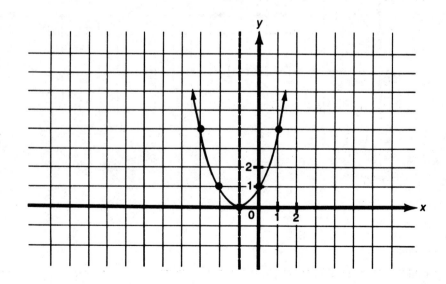

2.  $y = f(x) = -x^2 + 2x + 3$

    1.  This is a quadratic with $a = -1$, $b = 2$, $c = 3$.

    2.  The parabolic curve opens down.

    3.  The $y$-intercept is at 3.

    4.  The $x$-intercepts are $-1$ and 3 because

        $0 = -x^2 + 2x + 3$

        $0 = x^2 - 2x - 3$

        $0 = (x - 3)(x - 1)$

        $x = 3, \qquad x = -1$

    5.  The vertex is located at $(1, 4)$ because

        $$\frac{-b}{2a} = \frac{-2}{2(-1)} = \frac{-2}{-2} = 1$$

        and $f(1) = -1 + 2(1) + 3$

        $\qquad\qquad = -1 + 2 + 3$

        $\qquad\qquad = 4$

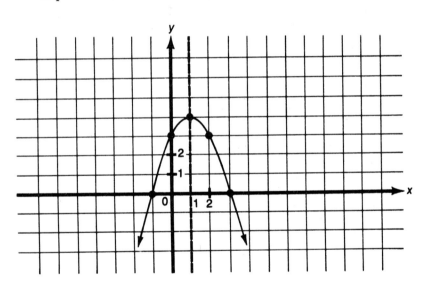

3.  $y = f(x) = -2x^2 + 4x$

    1.  This is a quadratic with $a = -2$, $b = 4$, $c = 0$.

    2.  The parabolic curve opens down.

    3.  The $y$-intercept is 0.

    4.  The $x$-intercepts are 0 and 2 because

        $0 = -2x^2 + 4x$

        $0 = 2x^2 - 4x$

        $0 = 2x(x - 2)$

        $x = 0 \qquad x = 2$

5. The vertex is located at $(1, 2)$ because

the axis of symmetry is $x = 1$ and

$$f(1) = -2(1)^2 + 4(1)$$
$$= -2 + 4$$
$$= 2$$

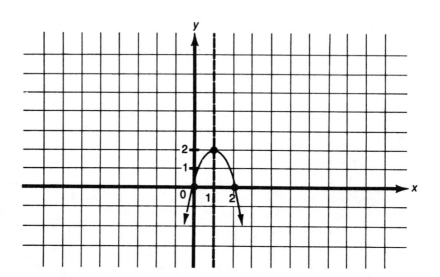

4. $y = f(x) = 2x^2 - x - 10$

1. This is a quadratic with $a = 2$, $b = -1$, $c = -10$.

2. The parabolic curve opens up.

3. The $y$-intercept is at $-10$.

4. The $x$-intercepts are $-2$ and $\dfrac{5}{2}$ because

$$0 = 2x^2 - x - 10$$
$$0 = (2x - 5)(x + 2)$$

$$2x - 5 = 0 \qquad x + 2 = 0$$
$$2x = 5 \qquad\qquad x = -2$$
$$x = \frac{5}{2}$$

5. The vertex is located at $\left(\dfrac{1}{4}, \dfrac{-81}{8}\right)$ or $(.25, -10.125)$

$$\frac{-b}{2a} = \frac{-(-1)}{2(2)} = \frac{1}{4}$$

$$\frac{4ac - b^2}{4a} = \frac{4(2)(-10) - (-1)^2}{4(2)} = \frac{-80 - 1}{8} = \frac{-81}{8}$$

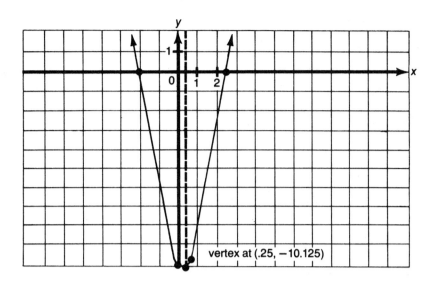

vertex at (.25, −10.125)

5.  $y = f(x) = -x^2 - 7$

    1.  This is a quadratic with $a = -1$, $b = 0$, $c = -7$.

    2.  The parabolic curve opens down.

    3.  The $y$-intercept is at $-7$.

    4.  There are no $x$-intercepts because

        $$0 = -x^2 - 7$$

        $x^2 = -7$ has no solution

    5.  The vertex is located at $(0, -7)$ because

        $$\frac{-b}{2a} = \frac{0}{2(-1)} = 0$$

        and $f(0) = (0)^2 - 7 = -7$

    6.  Locate one point in either side of vertex

        let $x = 1$, then $f(1) = -1 - 7 = -8$

        let $x = -1$, then $f(-1) = -(-1)^2 - 7 = -8$

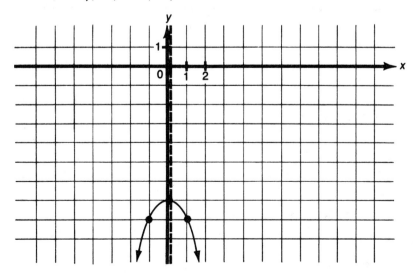

6.   $y = f(x) = x^2 - 4x + 2$

   1.   This is a quadratic with $a = 1$, $b = -4$, $c = 2$.

   2.   The parabolic curve opens up.

   3.   The $y$-intercept is at 2.

   4.   Use the quadratic formula to find $x$-intercepts.

$$x = \frac{-b \pm \sqrt{b^2 - 4ac}}{2a}$$

$$= \frac{-(-4) \pm \sqrt{(-4)^2 - 4(1)(2)}}{2(1)}$$

$$= \frac{4 \pm \sqrt{16 - 8}}{2} = \frac{4 \pm \sqrt{8}}{2} \approx \frac{4 \pm 2.8}{2}$$

$$x \approx (4 + 2.8)/2 \qquad x = (4 - 2.8)/2$$

$$\approx 3.4 \qquad\qquad \approx .6$$

The $x$-intercepts are approximately 3.4 and .6.

   5.   The vertex is located at $(2, -2)$ because

$$\frac{-b}{2a} = \frac{-(-4)}{2(1)} = \frac{4}{2} = 2$$

and

$$f(2) = (2)^2 - 4(2) + 2 = 4 - 8 + 2 = -2$$

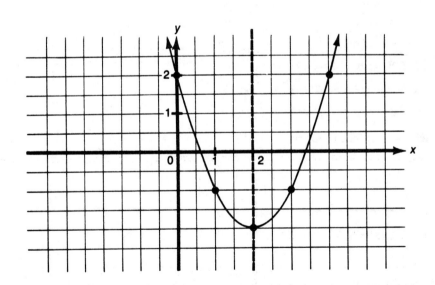

7.   $y = f(x) = 3x^2 - 9x$

   1.   This is a quadratic with $a = 3$, $b = -9$, $c = 0$.

   2.   The parabolic curve opens up.

   3.   The $y$-intercept is at 0.

4.  The $x$-intercepts are 0 and 3 because

    $0 = 3x^2 - 9x$

    $0 = 3x(x - 3)$

    $3x = 0 \qquad x - 3 = 0$

    $x = 0 \qquad\quad x = 3$

5.  The vertex is located at $(1.5, -6.75)$ because

    the axis of symmetry is $x = 1.5$

    and $f(1.5) = 3(1.5)^2 - 9(1.5)$

    $\qquad\qquad = 3(2.25) - 13.5$

    $\qquad\qquad = 6.75 - 13.5$

    $\qquad\qquad = -6.75$

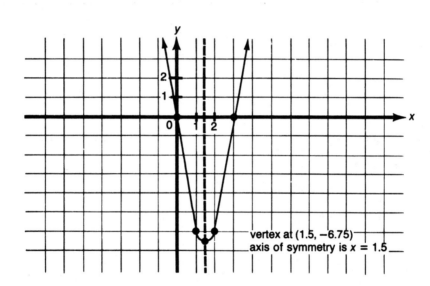

vertex at $(1.5, -6.75)$
axis of symmetry is $x = 1.5$

# UNIT 7

1.  $y = f(x) = .01x^2 - 8x$

    1.  This is a quadratic with $a = .01$, $b = -8$, $c = 0$.

    2.  The parabolic curve opens up.

    3.  The $y$-intercept is at 0.

    4.  The $x$-intercepts are at 0 and 800 because

        $0 = .01x^2 - 8x$

        $0 = x^2 - 800x$

        $0 = x(x - 800)$

        $x = 0 \qquad x = 800$

5.  The vertex is located at $(400, -1,600)$ because

    the axis of symmetry is $x = 400$

    and    $f(400) = .01(400)^2 - 8(400)$

    $\qquad\qquad = .01(160,000) - 3,200$

    $\qquad\qquad = 1,600 - 3,200$

    $\qquad\qquad = -1,600$

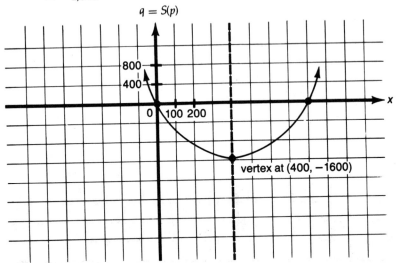

vertex at (400, -1600)

2.  a.  $q = S(p) = p^2 - 100$

    1.  This is a quadratic with $a = 1$, $b = 0$, $c = -100$.

    2.  The parabolic curve opens up.

    3.  The $y$-intercept is at $-100$.

    4.  The $x$-intercepts are at $10$ and $-10$ because

        $0 = p^2 - 100$

        $p^2 = 100$

        $p = \pm 10$

    5.  The vertex is located at $(0, -100)$

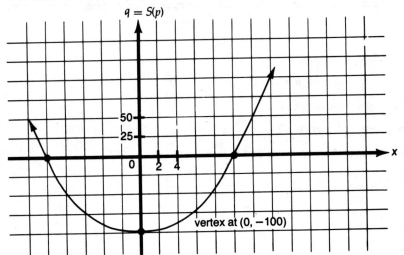

vertex at (0, -100)

b.  $S(18.50) = (18.50)^2 - 100 = 342.25 - 100 = 242.25$

At a price of $18.50, farmers are willing to supply 242.25 pounds of seed.

c.  If $D(p) = 800$, then

$$800 = p^2 - 100$$

$$p^2 = 900$$

$$p = \pm 30$$

The price per pound would have to be $30, before farmers would be willing to supply 800 pounds.

3.  Equilibrium occurs where supply and demand are equal. It does not change the definition just because the functions are quadratics.

Equilibrium is where $S(p) = D(p)$

$$2p^2 + 5p - 250 = 2p^2 - 297p + 10{,}960$$

$$297p + 5p = 10{,}960 + 250$$

$$302p = 11{,}210$$

$$p \approx 37.12$$

To find $q$ when $p = 37.12$

$$S(p) = 2p^2 + 5p - 250$$

$$S(37.12) = 2(37.12)^2 + 5(37.12) - 250$$

$$= 2(1{,}377.152) + 185.60 - 250$$

$$= 2{,}689.90$$

$$= 2{,}690$$

At a price of $37.12 per unit, the supply and demand for the item will be equal at approximately 2,690 units.

4.  a.  $H(t) = -16t^2 + 128t + 320$

1.  This is a quadratic with $a = -16$, $b = 128$, $c = 320$.

2.  The parabolic curve opens down.

3.  The $y$-intercept is at 320.

4.  The $x$-intercepts are at $-2$ and $10$ because

$$0 = -16t^2 + 128t + 320$$

$$0 = 16t^2 - 128t - 320$$

$$0 = t^2 - 8t - 20$$

$$0 = (t - 10)(t + 2)$$

$$t - 10 = 0 \qquad t + 2 = 0$$

$$t = 10 \qquad\quad t = -2$$

5.  The vertex is located at $(4, 576)$.

$$\frac{-b}{2a} = \frac{-128}{2(-16)} = \frac{-128}{-32} = 4$$

and

$$H(4) = -16(4)^2 + 128(4) + 320$$
$$= -16(16) + 512 + 320$$
$$= -256 + 512 + 320$$
$$= 576$$

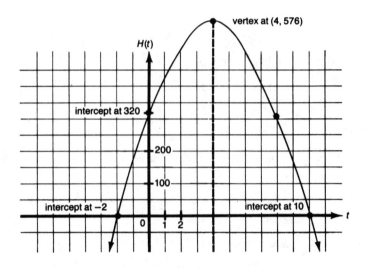

b. (2, 512)

Two seconds after the ball is thrown upward, it is 512 ft from the ground.

c. $H(t) = -16t^2 + 128t + 320$

$$H(9) = -16(9)^2 + 128(9) + 320$$
$$= -16(81) + 1,152 + 320$$
$$= -1,296 + 1,152 + 320$$
$$= 176$$

Nine seconds after the ball is thrown, it will be only 176 ft from the ground.

d. The ball will hit the ground in 10 seconds.

e. The ball went 576 ft high.

f. The building was 320 ft high.

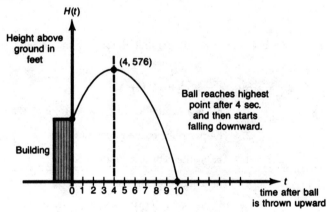

5. $P(x) = -x^2 + 140x - 4,000$

1. This is a quadratic with $a = -1$, $b = 140$, $c = -4,000$.

2. The parabolic curve opens down.

3. The $y$-intercept is at $-4,000$.

4. The $x$-intercepts are at 40 and 100 because

$$0 = -x^2 + 140x - 4,000$$
$$0 = x^2 - 140x + 4,000$$
$$0 = (x - 40)(x - 100)$$
$$x - 40 = 0 \qquad x - 100 = 0$$
$$x = 40 \qquad\quad x = 100$$

5. The vertex is located at (70, 900) because the axis of symmetry is $x = 70$

and $P(70) = -(70)^2 + 140(70) - 4,000$

$= -4,900 + 9,800 - 4,000$

$= 900$

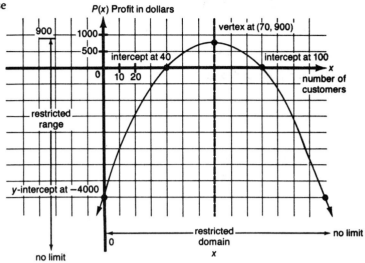

| | |
|---|---|
| $(0, -4,000)$ | If Jim has no customers, he will have losses of $4,000 per day. |
| $(40, 0)$ | If there are 40 customers per day, he will break even; that is, his profit will be 0, but no losses. |
| $(100, 0)$ | If there are 100 customers per day, he will also break even; that is, profit will be 0, but no losses. |
| $(70, 900)$ | If there are 70 customers, he will make a profit of $900. This is the best he can achieve in one day. |

If there are between 39 and 99 customers, per day, inclusive, Jim will make a profit.

If there are less than 40 customers per day, Jim will lose money.

If there are more than 100 customers per day, for whatever reason, Jim will also lose money.

Restricted domain:   $x \geq 0$ or if you prefer $0 \leq x$.

Restricted range:     $P(x) \leq 900$

This time $x$, number of people, must be non-negative; but $P(x)$, which is profit, could be a negative, denoting losses.

The relevant position of the graph is found in the first and fourth quadrants.

# UNIT 8

1.  $y = f(x) = x^3 - 9x^2 + 18x$

    1.  This is a cubic with $a = 1$, $d = 0$.

    2.  The curve opens up: ⌐⌐

    3.  The $y$-intercept is at 0.

    4.  The $x$-intercepts are 0, 3, and 6 because

        $0 = x^3 - 9x^2 + 18x$

        $0 = x(x^2 - 9x + 18)$

        $0 = x(x - 3)(x - 6)$

        $x = 0 \qquad x - 3 = 0 \qquad x - 6 = 0$

        $x = 0 \qquad\qquad x = 3 \qquad\qquad x = 6$

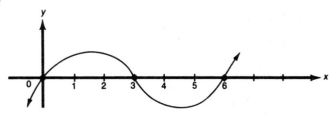

2.   $y = f(x) = 2x^3 + 6x^2 - 20x$

   1.   This is a cubic with $a = 2$, $d = 0$.

   2.   The curve opens up: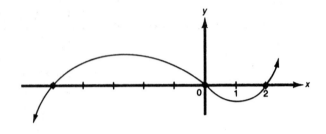

   3.   The $y$-intercept is at 0.

   4.   The $x$-intercepts are 0, $-5$, and 2 because

   $$0 = 2x^3 + 6x^2 - 20x$$
   $$0 = 2x(x^2 + 3x - 10)$$
   $$0 = 2x(x + 5)(x - 2)$$

   | $2x = 0$ | $x + 5 = 0$ | $x - 2 = 0$ |
   |---|---|---|
   | $x = 0$ | $x = -5$ | $x = 2$ |

3.   $y = f(x) = 12x^3 - 24x^2 + 12x$

   1.   This is a cubic with $a = 12$, $d = 0$.

   2.   The curve opens up: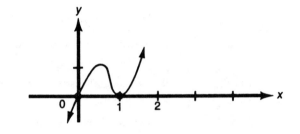

   3.   The $y$-intercept is at 0.

   4.   The $x$-intercepts are 0 and 1 because

   $$0 = 12x^3 - 24x^2 + 12x$$
   $$0 = 12x(x^2 - 2x + 1)$$
   $$0 = 12x(x - 1)^2$$

   | $12x = 0$ | $x - 1 = 0$ |
   |---|---|
   | $x = 0$ | $x = 1$ |

4.   $f(x) = (x - 2)(x + 3)(x - 1)$

   $$= x^3 + \cdots + 6$$

   1.   This is a cubic with $a = 1$, $d = 6$.

   2.   The curve opens up: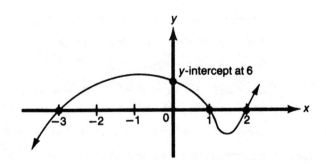

   3.   The $y$-intercept is at 6.

   y-intercept at 6

   4.   The $x$-intercepts are 2, $-3$, and 1 because

   $$0 = (x - 2)(x + 3)(x - 1)$$

   | $x - 2 = 0$ | $x + 3 = 0$ | $x - 1 = 0$ |
   |---|---|---|
   | $x = 2$ | $x = -3$ | $x = 1$ |

5.   $y = f(x) = -x^3 + 3x^2$

   1.   This is a cubic with $a = -1$, $d = 0$.

   2.   The curve opens down: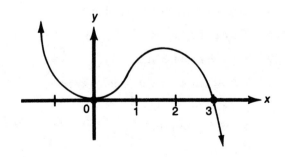

   3.   The $y$-intercept is at 0.

   4.   The $x$-intercepts are 0 and 3 because

   $$0 = -x^3 + 3x^2$$
   $$0 = -x^2(x - 3)$$

   | $-x^2 = 0$ | $x - 3 = 0$ |
   |---|---|
   | $x = 0$ | $x = 3$ |

6.   $y = f(x) = x^3 - 4x$

   1.   This is a cubic with $a = 1$, $d = 0$.

   2.   The curve opens up: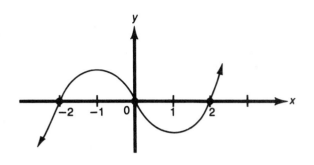

   3.   The $y$-intercept is at 0.

   4.   The $x$-intercepts are 0, 2, and $-2$ because

$$0 = x^3 - 4x$$

$$0 = x(x^2 - 4)$$

$$0 = x(x + 2)(x - 2)$$

   $x = 0$        $x + 2 = 0$              $x - 2 = 0$

   $x = 0$              $x = -2$              $x = 2$

7.

   $f(x) = 3$ if $5 < x$

   linear—half line—slope 0—horizontal

   lower limit        if $x = 5$, $f(5) = 3$        $(5, 3)$  open

   $f(x) = x^2$        if $x \leq 2$

   quadratic  ⌣          , vertex at $(0, 0)$

   upper limit        if $x = 2$, $f(2) = (2)^2 = 4$    $(2, 4)$

   $f(x) = x + 1$    if $3 \leq x \leq 5$

   lower limit        if $x = 3$, $f(3) = 3 + 1 = 4$  $(3, 4)$

   upper limit        if $x = 5$, $f(5) = 5 + 1 = 6$  $(5, 6)$

   linear—line segment

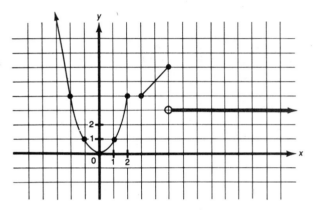

# UNIT 9

   1.   $f(x) = 3x^{15}$

   $f'(x) = 3(15x^{14})$

   $= 45x^{14}$

   2.   $f(x) = -7x^2$

   $f'(x) = -7(2x)$

   $= -14x$

   3.   $f(x) = 21x^7$

   $f'(x) = 21(7x^6)$

   $= 147x^6$

   4.   $f(x) = x$

   $f'(x) = 1$

   5.   $f(x) = -11x^9$

   $f'(x) = -11(9x^8)$

   $= -99x^8$

   6.   $f(x) = 10$

   $f'(x) = 0$

   7.   $f(x) = -8$

   $f'(x) = 0$

   8.   $f(x) = 4x + 3$

   $f'(x) = 4(1) + 0$

   $= 4$

9.　$f(x) = -3x + 17$

　　$f'(x) = -3(1) + 0$

　　　　$= -3$

10.　$f(x) = \dfrac{2}{5}x$

　　$f'(x) = \dfrac{2}{5}(1)$

　　　　$= \dfrac{2}{5}$

11.　$f(x) = 3x^2 + 2x - 6$

　　$f'(x) = 3(2x) + 2(1) + 0$

　　　　$= 6x + 2$

12.　$f(x) = -8x^3 - 7x^2 + 11x - 5$

　　$f'(x) = -8(3x^2) - 7(2x) + 11(1) + 0$

　　　　$= -24x^2 - 14x + 11$

13.　$f(x) = 5x^3 - 4x^2 + 3x + 20$

　　$f'(x) = 5(3x^2) - 4(2x) + 3(1) + 0$

　　　　$= 15x^2 - 8x + 3$

14.　$f(x) = 34x^2 + x^5$

　　$f'(x) = 34(2x) + 5x^4$

　　　　$= 68x + 5x^4$

15.　$f(x) = \dfrac{1}{3}x^3 + \dfrac{1}{2}x^2 - 2x + 13$

　　$f'(x) = \dfrac{1}{3}(3x^2) + \dfrac{1}{2}(2x) - 2(1) + 0$

　　　　$= x^2 + x - 2$

16.　$f(x) = ax^2 + bx + c$

　　$f'(x) = a(2x) + b(1) + 0$

　　　　$= 2ax + b$

17.　$f(x) = mx + b$

　　$f'(x) = m(1) + 0$

　　　　$= m$

18.　$f(x) = ax^3 + bx^2 + cx + d$

　　$f'(x) = a(3x^2) + b(2x) + c(1) + 0$

　　　　$= 3ax^2 + 2bx + c$

# UNIT 10

1.　$y = 11 + 2x - x^3$

　　$\dfrac{dy}{dx} = 2 - 3x^2$

2.　$D_x(7x + 2 - 3x^5) = 7 - 15x^4$

3.　$f(x) = 3 - 7x^2 + 21x + 2x^5$

　　$f'(x) = -14x + 21 + 10x^4$

　　$f'(1) = -14(1) + 21 + 10(1)^4$

　　　　$= -14 + 21 + 10$

　　　　$= 17$

4.　$f(x) = 11x - 2$

　　$f' = 11$

　　The slope of the line is 11.

5.　$\dfrac{d(1 - 3x^2)}{dx} = -6x$

6.　$y = x^6/3 - 2x$

　　　$= \dfrac{1}{3}x^6 - 2x$　　　rewrite first

　　$\dfrac{dy}{dx} = \dfrac{1}{3}(6x^5) - 2$

　　　$= 2x^5 - 2$

7.　$f(x) = x^4/4 - x^3/3 + 5x^2$

　　　$= \dfrac{1}{4}x^4 - \dfrac{1}{3}x^3 + 5x^2$　　　rewritten

　　$f'(x) = \dfrac{1}{4}(4x^3) - \dfrac{1}{3}(3x^2) + 5(2x)$

　　　$= x^3 - x^2 + 10x$

8.       $f(x) = 4x^2 - 2x + 9$

    a.  $f'(x) = 4(2x) - 2(1)$

        $= 8x - 2$

    b.  $f'(0) = 8(0) - 2$

        $= -2$

    c.  $f'(2) = 8(2) - 2$

        $= 16 - 2$

        $= 14$

    d.  Find $x$ such that $f'(x) = 0$ means solve

        $0 = 8x - 2$

        $8x = 2$

$$x = \frac{2}{8} = \frac{1}{4}$$

9.  $y = f(x) = x^9 - 5x^8 + x + 12$

       $f'(x) = 9x^8 - 40x^7 + 1$

    $f'(-1) = 9(-1)^8 - 40(-1)^7 + 1$

        $= 9(1) - 40(-1) + 1$

        $= 9 + 40 + 1$

        $= 50$

$$y - y_1 = m(x - x_1)$$
$$y - 5 = 50(x - (-1))$$
$$y - 5 = 50(x + 1)$$
$$y - 5 = 50x + 50$$

The equation of the line is $y = 50x + 55$.

10.      $f(x) = x^4 - 3x^3 + 2x^2 - 6$

    a.  $f'(x) = 4x^3 - 9x^2 + 4x$

    b.  Need coordinates of the point at $x = 2$

        $f(2) = (2)^4 - 3(2)^3 + 2(2)^2 - 6$

        $= 16 - 24 + 8 - 6$

        $= -6$, which is the point $(2, -6)$

    Need the slope at $x = 2$

    $f'(2) = 4(2)^3 - 9(2)^2 + 4(2)$

        $= 4(8) - 9(4) + 8$

        $= 32 - 36 + 8$

        $= 4$

To write equation:  $y - y_1 = m(x - x_1)$
$$y - (-6) = 4(x - 2)$$
$$y + 6 = 4x - 8$$

The equation of the line is $y = 4x - 14$.

---

# UNIT 11

1.       $N(t) = -t^2 - 5t - 750$

    a.   $N'(t) = -2t - 5$

    b.  $N'(10) = -2(10) - 5 = -20 - 5 = -25$

2.      $C(x) = 10 - 2.5x^2 + x^3$    $x \geq 2$

    a.  $MC = C'(x) = 0 - 2.5(2x) + 3x^2$

        $= -5x + 3x^2$

    b.  $MC = C'(10) = -5(10) + 3(10)^2$

        $= -50 + 3(100)$

        $= -50 + 300$

        $= 250$

    c.  The marginal cost of the 11th unit is $250.

3.           $C(q) = 25q^2 + q + 100$

    a.  $MC = $  $C'(q) = 25(2q) + 1$

          $= 50q + 1$

    To estimate the cost of 101st unit, use 100.

    $MC = C'(100) = 50(100) + 1$

        $= 5,000 + 1$

        $= 5,001$

The approximate cost of manufacturing the next unit (101st) is $5,001.

b.     $C(q) = 25q^2 + q + 100$

$C(101) = 25(101)^2 + (101) + 100$

$= 255{,}025 + 201$

$= \$255{,}226$

$C(100) = 25(100)^2 + 100 + 100$

$= 250{,}000 + 200$

$= 250{,}200$

Cost of 101st unit $= C(101) - C(100)$

$= 255{,}226 - 250{,}200$

$= \$5{,}026$

4.     $N = \dfrac{1}{3}t^3 - 6t^2 + t + 1{,}757$

a.     $N' = \dfrac{1}{3}(3t^2) - 6(2t) + 1$

$= t^2 - 12t + 1$

b.     $N'(6) = (6)^2 - 12(6) + 1$

$= 36 - 72 + 1$

$= -35$

Sales were decreasing at that point in time.

c.     $N'(12) = (12)^2 - 12(12) + 1$

$= 144 - 144 + 1$

$= 1$

Sales were increasing at that point in time.

5.     $Q(x) = .1x^3 + 3x^2 + 1$

a.     $Q'(x) = .3x^2 + 6x$

$Q'(10) = .3(10)^2 + 6(10)$

$= .3(100) + 60$

$= 30 + 60$

$= 90$

It is estimated that if the work force was increased by one additional employee, bringing the total to 11, daily output would increase by 90 units.

b.   Output if there are 11 workers is

$Q(11) = .1(11)^3 + 3(11)^2 + 1$

$= .1(1{,}331) + 3(121) + 1$

$= 133.1 + 363 + 1$

$= 497.1$ units

Output if there are 10 workers is

$Q(10) = .1(10)^3 + 3(10)^2 + 1$

$= .1(1{,}000) + 3(100) + 1$

$= 100 + 300 + 1$

$= 401$

Actual change in output $= Q(11) - Q(10)$

$= 497.1 - 401$

$= 96.1$ units

6.     $P(x) = -x^2 + 140x - 4{,}000$

a.   Marginal profit equals the derivative of the profit function; therefore

$MP = P'(x) = -2x + 140$

b.   $MP = P'(50) = -2(50) + 140$

$= -100 + 140$

$= 40$

c.   Profit is increasing.

The marginal profit of the 51st customer is $40. Or, at 50 customers per day, one more customer would increase profits by approximately $40.

7.     $y = H(t) = -16t^2 + 64t + 80$

a.   velocity $= H'(t) = -32t + 64$

$H'(1) = -32 + 64$

$= 32$ ft/sec

After one second, the ball's velocity is 32 feet per second.

b.   distance $= H(t) = -16t^2 + 64t + 80$

$H(4) = -16(4)^2 + 64(4) + 80$

$= -256 + 256 + 80$

$= 80$

The ball will be 80 feet above the ground after 4 seconds.

c. velocity $= H'(t) = -32t + 64$

$$H'(4) = -32(4) + 64$$
$$= -128 + 64$$
$$= -64$$

After 4 seconds, the ball's velocity will be $-64$ feet per second. Or, after 4 seconds, the ball will be falling at the rate of 64 feet per second.

d. Initial velocity occurs when $t = 0$.

velocity $= H'(t) = -32t + 64$

$$H'(0) = 0 + 64$$
$$= 64 \text{ ft/sec}$$

The ball was tossed upward with an initial velocity of 64 feet per second.

e. $y = H(t) = -16t^2 + 64t + 80$

The function is a quadratic with $a = -16$.

The parabola opens down

The $y$-intercept is at 80.

The $x$-intercepts are at $-1$ and 5 because

$$0 = -16t^2 + 64t + 80$$
$$0 = -16(t^2 - 4t - 5)$$
$$0 = -16(t - 5)(t + 1)$$
$$t - 5 = 0 \qquad t + 1 = 0$$
$$t = 5 \qquad t = -1$$

The vertex is located at $(2, 144)$ because the axis of symmetry is $x = 2$.

$$H(2) = -16(2)^2 + 64(2) + 80$$
$$= -64 + 128 + 80$$
$$= 144$$

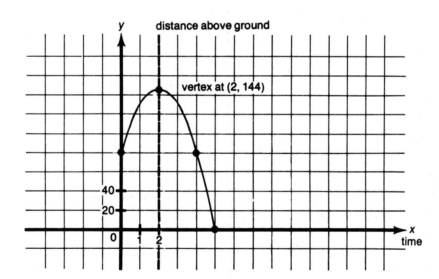

$y = H(t) = -16t^2 + 64t + 80$

From the graph, $t = 2$.

The ball will reach its maximum height after 2 seconds.

f. From the graph, $y = 144$.

The ball will rise to a maximum height of 144 feet.

g. From the graph, $t = 5$.

The ball will hit the ground in 5 seconds.

h. velocity $= H'(t) = -32t + 64$

$$H'(5) = -32(5) + 64$$
$$= -160 + 64$$
$$= -96 \text{ ft/sec}$$

The ball will hit the ground with a velocity of $-96$ feet per second.

# UNIT 12

1. $f(x) = (2x + 1)(x^2 - 3)$

   $f'(x) = (2x + 1)\dfrac{d}{dx}(x^2 - 3) + (x^2 - 3)\dfrac{d}{dx}(2x + 1)$

   $\quad = (2x + 1)(2x) + (x^2 - 3)(2)$

   $\quad = 4x^2 + 2x + 2x^2 - 6$

   $\quad = 6x^2 + 2x - 6$

2. $y = (x^2 + 1)(2 - x^3)$

   $\dfrac{dy}{dx} = (x^2 + 1)\dfrac{d}{dx}(2 - x^3) + (2 - x^3)\dfrac{d}{dx}(x^2 + 1)$

   $\quad = (x^2 + 1)(-3x^2) + (2 - x^3)(2x)$

   $\quad = -3x^4 - 3x^2 + 4x - 2x^4$

   $\quad = -5x^4 - 3x^2 + 4x$

3. $y = 3(x^7 + 2x^5 - x^3 + 6x - 1)$

   $y' = 3(7x^6 + 10x^4 - 3x^2 + 6)$

4. $f(u) = (u^3 + 5)^7$

   $f'(u) = 7(u^3 + 5)^6 \dfrac{d}{dx}(u^3 + 5)$

   $\quad = 7(u^3 + 5)^6(3u^2)$

   $\quad = 21u^2(u^3 + 5)^6$

5. $f(x) = -\dfrac{1}{x^2}$

   $f'(x) = \dfrac{x^2 \dfrac{d}{dx}(-1) - (-1)\dfrac{d}{dx}(x^2)}{x^4}$

   $\quad = \dfrac{x^2 \cdot 0 + 1(2x)}{x^4}$

   $\quad = \dfrac{2x}{x^4}$

   $\quad = \dfrac{2}{x^3}$

6. $D_x\left(\dfrac{3}{x + 4}\right) = \dfrac{(x + 4)\dfrac{d}{dx}(3) - 3\dfrac{d}{dx}(x + 4)}{(x + 4)^2}$

   $\quad = \dfrac{(x + 4) \cdot 0 - 3(1)}{(x + 4)^2}$

   $\quad = \dfrac{-3}{(x + 4)^2}$

7. $y = \dfrac{x^2 + 2}{x - 3}$

   $y' = \dfrac{(x - 3)\dfrac{d}{dx}(x^2 + 2) - (x^2 + 2)\dfrac{d}{dx}(x - 3)}{(x - 3)^2}$

   $\quad = \dfrac{(x - 3)(2x) - (x^2 + 2)(1)}{(x - 3)^2}$

   $\quad = \dfrac{2x^2 - 6x - x^2 - 2}{(x - 3)^2}$

   $\quad = \dfrac{x^2 - 6x - 2}{(x - 3)^2}$

8. $N = \dfrac{100t}{t - 9}$

   $\dfrac{dN}{dt} = \dfrac{(t - 9)\dfrac{d}{dt}(100t) - 100t\dfrac{d}{dt}(t - 9)}{(t - 9)^2}$

   $\quad = \dfrac{(t - 9)(100) - 100t(1)}{(t - 9)^2}$

   $\quad = \dfrac{100t - 900 - 100t}{(t - 9)^2}$

   $\quad = \dfrac{-900}{(t - 9)^2}$

9.  $f(x) = (6x^2 + 12x + 1)^5$

    a. $f(-2) = (6(-2)^2 + 12(-2) + 1)^5$

    $\quad = (24 - 24 + 1)^5$

    $\quad = (1)^5$

    $\quad = 1$

    The point $(-2, 1)$ is on the graph of the function. Or, when $x = -2$, $f(-2) = 1$.

    b. $f'(x) = 5(6x^2 + 12x + 1)^4 \dfrac{d}{dx}(6x^2 + 12x + 1)$

    $\quad = 5(6x^2 + 12x + 1)^4(12x + 12)$

    $f'(-2) = 5[6(-2)^2 + 12(-2) + 1]^4(12(-2) + 12)$

    $\quad = 5(1)^4(-12)$

    $\quad = -60$

    The slope of the tangent line at the point $(-2, 1)$ is $-60$.

10.     $R(x) = (x + 1)(2x^2 + x + 3)$     $x \geq 0$

a.      $MR = R'(x)$

$$= (x + 1)\frac{d}{dx}(2x^2 + x + 3) + (2x^2 + x + 3)\frac{d}{dx}(x + 1)$$

$$= (x + 1)(4x + 1) + (2x^2 + x + 3)(1)$$

$$= 4x^2 + x + 4x + 1 + 2x^2 + x + 3$$

$$= 6x^2 + 6x + 4$$

b.      $MR = 6x^2 + 6x + 4$

$R'(10) = 6(10)^2 + 6(10) + 4$

$$= 6(100) + 60 + 4$$

$$= 664$$

It is increasing because $R'(10)$ is positive.

11.     $D(p) = 6 + p - p^2$     $0 \leq p \leq 3$

a.      $D'(p) = 1 - 2p$

b.      $D(1.5) = 6 + (1.5) - (1.5)^2$

$$= 6 + 1.5 - 2.25$$

$$= 5.25$$

At a price of \$1.50 per pound, 5.25 pounds of jelly beans will be sold per day.

$D'(1.5) = 1 - 2(1.5)$

$$= 1 - 3$$

$$= -2$$

If the price of a pound of jelly beans is increased \$1 to \$2.50 per pound, the demand for jelly beans will drop by approximately 2 pounds.

12.     $P(t) = 15 - \dfrac{6}{t + 1}$ million

a.      $P'(t) = 0 - \dfrac{\left[(t + 1)\dfrac{d}{dt}(6) - 6\dfrac{d}{dt}(t + 1)\right]}{(t + 1)^2}$

$$= -\frac{[(t + 1)(0) - 6(1)]}{(t + 1)^2}$$

$$= +\frac{6}{(t + 1)^2}$$

b.      $P'(1) = \dfrac{6}{(1 + 1)^2}$

$$= \frac{6}{4}$$

$$= 1.5$$

The population will be growing at the rate of 1.5 million per year one year from now.

c.  $P(2) = 15 - \dfrac{6}{2+1}$

$\quad = 15 - \dfrac{6}{3}$

$\quad = 13$ million people 2 years from now

$P(1) = 15 - \dfrac{6}{1+1}$

$\quad = 15 - 3$

$\quad = 12$ million

$P(2) - P(1) = 13 - 12 = 1$

The population will actually increase by only 1 million during the second year.

d.  $P'(t) = \dfrac{6}{(t+1)^2}$

$P'(9) = \dfrac{6}{(9+1)^2}$

$\quad = \dfrac{6}{(10)^2}$

$\quad = \dfrac{6}{100}$

$\quad = .06$

Nine years from now the population will be growing at the rate of .06 million or 60,000 per year.

e.  In the long run, as $t$ gets very large, the rate of the population growth is going to get very small.

# UNIT 13

1.  $f(x) = 3x^{-4}$

$f'(x) = 3(-4x^{-4-1})$

$\quad = -12x^{-5}$

$\quad = \dfrac{-12}{x^5}$

2.  $f(x) = 2x^{\frac{3}{2}}$

$f'(x) = 2\left(\dfrac{3}{2}x^{\frac{3}{2}-1}\right)$

$\quad = \not{2} \cdot \dfrac{3}{\not{2}}x^{\frac{1}{2}}$

$\quad = 3\sqrt{x}$

3.  $f(x) = x^{-\frac{1}{3}}$

$f'(x) = -\dfrac{1}{3}x^{-\frac{1}{3}-1}$

$\quad = -\dfrac{1}{3}x^{-\frac{4}{3}}$

$\quad = -\dfrac{1}{3} \cdot \dfrac{1}{x^{\frac{4}{3}}}$

$\quad = \dfrac{-1}{3\sqrt[3]{x^4}}$

$\quad = \dfrac{-1}{3x\sqrt[3]{x}}$

4.  $f(x) = \dfrac{5}{x^2} - \dfrac{1}{x}$

$\quad = 5x^{-2} - x^{-1}$

$f'(x) = 5(-2x^{-2-1}) - (-1x^{-1-1})$

$\quad = -10x^{-3} + x^{-2}$

$\quad = \dfrac{-10}{x^3} + \dfrac{1}{x^2}$

5.  $y = 2\sqrt{x}$

$\quad = 2x^{\frac{1}{2}}$

$\dfrac{dy}{dx} = 2\left(\dfrac{1}{2}x^{\frac{1}{2}-1}\right)$

$\quad = \not{2} \cdot \dfrac{1}{\not{2}}x^{-\frac{1}{2}}$

$\quad = \dfrac{1}{x^{\frac{1}{2}}}$

$\quad = \dfrac{1}{\sqrt{x}}$

6.    $y = \sqrt{2x}$

$$= (2x)^{\frac{1}{2}}$$

$$\frac{dy}{dx} = \frac{1}{2}(2x)^{\frac{1}{2}-1}\frac{d}{dx}(2x) \qquad \text{Chain Rule}$$

$$= \frac{1}{2}(2x)^{-\frac{1}{2}} \cdot 2$$

$$= \frac{1}{\cancel{2}} \cdot \frac{1}{(2x)^{\frac{1}{2}}} \cdot \cancel{2}$$

$$= \frac{1}{\sqrt{2x}}$$

7.    $y = \dfrac{2}{\sqrt{x}}$

$$= 2x^{-\frac{1}{2}}$$

$$\frac{dy}{dx} = 2\left(-\frac{1}{2}x^{-\frac{1}{2}-1}\right)$$

$$= 2 \cdot -\frac{1}{2}x^{-\frac{3}{2}}$$

$$= \cancel{2} \cdot \frac{-1}{\cancel{2}} \cdot \frac{1}{x^{\frac{3}{2}}}$$

$$= \frac{-1}{\sqrt{x^3}} \qquad \text{or} \qquad \frac{-1}{x\sqrt{x}}$$

8.    $g(x) = \dfrac{1}{3x^2}$

$$= \frac{1}{3}x^{-2}$$

$$g'(x) = \frac{1}{3}(-2x^{-2-1})$$

$$= \frac{1}{3} \cdot -2x^{-3}$$

$$= \frac{1}{3} \cdot -2 \cdot \frac{1}{x^3}$$

$$= \frac{-2}{3x^3}$$

9.    $y = \sqrt{1-3x^2}$

$$= (1-3x^2)^{\frac{1}{2}}$$

$$\frac{dy}{dx} = \frac{1}{2}(1-3x^2)^{\frac{1}{2}-1}\frac{d}{dx}(1-3x^2) \text{ Chain Rule}$$

$$= \frac{1}{2}(1-3x^2)^{-\frac{1}{2}}(-6x)$$

$$= \frac{1}{\cancel{2}} \cdot \frac{1}{(1-3x^2)^{\frac{1}{2}}} \cdot -\overset{3}{\cancel{6}x}$$

$$= \frac{-3x}{\sqrt{1-3x^2}}$$

10.    $D(p) = \sqrt[3]{p^2+1}$

$$= (p^2+1)^{\frac{1}{3}}$$

$$D'(p) = \frac{1}{3}(p^2+1)^{\frac{1}{3}-1}\frac{d}{dp}(p^2+1) \text{ Chain Rule}$$

$$= \frac{1}{3}(p^2+1)^{-\frac{2}{3}}(2p)$$

$$= \frac{1}{3} \cdot \frac{1}{(p^2+1)^{\frac{2}{3}}} \cdot 2p$$

$$= \frac{2p}{3\sqrt[3]{(p^2+1)^2}}$$

11.    $f(x) = x^{\frac{1}{2}}$

$$f'(x) = \frac{1}{2}x^{\frac{1}{2}-1}$$

$$= \frac{1}{2}x^{-\frac{1}{2}}$$

$$= \frac{1}{2} \cdot \frac{1}{x^{\frac{1}{2}}}$$

$$= \frac{1}{2\sqrt{x}}$$

$$f'\left(\frac{1}{4}\right) = \frac{1}{2\sqrt{\dfrac{1}{4}}}$$

$$= \frac{1}{\cancel{2}\left(\dfrac{1}{\cancel{2}}\right)}$$

$$= 1$$

12.  $g(u) = 2u - 4u^{-1}$

$g'(u) = 2 - 4(-1u^{-1-1})$          $g'(2) = 2 + \dfrac{4}{(2)^2}$

$\qquad = 2 + 4u^{-2}$

$\qquad = 2 + 4\left(\dfrac{1}{u^2}\right)$          $\qquad = 2 + \dfrac{4}{4}$

$\qquad\qquad\qquad\qquad\qquad\qquad = 2 + 1$

$\qquad = 2 + \dfrac{4}{u^2}$          $\qquad = 3$

# UNIT 14

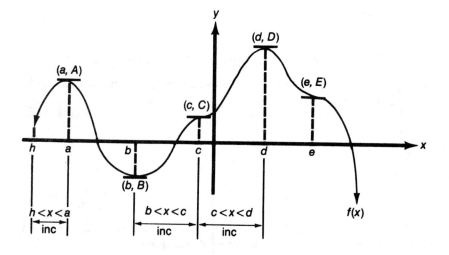

1. There are 5 critical points.

2. They are located as indicated in drawing.

3. See drawing.

4. $(a, A)$—relative maximum

   $(b, B)$—relative minimum

   $(c, C)$—stationary inflection point

   $(d, D)$—relative maximum, also global maximum

   $(e, E)$—stationary inflection point

5. Yes, the absolute maximum value for $y$ is $D$ and it occurs when $x = d$.

6. There is no global minimum.

7. Increasing when  $h \le x < a$

   $\qquad\qquad\qquad b < x < c$

   $\qquad\qquad\qquad c < x < d$

# UNIT 15

1.  $y = f(x) = 2x^2 - 4x$

    Find all critical points.

    $f'(x) = 4x - 4$

    $0 = 4x - 4$

    $4 = 4x$

    $x^* = 1$

    Use $f(x)$ to find $y$-coordinate.

    $f(1) = 2(1)^2 - 4(1) = 2 - 4 = -2$

    and $(1, -2)$ is a critical point.

    To classify:

    $f(x)$ is a quadratic function with $a = 2$. The parabola opens up:

    Answer:   $(1, -2)$ is an absolute minimum. Or stated another way, the minimum value for $y$ is $-2$ and
    it occurs when $x = 1$.

2.  $f(x) = -x^2 - 1$

    $f'(x) = -2x$

    $0 = -2x$

    $x^* = 0$

    $f(0) = -0 - 1 = -1$ and $(0, -1)$ is a critical point. The function is quadratic with $a = -1$, open-
    ing . The critical point $(0, -1)$ is an absolute maximum.

3.  $f(x) = 3x - 2$

    $f'(x) = 3$

    $0 = 3$ has no solution.

    There are no critical points.

4.      $f(x) = 6x^2 + x - 1$

    $f'(x) = 12x + 1$

    $0 = 12x + 1$

    $-12x = 1$

    $x^* = -\dfrac{1}{12}$

    $f\left(-\dfrac{1}{12}\right) = 6\left(-\dfrac{1}{12}\right)^2 + \left(\dfrac{1}{12}\right) - 1$

    $\qquad = \dfrac{6}{144} + \dfrac{1}{12} - 1$

    $\qquad = \dfrac{1}{24} + \dfrac{2}{24} - \dfrac{24}{24}$

$$= -\frac{21}{24}$$

$$= -\frac{7}{8} \text{ thus } \left(-\frac{1}{12}, -\frac{7}{8}\right) \text{ is a critical point.}$$

The function is quadratic with $a = 6$, opening . The critical point $\left(-\frac{1}{12}, -\frac{7}{8}\right)$ is an absolute minimum.

5.    $f(x) = x^3 + 3x^2$

$f'(x) = 3x^2 + 6x$

$0 = 3x^2 + 6x$

$\quad = 3x(x + 2)$

$3x = 0 \qquad\qquad\qquad\qquad x + 2 = 0$

$x^* = 0 \qquad\qquad\qquad\qquad x^* = -2$

$f(0) = 0 + 0 = 0 \qquad\qquad f(-2) = -8 + 12 = 4$

Critical points at $(0, 0)$ and $(-2, 4)$.

The function is a cubic with $a = 1$, opening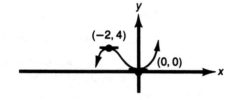

The critical point $(-2, 4)$ is a relative maximum and the critical point $(0, 0)$ is a relative minimum.

6.    $f(x) = -x^3 + 7$

$f'(x) = -3x^2$

$0 = -3x^2$

$x^* = 0$

$f(0) = 0 + 7 = 7$   thus $(0, 7)$ is a critical point.

The function is a cubic with one critical point; therefore it must be a stationary inflection point.

7.    $f(x) = x^3 - 3x^2 - 9x + 5$

$f'(x) = 3x^2 - 6x - 9$

$0 = 3x^2 - 6x - 9$

$\quad = x^2 - 2x - 3$

$\quad = (x - 3)(x + 1)$

$x^* = 3 \qquad\qquad\qquad\qquad x^* = -1$

$f(3) = 27 - 3(9) - 9(3) + 5 \qquad f(-1) = -1 - 3(1) - 9(-1) + 5$

$\quad = 27 - 27 - 27 + 5 \qquad\qquad\quad = -1 - 3 + 9 + 5$

$\quad = -22 \qquad\qquad\qquad\qquad\qquad = 10$

There are critical points at $(3, -22)$ and $(-1, 10)$.

The function is a cubic with $a = 1$, opening 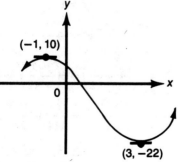.

The critical point $(-1, 10)$ is a relative maximum and $(3, -22)$ is a relative minimum.

8.  $f(x) = x^6 + 2$

$f'(x) = 6x^5$

$0 = 6x^5$

$x^* = 0$

$f(0) = 0 + 2 = 2,$   thus $(0, 2)$ is a critical point.

To classify, use the original function test.

On left:    Let $x = -1,$    $f(-1) = (-1)^6 + 2 = 1 + 2 = 3$

At the point:    $x = 0,$        $f(0) =$            $= 2$

On right:   Let $x = 1,$        $f(1) = 1 + 2$       $= 3$

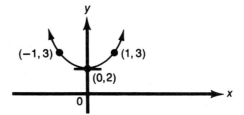

The critical point $(0, 2)$ is a relative minimum.

In fact, it is an absolute minimum.

9.  $f(x) = 5x^7 - 35x^6 + 63x^5$

$f'(x) = 35x^6 - 210x^5 + 315x^4$

$0 = 35x^6 - 210x^5 + 315x^4$

$= x^6 - 6x^5 + 9x^4$

$= x^4(x^2 - 6x + 9)$

$= x^4(x - 3)^2$

$x^4 = 0$                    $(x - 3)^2 = 0$

$x^* = 0$                    $x - 3 = 0$

$f(0) = 0 - 0 + 0 = 0$        $x^* = 3$

$f(3) = 5(2,187) - 35(729) + 63(243)$

$= 10,935 - 25,515 + 15,309$

$= 729$

There are critical points at $(0, 0)$ and $(3, 729)$.

For the point $(0, 0)$:

On the left:    Let $x = -1,$    $f(-1) = 5(-1) - 35(1) + 63(-1) = -103$

At the point:        $x = 0,$        $f(0) =$                            $=$    $0$

On the right:  Let $x = 1,$        $f(1) = 5 - 35 + 63$                $=$    $33$

For the point $(3, 729)$:

On the left:   Let $x = 2,$        $f(2) = 640 - 2,240 + 2,016$        $=$    $416$

At the point:        $x = 3,$        $f(0) =$                            $=$    $729$

On the right:  Let $x = 4,$        $f(4) = 81,920 - 143,360 + 64,512 = 3,072$

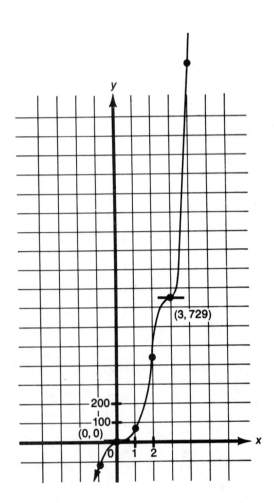

Both critical points, $(0, 0)$ and $(3, 729)$ are stationary inflection points.

10.   $f(x) = \dfrac{1}{3}x^3 - 3x^2 + 5x + 2$

$f'(x) = \dfrac{1}{3}(3x^2) - 6x + 5$

$\quad\quad = x^2 - 6x + 5$

$\quad 0 = (x - 1)(x - 5)$

$x^* = 1 \quad\quad\quad\quad\quad\quad x^* = 5$

$f(1) = \dfrac{1}{3} - 3 + 5 + 2 \quad\quad f(5) = \dfrac{1}{3}(125) - 3(25) + 5(5) + 2$

$\quad\quad = 4\dfrac{1}{3} \quad\quad\quad\quad\quad\quad\quad = 41\dfrac{2}{3} - 75 + 25 + 2$

$\quad\quad\quad\quad\quad\quad\quad\quad\quad\quad\quad\quad = -6\dfrac{1}{3}$

There are critical points at $\left(1, 4\dfrac{1}{3}\right)$ and $\left(5, -6\dfrac{1}{3}\right)$

The function is a cubic with $a = \dfrac{1}{3}$ and opening

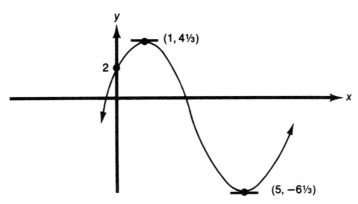

The critical point $\left(1, 4\dfrac{1}{3}\right)$ is a relative maximum and the critical point $\left(5, -6\dfrac{1}{3}\right)$ is a relative minimum.

# UNIT 16

1.  $f(x) = x^4 - 4x^3 + 10$

    $f'(x) = 4x^3 - 12x^2$

    $\qquad = 4x^2(x - 3)$

    Critical points occur where $f'(x) = 0$.

    $\qquad 0 = 4x^2(x - 3)$

    $4x^2 = 0 \qquad\qquad x - 3 = 0$

    $\quad x^* = 0 \qquad\qquad\quad x^* = 3$

    $f(0) = 10 \qquad\qquad f(3) = 81 - 108 + 10 = -17$

    Critical points at $(0, 10)$ and $(3, -17)$

    Use first-derivative test with $f'(x) = 4x^3 - 12x^2$.

    For the point $(0, 10)$                                        the curve is:

    On the left:    Let $x = -1$, $f'(-1) = -4 - 12 = -$      decreasing

    At the point:    $(0, 10)$                                  levels off

    On the right:    Let $x = 1$, $f'(1) = 4 - 12 = -$          decreasing

    The critical point is a stationary inflection point.

    For the point $(3, -17)$                                       the curve is:

    On the left:    Let $x = 2$, $f'(2) = 32 - 48 = -$        decreasing

    At the point:    $(3, -17)$                                 levels off

    On the right:    Let $x = 4$, $f'(4) = 256 - 192 = +$       increasing

    The critical point is a relative minimum.

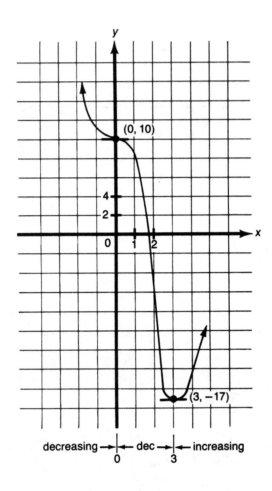

decreasing →|← dec →|← increasing
            0           3

2. a. The absolute minimum value of $f(x) = -17$ and occurs when $x = 3$. $f(x)$ does not have an absolute maximum.

   b. The function is increasing over the interval $3 < x$.

   c. The function is decreasing over the intervals $x < 0$ and $0 < x < 3$.

3. $$f(x) = \frac{1}{4}x^4 + x$$

$$f'(x) = x^3 + 1$$

$$0 = x^3 + 1$$

$$x^3 = -1$$

$$x^* = -1$$

$$f(-1) = \frac{1}{4}(1) - 1 = -\frac{3}{4}$$

There is one critical point at $\left(-1, -\frac{3}{4}\right)$.

Use the first-derivative test with $f'(x) = x^3 + 1$.

For the point $\left(-1, -\dfrac{3}{4}\right)$                                          the curve is:

On the left:    Let $x = -2$, $f'(-2) = -8 + 1 = -$          decreasing

At the point:    $\left(-1, -\dfrac{3}{4}\right)$                          levels off

On the right:   Let $x = 0$, $f'(0) = 0 + 1 = +$            increasing

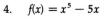

The critical point $\left(-1, -\dfrac{3}{4}\right)$ is a relative minimum.

4.   $f(x) = x^5 - 5x$

$f'(x) = 5x^4 - 5$

$\quad = 5(x^4 - 1)$

$\quad = 5(x^2 + 1)(x^2 - 1)$

$\quad = 5(x^2 + 1)(x + 1)(x - 1)$

$0 = 5(x^2 + 1)(x + 1)(x - 1)$

$\quad\quad x^2 + 1 = 0 \quad\quad\quad x + 1 = 0 \quad\quad\quad\quad\quad x - 1 = 0$

$\quad\quad$ has no solution $\quad\quad x^* = -1 \quad\quad\quad\quad\quad x^* = 1$

$\quad\quad\quad\quad\quad\quad\quad\quad f(-1) = -1 + 5 = 4 \quad\quad f(1) = 1 - 5 = -4$

The function has critical points at $(-1, 4)$ and $(1, -4)$.

For the point $(1, -4)$                                          the curve is:

On the left:    Let $x = 0$, $f'(0) = 0 - 5 = -$          decreasing

At the point:    $(1, -4)$                                       levels off

On the right:   Let $x = 2$, $f'(2) = 80 - 5 = +$          increasing

The critical point $(1, -4)$ is a relative minimum.

For the point $(-1, 4)$                                          the curve is:

On the left:    Let $x = -2$, $f'(-2) = 80 - 5 = +$       increasing

At the point:    $(-1, 4)$                                       levels off

On the right:   Let $x = 0$, $f'(0) = 0 - 5 = -$          decreasing

The critical point $(-1, 4)$ is a relative maximum.

5.   The function is a fifth-degree polynomial function.

The $y$-intercept is at 0.

The $x$-intercepts are 0, $\sqrt[4]{5}$, and $-\sqrt[4]{5}$.

$0 = x^5 - 5x$

$0 = x(x^4 - 5)$

$\quad x = 0 \quad\quad x^4 = 5$

$\quad\quad\quad\quad\quad x = \pm\sqrt[4]{5}$

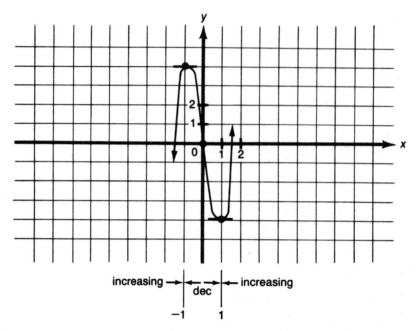

a.   $f(x)$ has neither an absolute maximum nor an absolute minimum.

b.   The function is increasing over the intervals $x < -1$ and $1 < x$.

c.   The function is decreasing over the interval $-1 < x < 1$.

6.   $f(x) = \dfrac{1}{x-2}$

$$f'(x) = \frac{(x-2)\dfrac{d}{dx}(1) - (1)\dfrac{d}{dx}(x-2)}{(x-2)^2}$$

$$= \frac{(x-2)(0) - (1)(1)}{(x-2)^2}$$

$$= \frac{-1}{(x-2)^2}$$

Critical points occur where $f'(x) = 0$.

$0 = \dfrac{-1}{(x-2)^2}$    Clear of fractions by multiplying

$0 = -1$ has no solution

There are no critical points for this function.

7.   $f(x) = \dfrac{x}{x^2+1}$

$$f'(x) = \frac{(x^2+1)\dfrac{d}{dx}(x) - x\dfrac{d}{dx}(x^2+1)}{(x^2+1)^2}$$

$$= \frac{(x^2 + 1)(1) - x(2x)}{(x^2 + 1)^2}$$

$$= \frac{x^2 + 1 - 2x^2}{(x^2 + 1)^2}$$

$$= \frac{1 - x^2}{(x^2 + 1)^2}$$

$$= \frac{(1 - x)(1 + x)}{(x^2 + 1)^2}$$

Critical points occur where $f'(x) = 0$.

$$0 = \frac{(1 - x)(1 + x)}{(x^2 + 1)^2} \quad \text{Clear of fractions}$$

$$0 = (1 - x)(1 + x)$$

$$x^* = 1 \qquad\qquad x^* = -1$$

$$f(1) = \frac{1}{1 + 1} = \frac{1}{2} \qquad f(-1) = \frac{-1}{1 + 1} = -\frac{1}{2}$$

The function has critical points at $\left(1, \frac{1}{2}\right)$ and $\left(-1, -\frac{1}{2}\right)$.

Use the first-derivative test with $f'(x) = \frac{1 - x^2}{(x^2 + 1)^2}$.

For the point $\left(1, \frac{1}{2}\right)$     the curve is:

On the left:   Let $x = 0$, $f'(0) = \frac{1}{(1)^2} = +$    increasing

At the point:   $\left(1, \frac{1}{2}\right)$    levels off

On the right:   Let $x = 2$, $f'(2) = \frac{1 - 4}{(5)^2} = -$    decreasing

The critical point $\left(1, \frac{1}{2}\right)$ is a relative maximum.

For the point $\left(-1, -\frac{1}{2}\right)$    the curve is:

On the left:   Let $x = -2$, $f'(-2) = \frac{1 - 4}{(5)^2} = -$    decreasing

At the point:   $\left(-1, -\frac{1}{2}\right)$    levels off

On the right:   Let $x = 0$, $f'(0) = \frac{1}{(1)^2} = +$    increasing

The critical point $\left(-1, -\frac{1}{2}\right)$ is a relative minimum.

8.  a.  $f(x) = 2x^3 - 21x^3 + 60x + 10$

    The function is a cubic with $a = 2$.

    The $y$-intercept is at 10.

    $f'(x) = 6x^2 - 42x + 60$

    $0 = 6x^2 - 42x + 60$

    $0 = 6(x^2 - 7x + 10)$

    $0 = 6(x - 2)(x - 5)$

    $x^* = 2$

    $f(2) = 16 - 84 + 120 + 10$

    $\phantom{f(2)} = 62$

    $x^* = 5$

    $f(5) = 250 - 525 + 300 + 10$

    $\phantom{f(5)} = 35$

    The function has critical points at
    (2, 62) and (5, 35).

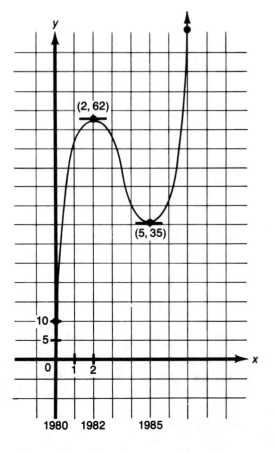

    b.  $0 \leq x$

    c.  $f(7) = 2(343) - 21(49) + 60(7) + 10 = 686 - 1{,}029 + 420 + 10 = 87$

        There were 87 paying guests in 1987.

    d.  The present

    e.  1985

    f.  35

    g.  The number of guests at the resort was decreasing.

    h.  It is estimated that the number of guests will continue to increase.

# UNIT 17

1.  $f(x) = x^4 - 18x^2$

    $f'(x) = 4x^3 - 36x$

    $f''(x) = 12x^2 - 36$

    Critical points are where $f'(x) = 0$.

    $0 = 4x^3 - 36x$

    $0 = 4x(x^2 - 9)$

    $0 = 4x(x + 3)(x - 3)$

$$x^* = 0 \qquad x^* = -3 \qquad x^* = 3$$

$$f(0) = 0 \qquad f(-3) = -81 \qquad f(3) = -81$$

There are critical points at $(0, 0)$, $(-3, -81)$, and $(3, -81)$.

At the point $(0, 0)$,

$f''(0) = 0 - 36 = -$, curve is concave down, ⌒

The critical point at $(0, 0)$ is a relative maximum.

At the point $(-3, -81)$,

$f''(-3) = 108 - 36 = +$, curve is concave up, ⌣

The critical point $(3, -81)$ is a relative minimum.

At the point $(3, -81)$,

$f''(3) = 108 - 36 = +$, curve is concave up, ⌣

The critical point $(3, -81)$, is a relative minimum.

2. The function is a fourth-degree polynomial function.

The $y$-intercept is at 0.

The $x$-intercepts are 0, $\sqrt{18}$, and $-\sqrt{18}$.

$$0 = x^4 - 18x^2$$

$$0 = x^2(x^2 - 18)$$

$$0 = x^2(x + \sqrt{18})(x - \sqrt{18})$$

From Exercise 1, $(0, 0)$ is a relative maximum,

$(3, -81)$ is a relative minimum,

$(-3, -81)$ is a relative minimum.

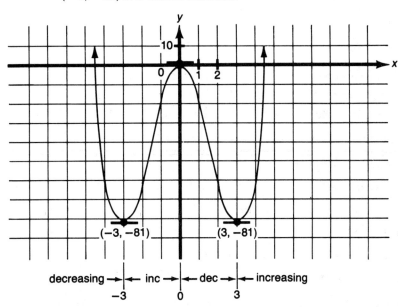

a. The absolute minimum value of $f(x)$ is $-81$ and occurs when $x = -3$ or $x = 3$.

b. The function is increasing over the intervals $-3 < x < 0$ and $3 < x$.

c. The function is decreasing over the intervals $x < -3$ and $0 < x < 3$.

3. $f(x) = 2 - x^5$

$f'(x) = -5x^4$

$f''(x) = -20x^3$

Critical points are where $f'(x) = 0$.

$0 = -5x^4$

$x^* = 0$

$f(0) = 2 - 0 = 2$, and $(0, 2)$ is a critical point.

Use the second-derivative test with $f''(x) = -20x^3$.

$f''(0) = 0$ and test fails.

Rework using first-derivative test with $f'(x) = -5x^4$.

| For the point $(0, 2)$ | | the curve is: |
|---|---|---|
| On the left: | Let $x = -1$, $f'(-1) = -5 = -$ | decreasing |
| At the point: | $(0, 2)$ | levels off |
| On the right: | Let $x = 1$, $f'(1) = -5 = -$ | decreasing |

The critical point $(0, 2)$ is a stationary inflection point.

4. $f(x) = 10x^6 + 24x^5 + 15x^4 + 2$

$f'(x) = 60x^5 + 120x^4 + 60x^3$

$f''(x) = 300x^4 + 480x^3 + 180x^2$

Critical points are where $f'(x) = 0$.

$0 = 60x^5 + 120x^4 + 60x^3$

$0 = 60x^3(x^2 + 2x + 1)$

$0 = 60x^3(x + 1)^2$

$x^* = 0 \qquad x^* = -1$

$f(0) = 2 \qquad f(-1) = 10 - 24 + 15 + 2 = 3$

There are critical points at $(0, 2)$ and $(-1, 3)$.

Use the second-derivative test with $f''(x) = 300x^4 + 480x^3 + 180x^2$.

At $(-1, 3)$, $f''(-1) = 300 - 480 + 180 = 0$, test fails

At $(0, 2)$, $f''(0) = 0$, test fails

Use first-derivative test with $f'(x) = 60x^5 + 120x^4 + 60x^3$

$$= 60x^3(x + 1)^2$$

| For the point $(-1, 3)$ | | the curve is: |
|---|---|---|
| On the left: | Let $x = -2$, $f'(-2) = (+)(-)^3(-)^2 = -$ | decreasing |
| At the point: | $(-1, 3)$ | levels off |
| On the right: | Let $x = -.5$, $f'(-.5) = (+)(-)^3(+)^2 = -$ | decreasing |

The critical point $(-1, 3)$ is a stationary inflection point.

| For the point $(0, 2)$ | the curve is: |

On the left:    Let $x = -.5$, $f'(-.5) = -$                          decreasing

At the point:   $(0, 2)$                                              levels off

On the right:   Let $x = 1$, $f'(1) = (+)(+)^3(+)^2 = +$              increasing

The critical point $(0, 2)$ is a relative minimum.

5.  The function is a sixth-degree polynomial function.

    The $y$-intercept is at 2.

    From Exercise 4, $(-1, 3)$ is a stationary inflection point with the curve decreasing both to the left and right of the point.

    $(0, 2)$ is a relative minimum.

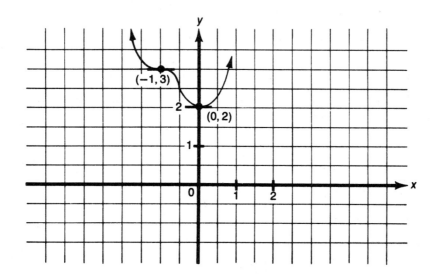

6.  $f(x) = \dfrac{x^2 + 1}{x}$

$$f'(x) = \frac{x \dfrac{d}{dx}(x^2 + 1) - (x^2 + 1)\dfrac{d}{dx}(x)}{x^2}$$

$$= \frac{x(2x) - (x^2 + 1)(1)}{x^2}$$

$$= \frac{2x^2 - x^2 - 1}{x^2}$$

$$= \frac{x^2 - 1}{x^2}$$

Critical points are where $f'(x) = 0$.

$0 = \dfrac{x^2 - 1}{x^2}$ \qquad Clear of fraction

$0 = x^2 - 1$

$0 = (x + 1)(x - 1)$

$$x^* = -1 \qquad\qquad x^* = 1$$

$$f(-1) = \frac{(-1)^2 + 1}{-1} \qquad f(1) = \frac{(1)^2 + 1}{1}$$

$$= \frac{1+1}{-1} \qquad\qquad = \frac{1+1}{1}$$

$$= -2 \qquad\qquad\qquad = 2$$

There are critical points at $(-1, -2)$ and $(1, 2)$.

7. Use the original-function test with $f(x) = \dfrac{x^2 + 1}{x}$.

At the point $(-1, -2)$

On the left:    Let $x = -2$,   $f(-2) = \dfrac{(-2)^2 + 1}{-2} = \dfrac{4 + 1}{-2} = \dfrac{5}{-2} = -2.5$

At the point:    $(-1, -2)$    $f(-1) = \qquad\qquad\qquad -2$

On the right:   Let $x = -.5$,   $f(-.5) = \dfrac{(-.5)^2 + 1}{-.5} = \dfrac{+.25 + 1}{-.5} = -2.5$

The critical point $(-1, -2)$ is a relative maximum.

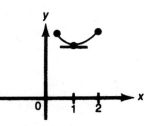

At the point $(1, 2)$

On the left:    Let $x = .5$,   $f(.5) = \dfrac{(.5)^2 + 1}{.5} = \dfrac{1.25}{.5} = 2.5$

At the point:    $(1, 2)$    $f(1) = \qquad\qquad = 2$

On the right:   Let $x = 2$,   $f(2) = \dfrac{(2)^2 + 1}{2} = \dfrac{5}{2} = 2.5$

The critical point $(1, 2)$ is a relative minimum.

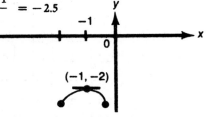

8. Use the first-derivative test with $f'(x) = \dfrac{x^2 - 1}{x^2} = \dfrac{(x + 1)(x - 1)}{x^2}$

For the point $(-1, -2)$                             the curve is:

On the left:    Let $x = -2, f'(-2) = \dfrac{(-)(-)}{(-)^2} = \dfrac{+}{+} = +$     increasing

At the point:    $(-1, -2)$                              levels off

On the right:   Let $x = -.5, f'(-.5) = \dfrac{(+)(-)}{+} = \dfrac{-}{+} = -$    decreasing

The curve is increasing, levels off, then decreasing.

The critical point $(-1, -2)$ is a relative maximum.

For the point    $(1, 2)$                                                    the curve is:

On the left:    Let $x = .5$, $f'(.5) = \dfrac{(+)(-)}{(+)^2} = \dfrac{-}{+} = -$          decreasing

At the point:    $(1, 2)$                                                  levels off

On the right:    Let $x = 2$, $f'(2) = \dfrac{(+)(+)}{(+)^2} = \dfrac{+}{+} = +$          increasing

The curve is decreasing, levels off, then increasing.

The critical point $(1, 2)$ is a relative minimum

9.  $f'(x) = \dfrac{x^2 - 1}{x^2}$

$$f''(x) = \dfrac{x^2 \dfrac{d}{dx}(x^2 - 1) - (x^2 - 1)\dfrac{d}{dx}(x^2)}{x^4}$$

$$= \dfrac{x^2(2x) - (x^2 - 1)(2x)}{x^4}$$

$$= \dfrac{2x^3 - 2x^3 + 2x}{x^4}$$

$$= \dfrac{2x}{x^4}$$

$$= \dfrac{2}{x^3}$$

At the point $(-1, -2)$,

$f''(-1) = \dfrac{2}{(-1)^3} = -$ , curve is concave down,

The critical point $(-1, -2)$ is a relative maximum.

At the point $(1, 2)$,

$f''(1) = \dfrac{2}{1} = +$ , curve is concave up,

The critical point $(1, 2)$ is a relative minimum.

10.  The critical points are $(-1, -2)$, a relative maximum, and $(1, 2)$, a relative minimum, no matter which method was used.

# UNIT 18

1.  $y = f(x) = x^2 - 12x + 1$        $1 \le x \le 10$

See Example 4, Unit 15, for the graph and determination of the critical points.

$$f(6) = 36 - 12(6) + 1 = -35$$

$$f(1) = 1 - 12 + 1 = -10$$

$$f(10) = 100 - 120 + 1 = -19$$

The absolute maximum is $-10$ and occurs when $x = 1$.

The absolute minimum is $-35$ and occurs when $x = 6$.

2.  $y = f(x) = 5x - 1 \qquad 0 \le x \le 5$

There are no critical points, $f(x)$ is linear.

$$f(0) = 0 - 1 = -1$$
$$f(5) = 25 - 1 = 24$$

The absolute maximum is 24 and occurs when $x = 5$.

The absolute minimum is $-1$ and occurs when $x = 0$.

3.  $y = f(x) = \dfrac{1}{3}x^3 - x^2 - 3x + 1 \qquad 0 \le x \le 3$

See Example 7, Unit 15, for the graph and determination of critical points.

$$f(0) = 0 - 0 - 0 + 1 = 1$$
$$f(3) = 9 - 9 - 9 + 1 = -8$$

The absolute maximum is 1 and occurs when $x = 0$.

The absolute minimum is $-8$ and occurs when $x = 3$.

4.  $y = f(x) = -x^3 - 3x^2 + 1 \qquad -2 \le x \le 2$

See Example 8, Unit 15, for graph and determination of critical points.

$$f(-2) = 8 - 12 + 1 = -3$$
$$f(0) = 0 - 0 + 1 = 1$$
$$f(2) = -8 - 12 + 1 = -19$$

The absolute maximum is 1 and occurs when $x = 0$.

The absolute minimum is $-19$ and occurs when $x = 2$.

5.  $y = f(x) = x^4 - 8x^3 + 18x^2 - 27 \qquad 1 \le x \le 2$

See Example 5, Unit 16, for the graph and determination of critical points.

$$f(1) = 1 - 8 + 18 - 27 = -16$$
$$f(2) = 16 - 64 + 72 - 27 = -3$$

The absolute maximum is $-3$ and occurs when $x = 2$.

The absolute minimum is $-16$ and occurs when $x = 1$.

6.  $y = f(x) = 3x^4 - 16x^3 + 18x^2 \qquad 0 \le x \le 2$

See Example 4, Unit 17, for the graph and determination of critical points.

$$f(0) = 0 - 0 + 0 = 0$$
$$f(1) = 3 - 16 + 18 = 5$$
$$f(2) = 48 - 128 + 72 = -8$$

The absolute maximum is 5 and occurs when $x = 1$.

The absolute minimum is $-8$ and occurs when $x = 2$.

# UNIT 19

1.  $P(x) = 50x - 0.01x^2 - 1,000$ with $x \geq 0$

    $P'(x) = 50 - 0.02x$

    $0 = 50 - 0.02x$

    $.02x = 50$

    $2x = 5,000$

    $x = 2,500$

    $P(2,500) = 50(2,500) - 0.01(2,500)^2 - 1,000$

    $\qquad = 125,000 - 62,500 - 1,000$

    $\qquad = 61,500$

    The function, $P(x)$, is quadratic, opening down, absolute maximum.

    Answer:   Maximum profit of \$61,500 would be achieved when 2,500 units are produced.

2.  $C(x) = x^2 + 4x + 16 \qquad x > 0$

    $A(x) = \dfrac{x^2 + 4x + 16}{x}$

    a.  $C(10) = 100 + 40 + 16 = 156$

        The total cost of producing 10 units is \$156.

    b.  The average cost per unit for producing 10 units is \$15.60.

    c.  $A'(x) = \dfrac{x\dfrac{d}{dx}(x^2 + 4x + 16) - (x^2 + 4x + 16)\dfrac{d}{dx}(x)}{x^2}$

    $\qquad = \dfrac{x(2x + 4) - (x^2 + 4x + 16)(1)}{x^2}$

    $\qquad = \dfrac{2x^2 + 4x - x^2 - 4x - 16}{x^2}$

    $\qquad = \dfrac{x^2 - 16}{x^2}$

    $0 = \dfrac{x^2 - 16}{x^2}$

    $0 = x^2 - 16$

    $0 = (x + 4)(x - 4)$

    $\qquad x^* = 4 \qquad x^* = -4$ not in domain

    $\qquad A(4) = \dfrac{16 + 16 + 16}{4} = \dfrac{48}{4} = 12$

    d.  The average cost per unit is mimimized if 4 units are produced, in which case the average cost is **\$12** per unit.

3.    Revenue = price · quantity

$$R(p) = p(1,152 - 5p)$$

$$= 1,152p - 5p^2$$

$$R'(p) = 1,152 - 10p$$

$$0 = 1,152 - 10p$$

$$10p = 1,152$$

$$p = 115.2$$

If $p = \$115.20$ and $q = 1,152 - 5p$

then $q = 1,152 - 5(115.20) = 576$

$$R(115.20) = (115.20)(576) = \$66,355.20$$

a.    A price of $115.20 will result in maximum total revenue.

b.    At a price of $115.20, maximum revenue of $66,355.20 can be expected.

c.    At a price of $115.20 per unit, revenue will be maximized, and 576 units will be sold.

d.    parabola

$x$-intercepts

$$0 = p(1,152 - 5p)$$

$$p = 0 \qquad 1,152 - 5p = 0$$

$$p = 230.4$$

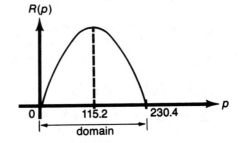

e.    Restricted domain $0 \le p \le 230.40$

4.    Profit = Total Revenue − Total Cost

$$= (\text{price})(\text{quantity}) - (\text{total cost})$$

$$P(x) = 50x - (110 - 2.5x^2 + x^3) \qquad x \ge 0$$

$$= 50x - 110 + 2.5x^2 - x^3$$

$$P'(x) = 50 + 5x - 3x^2$$

$$0 = -3x^2 + 5x + 50$$

$$0 = 3x^2 - 5x - 50$$

$$= (3x + 10)(x - 5)$$

$$x^* = 5 \qquad 3x + 10 = 0$$

$$3x = -10$$

Test:   $P(x)$ is a cubic, ↰↷↓

$x^* = 5$ is absolute maximum on domain

$$P(5) = 50(5) - 110 + 2.5(25) - (125)$$

$$= 250 - 110 + 62.5 - 125$$

$$= 77.5$$

The company should manufacture 5 units which would lead to maximum profit of $77.50.

5.    $C(x) = 20x + \dfrac{1,280}{x}$

$= 20x + 1,280x^{-1}$

$C'(x) = 20 + 1,280(-x^{-1-1})$

$= 20 + 1,280x^{-2}$

$0 = 20 - \dfrac{1,280}{x^2}$

$0 = 20x^2 - 1,280$

$20x^2 = 1,280$

$x^2 = 64$

$x = \pm 8$

$x = 8$

Test:   $C''(x) = -1,280(-2x^{-2-1})$

$= 2,560x^{-3}$

$= \dfrac{2,560}{x^3}$

$C''(8) = +$, concave up, minimum

To minimize costs the company should operate 8 machines.

6.    Maximize area

Area = length · width

$= y \cdot x$

$A(x) = (50 - 1.25x)x$          ends \$1.25 per foot

$= 50x - 1.25x^2$          front \$2.00 per foot

$A'(x) = 50 - 2.5x$          \$100 for fencing

$0 = 50 - 2.5x$

$2.5x = 50$          constraint

$25x = 500$          $100 = 2.0(y) + 1.25(2x)$

$x = 20$          $100 = 2y + 2.5x$

$2y = 100 - 2.5x$

$y = 50 - 1.25x$

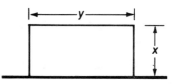

If $x = 20$,

then $y = 50 - 1.25(20)$

$= 50 - 25$

$= 25$

To maximize area the rectangular garden should be 20′ × 25′ with the long side being the front. The area would be 500 sq ft.

# UNIT 20

**1.**     $$x^2 + 5y = 10$$

$$\frac{d}{dx}(x^2) + \frac{d}{dx}(5y) = \frac{d}{dx}(10)$$

$$2x + 5\frac{dy}{dx} = 0$$

$$5\frac{dy}{dx} = -2x$$

$$\frac{dy}{dx} = \frac{-2x}{5}$$

**2.**     $$y^2 = x^3$$

$$\frac{d}{dx}(y^2) = \frac{d}{dx}(x^3)$$

$$2y\frac{dy}{dx} = 3x^2$$

$$\frac{dy}{dx} = \frac{3x^2}{2y}$$

**3.**     $$x^2 + y^2 = 9$$

$$\frac{d}{dx}(x^2) + \frac{d}{dx}(y^2) = \frac{d}{dx}(9)$$

$$2x + 2y\frac{dy}{dx} = 0$$

$$2y\frac{dy}{dx} = -2x$$

$$\frac{dy}{dx} = \frac{-2x}{2y}$$

$$\frac{dy}{dx} = \frac{-x}{y}$$

**4.**     $$x^2 - y^2 + 3y = 7$$

$$\frac{d}{dx}(x^2) - \frac{d}{dx}(y^2) + \frac{d}{dx}(3y) = \frac{d}{dx}(7)$$

$$2x - 2y\frac{dy}{dx} + 3\frac{dy}{dx} = 0$$

$$3\frac{dy}{dx} - 2y\frac{dy}{dx} = -2x$$

$$\frac{dy}{dx}(3 - 2y) = -2x$$

$$\frac{dy}{dx} = \frac{-2x}{3 - 2y}$$

**5.**     $$x^2 - y^2 - x = 1 \qquad \text{at } (2,\ 1)$$

$$\frac{d}{dx}(x^2) - \frac{d}{dx}(y^2) - \frac{d}{dx}(x) = \frac{d}{dx}(1)$$

$$2x - 2y\frac{dy}{dx} - 1 = 0$$

$$-2y\frac{dy}{dx} = 1 - 2x$$

$$\frac{dy}{dx} = \frac{1 - 2x}{-2y}$$

at $(2,\ 1)$     $$\frac{dy}{dx} = \frac{1 - 2(2)}{-2(1)}$$

$$= \frac{1 - 4}{-2}$$

$$= \frac{3}{2}$$

**6.**     $$x^3y = 2$$

$$\frac{d}{dx}(x^3y) = \frac{d}{dx}(2)$$

$$x^3\frac{d}{dx}(y) + y\frac{d}{dx}(x^3) = 0$$

$$x^3\frac{dy}{dx} + y(3x^2) = 0$$

$$x^3\frac{dy}{dx} = -3x^2y$$

$$\frac{dy}{dx} = \frac{-3x^2y}{x^3}$$

$$= \frac{-3y}{x}$$

**7.**     $$xy = 15 \qquad \text{at } (3,\ 5)$$

$$\frac{d}{dx}(xy) = \frac{d}{dx}(15)$$

$$x\frac{d}{dx}(y) + y\frac{d}{dx}(x) = 0$$

$$x\frac{dy}{dx} + y = 0$$

$$x\frac{dy}{dx} = -y$$

$$\frac{dy}{dx} = \frac{-y}{x}$$

at $(3, 5)$ $\dfrac{dy}{dx} = \dfrac{-5}{3}$

8. $y = \sqrt{x}$

$y^2 = x$

$\dfrac{d}{dx}(y^2) = \dfrac{d}{dx}(x)$

$2y\dfrac{dy}{dx} = 1$

$\dfrac{dy}{dx} = \dfrac{1}{2y}$

9. $q = p^2 - 40p + 2{,}150$ $\quad 0 \le p \le 20$

a. $\dfrac{d}{dq}(q) = \dfrac{d}{dq}(p^2) - \dfrac{d}{dq}(40p) + \dfrac{d}{dq}(2{,}150)$

$1 = 2p\dfrac{dp}{dq} - 40\dfrac{dp}{dq} + 0$

$1 = \dfrac{dp}{dq}(2p - 40)$

$\dfrac{dp}{dq}(2p - 40) = 1$

$\dfrac{dp}{dq} = \dfrac{1}{2p - 40}$

b. $q = 100 - 400 + 2{,}150 = 1{,}850$

c. At $p = 10$, $\dfrac{dp}{dq} = \dfrac{1}{20 - 40} = \dfrac{1}{-20} = -.05$

d. At a price of \$10, 1,850 units will be demanded. And if demand increases by one additional unit, the price will decrease by 5 cents to \$9.95.

# UNIT 21

1. $f(x) = e^{x^2}$

$f'(x) = \dfrac{d}{dx}(x^2) \cdot e^{x^2}$

$= 2xe^{x^2}$

Critical points occur where $f'(x) = 0$.

$0 = 2xe^{x^2}$

$2x = 0 \qquad e^{x^2} = 0$ has no solution

$x^* = 0$

If $x^* = 0$, $f(0) = e^0 = 1$,

thus $(0, 1)$ is the only critical point.

Classify critical point    First-derivative test

At $(0, 1)$ with $f'(x) = 2xe^{x^2}$ — the curve is

On the left:   Let $x = -1$, $f'(-1) = (+)(-)(+) = -$   decreasing

At the point:   $(0, 1)$ curve — levels off

On the right:   Let $x = 1$, $f'(1) = (+)(+)(+) = +$   increasing

Answer:   The critical point $(0,1)$ is a relative minimum.

The minimum value for $f(x)$ is 1 and it occurs when $x = 0$.

2.  $f(x) = \dfrac{e^x}{x^2}$

$$f'(x) = \frac{x^2 \dfrac{d}{dx}(e^x) - e^x \dfrac{d}{dx}(x^2)}{(x^2)^2} \qquad \text{Quotient Rule}$$

$$= \frac{x^2 e^x - e^x(2x)}{x^4}$$

$$= \frac{xe^x(x-2)}{x^{4\ 3}}$$

$$= \frac{e^x(x-2)}{x^3}$$

Find critical points.

$$0 = \frac{e^x(x-2)}{x^3}$$

$$0 = e^x(x-2)$$

$\qquad x^* = 2 \qquad e^x = 0$ has no solution.

If $x^* = 2$, $f(2) = \dfrac{e^2}{4}$ and $\left(2, \dfrac{e^2}{4}\right)$ is a critical point.

Classify critical point      First-derivative test

For the point $\left(2, \dfrac{e^2}{4}\right)$ with $f'(x) = \dfrac{e^x(x-2)}{x^3}$              the curve is:

On the left:   Let $x = 1$, $f'(1) = \dfrac{(+)(-)}{(+)} = -$      decreasing

At the point:  $\left(2, \dfrac{e^2}{4}\right)$                                          levels off

On the right:  Let $x = 3$, $f'(3) = \dfrac{(+)(+)}{(+)} = +$      increasing

Answer:  The critical point $\left(2, \dfrac{e^2}{4}\right)$ is a relative minimum.

3.  $f(x) = x^2 e^x$

$$f'(x) = x^2 \frac{d}{dx}(e^x) + (e^x) \frac{d}{dx}(x^2) \quad \text{Product Rule}$$

$$= x^2 e^x + e^x(2x)$$

$$= x^2 e^x + 2xe^x$$

$$= xe^x(x+2)$$

Find all critical points.

$$0 = xe^x(x+2)$$

$\qquad x^* = 0 \qquad x + 2 = 0 \qquad e^x = 0$ has no solution

$\qquad\qquad\qquad\quad x = -2$

If $x^* = 0$, $f(0) = 0e^0 = 0$,

thus $(0, 0)$ is a critical point.

If $x^* = -2$, $f(-2) = (-2)^2 e^{-2} = 4e^{-2}$,

thus $(-2, 4e^{-2})$ or $\left(-2, \dfrac{4}{e^2}\right)$ is a critical point.

Classify the critical points    First-derivative test

For the point $(0, 0)$ with $f'(x) = xe^x(x + 2)$                    the curve is:

$$(0,0) \begin{cases} \text{On the left:} & \text{Let } x = -1, f'(-1) = (-)(+)(+) = - & \text{decreasing} \\ \text{At the point:} & (0, 0) & \text{levels off} \\ \text{On the right:} & \text{Let } x = 1, f'(1) = (+)(+)(+) = \quad + & \text{increasing} \end{cases}$$

$$(-2, 4e^{-2}) \begin{cases} \text{On the left:} & \text{Let } x = -3, f'(-3) = (-)(+)(-) = + & \text{increasing} \\ \text{At the point:} & (-2, 4e^{-2}) & \text{levels off} \\ \text{On the right:} & \text{Let } x = -1, f'(-1) = (-)(+)(+) = - & \text{decreasing} \end{cases}$$

Answer:   The critical point $(0, 0)$ is a relative minimum.

The critical point $(-2, 4e^{-2})$ is a relative maximum.

4.   $p = 5e^{-x}$      $0 \leq x \leq 2$

where $p$ is price in dollars

$x$ is quantity in thousands

Revenue is to be maximized on restricted domain

Revenue = price · quantity

$$= 5e^{-x} \cdot x$$

$$R(x) = 5xe^{-x}$$

$$R'(x) = 5x \frac{d}{dx}(e^{-x}) + (e^{-x}) \frac{d}{dx}(5x)$$

$$= 5x(-e^{-x}) + (e^{-x})(5)$$

$$= 5e^{-x}(-x + 1)$$

$$-x + 1 = 0 \qquad 5e^{-x} = 0 \text{ has no solution}$$

$$x = 1$$

Looking for maximum value on restricted domain

critical point   If $x = 1$, $R(1) = 5e^{-1} = 5(.36788) = 1.8394$

lower limit     If $x = 0$, $R(0) = 5(0)e^0 = 5 \cdot 0 \cdot 1 = 0.000$

upper limit     If $x = 2$, $R(2) = 5(2)e^{-2} = 10(.13534) = 1.3534$

Maximum value is 1.8394 and occurs when $x = 1$.

When $x = 1$, $p = 5e^{-1} = 5(.36788) = 1.8394 \approx 1.84$.

The weekly revenue will be maximized at a price of **$1.84** per box of popcorn.

# UNIT 22

1. $f(x) = \ln 3x$

$$f'(x) = \frac{\dfrac{d(3x)}{dx}}{3x}$$

$$= \frac{3}{3x}$$

$$= \frac{1}{x}$$

2. $f(x) = \ln (x^2 - 5)$

$$f'(x) = \frac{\dfrac{d(x^2 - 5)}{dx}}{(x^2 - 5)}$$

$$= \frac{2x}{x^2 - 5}$$

3. $f(x) = \ln x^2 - 5$

$$f'(x) = \frac{\dfrac{d(x^2)}{dx}}{x^2} - 0$$

$$= \frac{2x}{x^2}$$

$$= \frac{2}{x}$$

4. $f(x) = \dfrac{2x}{\ln x}$

$$f'(x) = \frac{(\ln x) \dfrac{d}{dx}(2x) - (2x) \dfrac{d}{dx}(\ln x)}{(\ln x)^2}$$

$$= \frac{(\ln x)(2) - 2x\left(\dfrac{1}{x}\right)}{(\ln x)^2}$$

$$= \frac{2 \ln x - 2}{(\ln x)^2}$$

$$= \frac{2(\ln x - 1)}{(\ln x)^2},$$

if you prefer the answer factored

5. $f(x) = e^x \ln x$

$$f'(x) = e^x \frac{d}{dx}(\ln x) + (\ln x) \frac{d}{dx}(e^x)$$

$$= e^x \left(\frac{1}{x}\right) + (\ln x)(e^x)$$

$$= e^x \left(\frac{1}{x}\right) + e^x \ln x$$

$$= e^x \left(\frac{1}{x} + \ln x\right), \quad \text{if factored}$$

6. $f(x) = \ln (x^2 - 6x + 25)$

$$f'(x) = \frac{\dfrac{d}{dx}(x^2 - 6x + 25)}{(x^2 - 6x + 25)}$$

$$= \frac{2x - 6}{x^2 - 6x + 25}$$

Critical points occur where $f'(x) = 0$.

$$0 = \frac{2x - 6}{x^2 - 6x + 25}$$

$0 = 2x - 6$       cleared of fractions

$x^* = 3$

If $x^* = 3$ and $f(x) = \ln (x^2 - 6x + 25)$,

then $f(3) = \ln (9 - 18 + 25)$

$$= \ln 16$$

Answer:    The function has one critical point at $(3, \ln 16)$.

7. $f(x) = 4 \ln x - 2x$

$$f'(x) = 4\left(\frac{1}{x}\right) - 2$$

$$= \frac{4}{x} - 2$$

Find all critical points.

$$0 = \frac{4}{x} - 2$$

$0 = 4 - 2x$       clear of fractions

$2x = 4$

$x^* = 2$

If $x^* = 2$, then $f(2) = 4 \ln 2 - 4$ and $(2, 4 \ln 2 - 4)$ is the critical point.

If you wish $4 \ln 2 - 4$ can be rewritten as a decimal approximation of $4(0.69315) - 4 \approx -1.2274$

Classify critical points. Use first-derivative test.

For the point $(2, 4 \ln 2 - 4)$ with $f'(x) = \dfrac{4}{x} - 2$          the curve is:

On the left:     Let $x = 1$, then $f'(1) = 4 - 2 = +$          increasing

At the point:    $(2, 4 \ln 2 - 4)$                                          levels off

On the right:    Let $x = 3$, then $f'(3) = \dfrac{4}{3} - 2 = -$          decreasing

Answer:   The critical point $(2, 4 \ln 2 - 4)$ is a relative maximum. Or, the maximum value of $f(x)$ is $4 \ln 2 - 4$ and occurs when $x = 2$.

# UNIT 23

1.  $\displaystyle\int (2x^4 + x^2)\, dx = 2 \int x^4\, dx + \int x^2\, dx$

$$= \frac{2x^5}{5} + \frac{x^3}{3} + C$$

2.  $\displaystyle\int \left(x + \frac{1}{x}\right) dx = \int x\, dx + \int \frac{1}{x}\, dx$

$$= \frac{x^2}{2} + \ln |x| + C$$

3.  $\displaystyle\int x^{\frac{1}{3}}\, dx = \frac{x^{\frac{1}{3}+1}}{\frac{1}{3}+1} + C$

$$= \frac{x^{\frac{4}{3}}}{\frac{4}{3}} + C$$

$$= \frac{3x^{\frac{4}{3}}}{4} + C$$

4.  $\displaystyle\int 2x^{-5}\, dx = 2 \int x^{-5}\, dx$

$$= 2\frac{x^{-5+1}}{-5+1} + C$$

$$= \frac{2x^{-4}}{-4} + C$$

$$= \frac{2}{-4x^4} + C$$

$$= \frac{-1}{2x^4} + C$$

5.  $\displaystyle\int (15x^2 - \sqrt{2})\, dx = 15 \int x^2\, dx - \sqrt{2} \int dx$

$$= 15\frac{x^3}{3} - \sqrt{2}x + C$$

$$= 5x^3 - \sqrt{2}x + C$$

6.  $\displaystyle\int 3e^x\, dx = 3 \int e^x\, dx$

$$= 3e^x + C$$

7.  $\displaystyle\int \ln 3\, dx = \ln 3 \int dx$

$$= (\ln 3)x + C$$
Remember the ln 3 is a constant.

8.  $\displaystyle\int 2 \ln x\, dx = 2 \int \ln x\, dx$

$$= 2(x \ln x - x) + C$$

$$= 2x \ln x - 2x + C$$

9.  $\displaystyle\int 5ax^4 q\, dq = 5ax^4 \int q\, dq$

$$= \frac{5ax^4 q^2}{2} + C$$

10. $\int \left(x^3 - \dfrac{2}{x^2} + 2\right) dx = \int x^3\,dx - 2\int x^{-2}\,dx + 2\int dx$

$= \dfrac{x^4}{4} - 2\,\dfrac{x^{-2+1}}{-2+1} + 2x + C$

$= \dfrac{x^4}{4} - \dfrac{2x^{-1}}{-1} + 2x + C$

$= \dfrac{x^4}{4} + \dfrac{2}{x} + 2x + C$

or, if you prefer $= \dfrac{1}{4}x^4 + 2x^{-1} + 2x + C$

11. $\int (1 - \ln x)\,dx = \int dx - \int \ln x\,dx$

$= x - (x\ln x - x) + C$

$= x - x\ln x + x + C$

$= 2x - x\ln x + C$

12. $\int \dfrac{e^x - 2x}{2}\,dx = \dfrac{1}{2}\int (e^x - 2x)\,dx$

$= \dfrac{1}{2}\left[\int e^x\,dx - 2\int x\,dx\right]$

$= \dfrac{1}{2}\left(e^x - 2\dfrac{x^2}{2}\right) + C$

$= \dfrac{1}{2}(e^x - x^2) + C$

or, $= \dfrac{e^x - x^2}{2} + C$

13. $\int (3x^{-2} + x^{-1})\,dx = 3\int x^{-2}\,dx + \int x^{-1}\,dx$

$= \dfrac{3x^{-2+1}}{-2+1} + \ln|x| + C$

$= -3x^{-1} + \ln|x| + C$

$= \dfrac{-3}{x} + \ln|x| + C$

14. $\int \dfrac{\sqrt{x}}{2}\,dx = \dfrac{1}{2}\int \sqrt{x}\,dx$

$= \dfrac{1}{2}\int x^{\frac{1}{2}}\,dx$

$= \dfrac{1}{2}\cdot\dfrac{x^{\frac{1}{2}+1}}{\frac{1}{2}+1} + C$

$= \dfrac{1}{2}\cdot\dfrac{x^{\frac{3}{2}}}{\frac{3}{2}} + C$

$= \dfrac{1}{2}\cdot\dfrac{2}{3}x^{\frac{3}{2}} + C$

$= \dfrac{x^{\frac{3}{2}}}{3} + C$

$= \dfrac{\sqrt{x^3}}{3} + C$

15. $\int x^3\left(\dfrac{2}{x} + 4x\right) dx = \int (2x^2 + 4x^4)\,dx$

$= 2\int x^2\,dx + 4\int x^4\,dx$

$= \dfrac{2x^3}{3} + \dfrac{4x^5}{5} + C$

16. If $f'(x) = 2x^3 - x + 5$

then $f(x) = \int (2x^3 - x + 5)\,dx$

$= 2\int x^3\,dx - \int x\,dx + 5\int dx$

$= \dfrac{2x^4}{4} - \dfrac{x^2}{2} + 5x + C$

The point (3, 38) satisfies the function;

$f(x) = \dfrac{1}{2}x^4 - \dfrac{1}{2}x^2 + 5x + C$

$38 = \dfrac{1}{2}(3)^4 - \dfrac{1}{2}(3)^2 + 5(3) + C$

$38 = 40.5 - 4.5 + 15 + C$

$C = -13$

The equation of the curve is
$f(x) = \dfrac{1}{2}x^4 - \dfrac{1}{2}x^2 + 5x - 13.$

17. If $f'(x) = 2x - 5$

then $f(x) = \int (2x - 5)\,dx$

$= 2\int x\,dx - 5\int dx$

$= \dfrac{2x^2}{2} - 5x + C$

$= x^2 - 5x + C$

Given the point $(5, 4)$ as an initial condition

$$f(x) = x^2 - 5x + C$$

$$4 = (5)^2 - 5(5) + C$$

$$4 = C$$

The function is $f(x) = x^2 - 5x + 4$.

18.    $MC = 6x + 1$

$$TC = \int (6x + 1)\, dx$$

$$= \frac{6x^2}{2} + x + C$$

$$= 3x^2 + x + C \quad \text{if fixed cost is \$50}$$

$$TC(x) = 3x^2 + x + 50$$

$$TC(10) = 3(100) + 10 + 50$$

$$= 360$$

The total cost of producing 10 units is $360.

19.    $MP = 100 - 2x$

$$TP = \int (100 - 2x)\, dx$$

$$= 100x - \frac{2x^2}{2} + C$$

$$= 100x - x^2 + C$$

Given that if 10 units are produced and sold yields a profit of 150, then

$$\text{Profit} = 100x - x^2 + C$$

$$150 = 100(10) - (10)^2 + C$$

$$150 = 1{,}000 - 100 + C$$

$$C = -750$$

The profit function is $P(x) = 100x - x^2 - 750$.

---

# UNIT 24

1.

$$\int e^{3 - 5x}\, dx = \int e^u \frac{du}{-5}$$

Let $u = 3 - 5x$

$$= -\frac{1}{5} \int e^u\, du$$

$$\frac{du}{dx} = -5$$

$$= -\frac{1}{5} e^u + C$$

$$du = -5dx$$

$$= -\frac{1}{5} e^{3 - 5x} + C$$

$$\frac{du}{-5} = dx$$

2.

$$\int \frac{x}{(5 + x^2)^4}\, dx = \int \frac{x}{u^4} \frac{du}{2x}$$

Let $u = 5 + x^2$

$$= \frac{1}{2} \int \frac{du}{u^4}$$

$$\frac{du}{dx} = +2x$$

$$= \frac{1}{2} \int u^{-4}\, du$$

$$du = 2x\, dx$$

$$= \frac{1}{2} \cdot \frac{u^{-3}}{-3} + C$$

$$\frac{du}{2x} = dx$$

$$= \frac{u^{-3}}{-6} + C$$

$$= \frac{1}{-6u^3} + C$$

$$= \frac{1}{-6(5 + x^2)^3} + C$$

3.

$$\int \frac{2x^3}{x^4 + 1}\, dx = \int \frac{2x^3}{u} \frac{du}{4x^3}$$

Let $u = x^4 + 1$

$$= \frac{2}{4} \int \frac{du}{u}$$

$$\frac{du}{dx} = 4x^3$$

$$= \frac{1}{2} \ln |u| + C$$

$$du = 4x^3\, dx$$

$$= \frac{1}{2} \ln |x^4 + 1| + C$$

$$\frac{du}{4x^3} = dx$$

$$= \frac{1}{2} \ln (x^4 + 1) + C$$

Note:   The absolute value sign is unnecessary since $x^4 + 1$ is a positive number.

4.
$$\int (2x - 2)e^{x^2 - 2x + 1}\, dx = \int (2x - 2)\, e^u \frac{du}{(2x - 2)}$$

Let $u = x^2 - 2x + 1$            $= \int e^u\, du$

$\dfrac{du}{dx} = 2x - 2$            $= e^u + C$

$du = (2x - 2)\, dx$            $= e^{x^2 - 2x + 1} + C$

$\dfrac{du}{(2x - 2)} = dx$

5.
$$\int 3x^2 \ln (x^3 - 2)\, dx = \int 3x^2 \ln u \frac{du}{3x^2}$$

Let $u = x^3 - 2$            $= \int \ln u\, du$

$\dfrac{du}{dx} = 3x^2$            $= u \ln u - u + C$

$du = 3x^2\, dx$            $= (x^3 - 2) \ln (x^3 - 2) - (x^3 - 2) + C$

$\dfrac{du}{3x^2} = dx$            $= (x^3 - 2) \ln (x^3 - 2) - x^3 + 2 + C$

                                        $= (x^3 - 2) \ln (x^3 - 2) - x^3 + C$

6.
$$\int 36x^2 \sqrt{6x^3 + 1}\, dx = \int 36x^2 (6x^3 + 1)^{\frac{1}{2}}\, dx$$

Let $u = 6x^3 + 1$            $= \int 36x^2 u^{\frac{1}{2}} \frac{du}{18x^2}$

$\dfrac{du}{dx} = 18x^2$            $= \dfrac{36}{18} \int u^{\frac{1}{2}}\, du$

$du = 18x^2\, dx$            $= 2\dfrac{u^{\frac{3}{2}}}{\frac{3}{2}} + C$

$\dfrac{du}{18x^2} = dx$            $= \dfrac{4}{3} u^{\frac{3}{2}} + C$

                                        $= \dfrac{4}{3} (6x^3 + 1)^{\frac{3}{2}} + C$

7.
$$\int 2x(5x^3 - 2x)^7\, dx = \int 2x\, u^7 \frac{du}{15x^2 - 2}$$

Let $u = 5x^3 - 2x$         This problem cannot be integrated using substitution.

$\dfrac{du}{dx} = 15x^2 - 2$

$du = (15x^2 - 2)\, dx$

$\dfrac{du}{15x^2 - 2} = dx$

8.
$$\int \frac{4x^2}{\sqrt{x^3 - 7}}\, dx = \int \frac{4x^2}{(x^3 - 7)^{\frac{1}{2}}}\, dx$$

Let $u = x^3 - 7$
$$= \int \frac{4x^2\, du}{u^{\frac{1}{2}}\, 3x^2}$$

$$\frac{du}{dx} = 3x^2 \qquad\qquad = \frac{4}{3} \int u^{-\frac{1}{2}}\, du$$

$$du = 3x^2\, dx \qquad\qquad = \frac{4}{3} \cdot \frac{u^{\frac{1}{2}}}{\frac{1}{2}} + C$$

$$\frac{du}{3x^2} = dx \qquad\qquad = \frac{8}{3} u^{\frac{1}{2}} + C$$

$$= \frac{8}{3}(x^3 - 7)^{\frac{1}{2}} + C$$

$$= \frac{8}{3}\sqrt{x^3 - 7} + C$$

# UNIT 25

1. $\int x^n\, dx = (x + 1)e^x + C$     Rule 9

2. $\int \frac{dx}{x^2 - a^2} = \frac{1}{2a} \ln\left|\frac{x - a}{x + a}\right|$

$\int \frac{dx}{x^2 - 1} = \frac{1}{2} \ln\left|\frac{x - 1}{x + 1}\right| + C$

Rule 11 with $a = 1$

3. $\int (ax + b)^n\, dx = \frac{(ax + b)^{n+1}}{(n + 1)a}$

$\int (2x - 3)^3\, dx = \frac{(2x - 3)^{3+1}}{(3 + 1)(2)} + C$

$= \frac{(2x - 3)^4}{8} + C$

Rule 25 with $a = 2$, $b = -3$, and $n = 3$

4. $\int \frac{dx}{x \ln x} = \ln|\ln x| + C$     Rule 8

5. $\int \frac{dx}{\sqrt{x^2 - a^2}} = \ln\left|x + \sqrt{x^2 - a^2}\right|$

$\int \frac{dx}{\sqrt{x^2 - 100}} = \ln\left|x + \sqrt{x^2 - 100}\right| + C$

Rule 12 with $a^2 = 100$

6. $\int x^n \ln x\, dx = x^{n+1}\left(\frac{\ln x}{n + 1} - \frac{1}{(n + 1)^2}\right)$

$\int x^5 \ln x\, dx = x^6\left(\frac{\ln x}{6} - \frac{1}{36}\right) + C$

Rule 16 with $n = 5$

7. $\int x^n e^x\, dx = x^n e^x - n \int x^{n-1} e^x\, dx$

$\int x^2 e^x\, dx = x^2 e^x - 2 \int x^{2-1} e^x\, dx$

Rule 20 with $n = 2$

$= x^2 e^x - 2 \int x e^x\, dx$

Rule 9 $\quad = x^2 e^x - 2[(x - 1)e^x] + C$

$= x^2 e^x - 2[xe^x - e^x] + C$

$= x^2 e^x - 2xe^x + 2e^x + C$

$= e^x(x^2 - 2x + 2) + C$

8. $$\int x(ax+b)^{\frac{m}{2}}dx = \frac{2(ax+b)^{\frac{m+4}{2}}}{a^2(m+4)} - \frac{2b(ax+b)^{\frac{m+2}{2}}}{a^2(m+2)}$$

$$\int x(x-6)^{\frac{1}{2}}dx = \frac{2(x-6)^{\frac{1+4}{2}}}{(1)(1+4)} - \frac{2(-6)(x-6)^{\frac{3}{2}}}{(1)(3)} + C$$

Rule 26 with

$a = 1, b = -6, m = 1$

$$= \frac{2(x-6)^{\frac{5}{2}}}{5} - \frac{-12(x-6)^{\frac{3}{2}}}{3} + C$$

9. $$\int x^n(\ln x)^m dx = \frac{x^{n+1}(\ln x)^m}{n+1} - \frac{m}{n+1}\int x^n(\ln x)^{m-1}dx$$

$$\int x^3(\ln x)^2 dx = \frac{x^4(\ln x)^2}{4} - \frac{2}{4}\int x^3(\ln x)dx$$

Rule 22 with $n = 3$ and $m = 2$

Rule 16 with $n = 3$

$$= \frac{x^4(\ln x)^2}{4} - \frac{2}{4}\left[x^4\left(\frac{\ln x}{4} - \frac{1}{(4)^2}\right)\right] + C$$

$$= \frac{x^4(\ln x)^2}{4} - \frac{1}{2}\left[\frac{x^4(\ln x)}{4} - \frac{x^4}{16}\right] + C$$

$$= \frac{x^4(\ln x)^2}{4} - \frac{x^4(\ln x)}{8} - \frac{x^4}{32} + C$$

10. $$\int x^n e^{ax}\, dx = \frac{x^n e^{ax}}{a} - \frac{n}{a}\int x^{n-1}e^{ax}\,dx$$

$$\int x^2 e^{3x}\, dx = \frac{x^2 e^{3x}}{3} - \frac{2}{3}\int x^{2-1}e^{3x}\,dx$$

Rule 24 with $n = 2,\ a = 3$

$$= \frac{x^2 e^{3x}}{3} - \frac{2}{3}\int xe^{3x}\,dx$$

Rule 24 with $n = 1,\ a = 3$

$$= \frac{x^2 e^{3x}}{3} - \frac{2}{3}\left[\frac{xe^{3x}}{3} - \frac{1}{3}\int x^{1-1}e^{3x}\,dx\right]$$

$$= \frac{x^2 e^{3x}}{3} - \frac{2}{3}\left[\frac{xe^{3x}}{3} - \frac{1}{3}\int x^0 e^{3x}\,dx\right]$$

$$= \frac{x^2 e^{3x}}{3} - \frac{2xe^{3x}}{9} + \frac{2}{9}\int e^{3x}\,dx$$

Rule 23 with $a = 23$

$$= \frac{x^2 e^{3x}}{3} - \frac{2xe^{3x}}{9} + \frac{2}{9}\left[\frac{e^{3x}}{3}\right] + C$$

$$= \frac{x^2 e^{3x}}{3} - \frac{2xe^{3x}}{9} + \frac{2e^{3x}}{27} + C$$

# UNIT 26

1. $$\int_1^2 x^2\, dx = \frac{x^3}{3}\Big|_1^2$$

$$= \frac{8}{3} - \frac{1}{3}$$

$$= \frac{7}{3}$$

2. $$\int_2^{10}(-x+10)\, dx = \frac{-x^2}{2} + 10x\Big|_2^{10}$$

$$= \left(\frac{-100}{2} + 100\right) - \left(\frac{-4}{2} + 20\right)$$

$$= -50 + 100 + 2 - 20$$

$$= 32$$

3. $$\int_0^1 (x - x^2)\, dx = \frac{x^2}{2} - \frac{x^3}{3}\Big|_0^1$$

$$= \left(\frac{1}{2} - \frac{1}{3}\right) - 0$$

$$= \frac{3-2}{6}$$

$$= \frac{1}{6}$$

4. $$\int_1^e \ln x\, dx = x\ln x - x\Big|_1^e$$

$$= (e\ln e - e) - (\ln 1 - 1)$$

$$= e\ln e - e - \ln 1 + 1$$

Reminder

$\ln e = 1$

$\ln 1 = 0$

$$= e - e - 0 + 1$$

$$= 1$$

5.  Area $= \displaystyle\int_{-1}^{4} (x^3 + 8)\, dx$

$$= \frac{x^4}{4} + 8x \bigg|_{-1}^{4}$$

$$= \left( \frac{256}{4} + 32 \right) - \left( \frac{1}{4} - 8 \right)$$

$$= 64 + 32 - .25 + 8$$

$$= 103.75 \text{ square units}$$

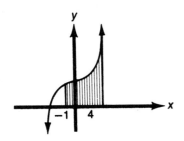

6.  Area $= \displaystyle\left| \int_{0}^{4} (x^2 - 4x)\, dx \right|$

$$= \left| \frac{x^3}{3} - \frac{\cancel{4}x^2}{\cancel{2}_1} \bigg|_{0}^{4} \right|$$

$$= \left| \frac{x^3}{3} - \cancel{2}x^2 \bigg|_{0}^{4} \right|$$

$$= \left| \left( \frac{64}{3} - 32 \right) - 0 \right|$$

$$= \left| 21\frac{1}{3} - 32 \right|$$

$$= \left| -10\frac{2}{3} \right|$$

$$= 10\frac{2}{3} \text{ square units}$$

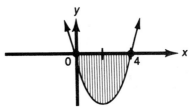

7.  Graph $f(x) = x^3 - 4x$

Cubic with $a = 1$

Curve opens up:

$y$-intercept is at 0

$x$-intercepts at 0, 2, $-2$

 because $0 = x^3 - 4x$

$$= x(x^2 - 4)$$

$$= x(x + 2)(x - 2)$$

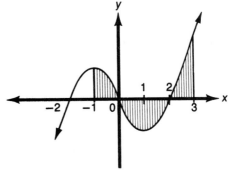

Area $= \displaystyle\int_{-1}^{0} (x^3 - 4x)\, dx - \int_{0}^{2} (x^3 - 4x)\, dx + \int_{2}^{3} (x^3 - 4x)\, dx$

$$= \frac{x^4}{4} - 2x^2 \bigg|_{-1}^{0} - \left( \frac{x^4}{4} - 2x^2 \bigg|_{0}^{2} \right) + \frac{x^4}{4} - 2x^2 \bigg|_{2}^{3}$$

$$= \left[ (0 - 0) - \left( \frac{1}{4} - 2 \right) \right] - \left[ \left( \frac{16}{4} - 8 \right) - (0) \right] + \left[ \left( \frac{81}{4} - 18 \right) - \left( \frac{16}{4} - 8 \right) \right]$$

$$= \left[ -\frac{1}{4} + 2 \right] - [4 - 8] + \left[ \frac{81}{4} - 18 - 4 + 8 \right]$$

$$= -\frac{1}{4} + 2 - 4 + 8 + 20\frac{1}{4} - 18 - 4 + 8$$

$$= 12 \text{ square units}$$

8.  $f(x) = 4$

$g(x) = x^2$

$\text{Area} = \int_{-2}^{2} [f(x) - g(x)]\,dx$

$= \int_{-2}^{2} (4 - x^2)\,dx$

$= 4x - \dfrac{x^3}{3}\Big|_{-2}^{2}$

$= \left(8 - \dfrac{8}{3}\right) - \left(-8 - \dfrac{-8}{3}\right)$

$= 8 - \dfrac{8}{3} + 8 - \dfrac{8}{3}$

$= 8 - 2.67 + 8 - 2.67$

$= 10.67$ square units

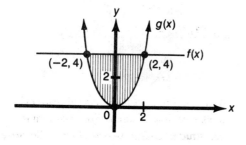

9.  Use the graph from Example 14.
    Notice from $x = -2$ to $x = 0$,
    the top function is $g(x) = x^2 - 2x + 1$.

$\text{Area} = \int_{-2}^{0} [g(x) - f(x)]\,dx$

$= \int_{-2}^{0} [(x^2 - 2x + 1) - (x + 1)]\,dx$

$= \int_{-2}^{0} (x^2 - 2x + 1 - x - 1)\,dx$

$= \int_{-2}^{0} (x^2 - 3x)\,dx$

$= \dfrac{x^3}{3} - \dfrac{3x^2}{2}\Big|_{-2}^{0}$

$= (0) - \left(\dfrac{-8}{3} - \dfrac{12}{2}\right)$

$= 2.67 + 6$

$= 8.67$ square units

10.  $M'(x) = 9x^2 + 50$

Total maintenance from 5 to 10 years after
the house was built.

$M(x) = \int_{5}^{10} M'(x)\,dx$

$= \int_{5}^{10} (9x^2 + 50)\,dx$

$= \dfrac{9x^3}{3} + 50x\Big|_{5}^{10}$

$3x^3 + 50x\Big|_{5}^{10}$

$= [3(10)^3 + 50(10)] - [3(5)^3 + 50(5)]$

$= 3{,}000 + 500 - 375 - 250$

$= \$2{,}875$

It is estimated that the total maintenance costs
for a small house from 5 to 10 years after the
house was built would be \$2,875.

# UNIT 27

1.  $f(x, y) = (3x - 7y)^5$

$f_x = 5(3x - 7y)^4(3)$

$= 15(3x - 7y)^4$

$f_y = 5(3x - 7y)^4(-7)$

$= -35(3x - 7y)^4$

2.  $f(w, x, y) = 2xy + 3x^2yw + w^2 + 10$

$f_w = \dfrac{\partial(2xy)}{\partial w} + 3x^2y\dfrac{\partial(w)}{\partial w} + \dfrac{\partial(w^2)}{\partial w} + \dfrac{\partial(10)}{\partial w}$

$= 0 + 3x^2y(1) + 2w + 0$

$= 3x^2y + 2w$

$f_x = 2y\dfrac{\partial(x)}{\partial x} + 3yw\dfrac{\partial(x^2)}{\partial x} + \dfrac{\partial(w^2)}{\partial x} + \dfrac{\partial(10)}{\partial x}$

$= 2y(1) + 3yw(2x) + 0 + 0$

$= 2y + 6wxy$

$f_y = 2x\dfrac{\partial(y)}{\partial y} + 3x^2w\dfrac{\partial(y)}{\partial y} + \dfrac{\partial(w^2)}{\partial y} + \dfrac{\partial(10)}{\partial y}$

$= 2x(1) + 3x^2w(1) + 0 + 0$

$= 2x + 3wx^2$

3. $f(x, y) = 3x^4y^3 + 2xy$

$$f_x = 3y^3 \frac{\partial(x^4)}{\partial x} + 2y \frac{\partial(x)}{\partial x}$$

$$= 3y^3(4x^3) + 2y(1)$$

$$= 12x^3y^3 + 2y$$

$$f_y = 3x^4 \frac{\partial(y^3)}{\partial y} + 2x \frac{\partial(y)}{\partial y}$$

$$= 3x^4(3y^2) + 2x(1)$$

$$= 9x^4y^2 + 2x$$

$$f_{xx} = 12y^3 \frac{\partial(x^3)}{\partial x} + \frac{\partial(2y)}{\partial x}$$

$$= 12y^3(3x^2) + 0$$

$$= 36x^2y^3$$

$$f_{yy} = 9x^4 \frac{\partial(y^2)}{\partial y} + \frac{\partial(2x)}{\partial y}$$

$$= 9x^4(2y) + 0$$

$$= 18x^4y$$

$$f_{xy} = 12x^3 \frac{\partial(y^3)}{\partial y} + 2 \frac{\partial(y)}{\partial y}$$

$$= 12x^3(3y^2) + 2(1)$$

$$= 36x^3y^2 + 2$$

$$f_{yx} = 9y^2 \frac{\partial(x^4)}{\partial x} + 2 \frac{\partial(x)}{\partial x}$$

$$= 9y^2(4x^3) + 2(1)$$

$$= 36x^3y^2 + 2$$

4. $f(x, y) = 3 - x^2 - y^2$

$$f_x = -2x$$

$$f_y = -2y$$

Critical points occur where $f_x = 0$ and $f_y = 0$.

$$-2x = 0 \qquad x = 0$$

$$-2y = 0 \qquad y = 0$$

Critical point is located at $(0, 0, 3)$

To classify, use second-derivative test with

$$f_{xx} = -2$$

$$f_{yy} = -2$$

$$f_{xy} = 0$$

Let $D = f_{xx} \cdot f_{yy} - (f_{xy})^2$

$$= (-2)(-2) - 0$$

$$= 4$$

Since $D$ is positive, critical point is either a relative maximum or minimum. Look at $f_{xx}$. Since $f_{xx}$ is negative, the critical point $(0, 0, 3)$ is a relative maximum.

5.     $f(x, y) = 10x^2 + 20y^2 - 10xy$

a.     $f_y(x, y) = 40y - 10x$

$$f_y(100, 10) = 40(10) - 10(100)$$

$$= 400 - 1,000$$

$$= -600$$

Currently there are 10 rooms and a weekly budget of $100. If one additional room is constructed, profits are expected to decrease by $600.

b.     $f_x(x, y) = 20x - 10y$

$$f_x(100, 10) = 20(100) - 10(10)$$

$$= 2,000 - 100$$

$$= 1,900$$

If an additional dollar per week is spent on advertising, annual profits are expected to increase by $1,900.

# Appendix

## Table I   Exponential Functions

| $x$ | $e^x$ | $e^{-x}$ | $x$ | $e^x$ | $e^{-x}$ | $x$ | $e^x$ | $e^{-x}$ |
|------|--------|----------|------|--------|----------|-------|---------|----------|
| 0.00 | 1.0000 | 1.00000 | 0.45 | 1.5683 | .63763 | 0.90 | 2.4596 | .40657 |
| 0.01 | 1.0101 | 0.99005 | 0.46 | 1.5841 | .63128 | 0.91 | 2.4843 | .40252 |
| 0.02 | 1.0202 | .98020 | 0.47 | 1.6000 | .62500 | 0.92 | 2.5093 | .39852 |
| 0.03 | 1.0305 | .97045 | 0.48 | 1.6161 | .61878 | 0.93 | 2.5345 | .39455 |
| 0.04 | 1.0408 | .96079 | 0.49 | 1.6323 | .61263 | 0.94 | 2.5600 | .39063 |
| 0.05 | 1.0513 | .95123 | 0.50 | 1.6487 | .60653 | 0.95 | 2.5857 | .38674 |
| 0.06 | 1.0618 | .94176 | 0.51 | 1.6653 | .60050 | 0.96 | 2.6117 | .38298 |
| 0.07 | 1.0725 | .93239 | 0.52 | 1.6820 | .59452 | 0.97 | 2.6379 | .37908 |
| 0.08 | 1.0833 | .92312 | 0.53 | 1.6989 | .58860 | 0.98 | 2.6645 | .37531 |
| 0.09 | 1.0942 | .91393 | 0.54 | 1.7160 | .58275 | 0.99 | 2.6912 | .37158 |
| 0.10 | 1.1052 | .90484 | 0.55 | 1.7333 | .57695 | 1.00 | 2.7183 | .36788 |
| 0.11 | 1.1163 | .89583 | 0.56 | 1.7507 | .57121 | 1.20 | 3.3201 | .30119 |
| 0.12 | 1.1275 | .88692 | 0.57 | 1.7683 | .56553 | 1.30 | 3.6693 | .27253 |
| 0.13 | 1.1388 | .87809 | 0.58 | 1.7860 | .55990 | 1.40 | 4.0552 | .24660 |
| 0.14 | 1.1503 | .86936 | 0.59 | 1.8040 | .55433 | 1.50 | 4.4817 | .22313 |
| 0.15 | 1.1618 | .86071 | 0.60 | 1.8221 | .54881 | 1.60 | 4.9530 | .20190 |
| 0.16 | 1.1735 | .85214 | 0.61 | 1.8404 | .54335 | 1.70 | 5.4739 | .18268 |
| 0.17 | 1.1853 | .84366 | 0.62 | 1.8589 | .53794 | 1.80 | 6.0496 | .16530 |
| 0.18 | 1.1972 | .83527 | 0.63 | 1.8776 | .53259 | 1.90 | 6.6859 | .14957 |
| 0.19 | 1.2092 | .82696 | 0.64 | 1.8965 | .52729 | 2.00 | 7.3891 | .13534 |
| 0.20 | 1.2214 | .81873 | 0.65 | 1.9155 | .52205 | 3.00 | 20.086 | .04979 |
| 0.21 | 1.2337 | .81058 | 0.66 | 1.9348 | .51685 | 4.00 | 54.598 | .01832 |
| 0.22 | 1.2461 | .80252 | 0.67 | 1.9542 | .51171 | 5.00 | 148.41 | .00674 |
| 0.23 | 1.2586 | .79453 | 0.68 | 1.9739 | .50662 | 6.00 | 403.43 | .00248 |
| 0.24 | 1.2712 | .78663 | 0.69 | 1.9937 | .50158 | 7.00 | 1096.6 | .00091 |
| 0.25 | 1.2840 | .77880 | 0.70 | 2.0138 | .49659 | 8.00 | 2981.0 | .00034 |
| 0.26 | 1.2969 | .77105 | 0.71 | 2.0340 | .49164 | 9.00 | 8103.1 | .00012 |
| 0.27 | 1.3100 | .76338 | 0.72 | 2.0544 | .48675 | 10.00 | 22026.5 | .00005 |
| 0.28 | 1.3231 | .75578 | 0.73 | 2.0751 | .48191 | | | |
| 0.29 | 1.3364 | .74826 | 0.74 | 2.0959 | .47711 | | | |
| 0.30 | 1.3499 | .74082 | 0.75 | 2.1170 | .47237 | | | |
| 0.31 | 1.3634 | .73345 | 0.76 | 2.1383 | .46767 | | | |
| 0.32 | 1.3771 | .72615 | 0.77 | 2.1598 | .46301 | | | |
| 0.33 | 1.3910 | .71892 | 0.78 | 2.1815 | .45841 | | | |
| 0.34 | 1.4049 | .71177 | 0.79 | 2.2034 | .45384 | | | |
| 0.35 | 1.4191 | .70469 | 0.80 | 2.2255 | .44933 | | | |
| 0.36 | 1.4333 | .69768 | 0.81 | 2.2479 | .44486 | | | |
| 0.37 | 1.4477 | .69073 | 0.82 | 2.2705 | .44043 | | | |
| 0.38 | 1.4623 | .68386 | 0.83 | 2.2933 | .43605 | | | |
| 0.39 | 1.4770 | .67706 | 0.84 | 2.3164 | .43171 | | | |
| 0.40 | 1.4918 | .67032 | 0.85 | 2.3396 | .42741 | | | |
| 0.41 | 1.5068 | .66365 | 0.86 | 2.3632 | .43216 | | | |
| 0.42 | 1.5220 | .65705 | 0.87 | 2.3869 | .41895 | | | |
| 0.43 | 1.5373 | .65051 | 0.88 | 2.4109 | .41478 | | | |
| 0.44 | 1.5527 | .64404 | 0.89 | 2.4351 | .41066 | | | |

## Table II    Natural Logarithms

| $x$ | $\ln x$ | $x$ | $\ln x$ | $x$ | $\ln x$ | $x$ | $\ln x$ |
|---|---|---|---|---|---|---|---|
| .01 | −4.60517 | .45 | −0.79851 | 0.90 | −0.10536 | 1.35 | 0.30010 |
| .02 | −3.91202 | .46 | −0.77653 | .91 | −0.09431 | 1.36 | 0.30748 |
| .03 | −3.50656 | .47 | −0.75502 | .92 | −0.08338 | 1.37 | 0.31481 |
| .04 | −3.21888 | .48 | −0.73397 | .93 | −0.07257 | 1.38 | 0.32208 |
| | | .49 | −0.71335 | .94 | −0.06188 | 1.39 | 0.32930 |
| .05 | −2.99573 | 0.50 | −0.69315 | .95 | −0.05129 | 1.40 | 0.33647 |
| .06 | −2.81341 | .51 | −0.67334 | .96 | −0.04082 | 1.41 | 0.34359 |
| .07 | −2.65926 | .52 | −0.65393 | .97 | −0.03046 | 1.42 | 0.35066 |
| .08 | −2.52573 | .53 | −0.63488 | .98 | −0.02020 | 1.43 | 0.35767 |
| .09 | −2.40795 | .54 | −0.61619 | .99 | −0.01005 | 1.44 | 0.36464 |
| 0.10 | −2.30259 | .55 | −0.59784 | 1.00 | 0.00000 | 1.45 | 0.37156 |
| .11 | −2.20727 | .56 | −0.57982 | 1.01 | 0.00995 | 1.46 | 0.37844 |
| .12 | −2.12026 | .57 | −0.56212 | 1.02 | 0.01980 | 1.47 | 0.38526 |
| .13 | −2.04022 | .58 | −0.54473 | 1.03 | 0.02956 | 1.48 | 0.39204 |
| .14 | −1.96611 | .59 | −0.52763 | 1.04 | 0.03922 | 1.49 | 0.39878 |
| .15 | −1.89712 | 0.60 | −0.51083 | 1.05 | 0.04879 | 1.5 | 0.40547 |
| .16 | −1.83258 | .61 | −0.49430 | 1.06 | 0.05827 | 1.6 | 0.47000 |
| .17 | −1.77196 | .62 | −0.47804 | 1.07 | 0.06766 | 1.7 | 0.53063 |
| .18 | −1.71480 | .63 | −0.46204 | 1.08 | 0.07696 | 1.8 | 0.58779 |
| .19 | −1.66073 | .64 | −0.44629 | 1.09 | 0.08618 | 1.9 | 0.64185 |
| 0.20 | −1.60944 | .65 | −0.43078 | 1.10 | 0.09531 | 2.0 | 0.69315 |
| .21 | −1.56065 | .66 | −0.41552 | 1.11 | 0.10436 | 2.1 | 0.74194 |
| .22 | −1.51413 | .67 | −0.40048 | 1.12 | 0.11333 | 2.2 | 0.78846 |
| .23 | −1.46968 | .68 | −0.38566 | 1.13 | 0.12222 | 2.3 | 0.83291 |
| .24 | −1.42712 | .69 | −0.37106 | 1.14 | 0.13103 | 2.4 | 0.87547 |
| .25 | −1.38629 | 0.70 | −0.35667 | 1.15 | 0.13976 | 2.5 | 0.91629 |
| .26 | −1.34707 | .71 | −0.34249 | 1.16 | 0.14842 | 2.6 | 0.95551 |
| .27 | −1.30933 | .72 | −0.32850 | 1.17 | 0.15700 | 2.7 | 0.99325 |
| .28 | −1.27297 | .73 | −0.31471 | 1.18 | 0.16551 | 2.8 | 1.02962 |
| .29 | −1.23787 | .74 | −0.30111 | 1.19 | 0.17395 | 2.9 | 1.06471 |
| 0.30 | −1.20397 | .75 | −0.28768 | 1.20 | 0.18232 | 3.0 | 1.09861 |
| .31 | −1.17118 | .76 | −0.27444 | 1.21 | 0.19062 | 4.0 | 1.38629 |
| .32 | −1.13943 | .77 | −0.26136 | 1.22 | 0.19885 | 5.0 | 1.60944 |
| .33 | −1.10866 | .78 | −0.24846 | 1.23 | 0.20701 | 10.0 | 2.30258 |
| .34 | −1.07881 | .79 | −0.23572 | 1.24 | 0.21511 | | |
| .35 | −1.04982 | 0.80 | −0.22314 | 1.25 | 0.22314 | | |
| .36 | −1.02165 | .81 | −0.21072 | 1.26 | 0.23111 | | |
| .37 | −0.99425 | .82 | −0.19845 | 1.27 | 0.23902 | | |
| .38 | −0.96758 | .83 | −0.18633 | 1.28 | 0.24686 | | |
| .39 | −0.94161 | .84 | −0.17435 | 1.29 | 0.25464 | | |
| 0.40 | −0.91629 | .85 | −0.16252 | 1.30 | 0.26236 | | |
| .41 | −0.89160 | .86 | −0.15032 | 1.31 | 0.27003 | | |
| .42 | −0.86750 | .87 | −0.13926 | 1.32 | 0.27763 | | |
| .43 | −0.84397 | .88 | −0.12783 | 1.33 | 0.28518 | | |
| .44 | −0.82098 | .89 | −0.11653 | 1.34 | 0.29267 | | |

# Index

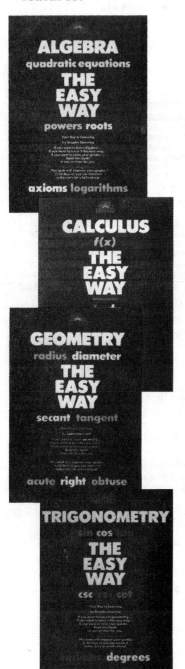